CLASSIC TRAINS

AND

RAILROAD ENGINEERING

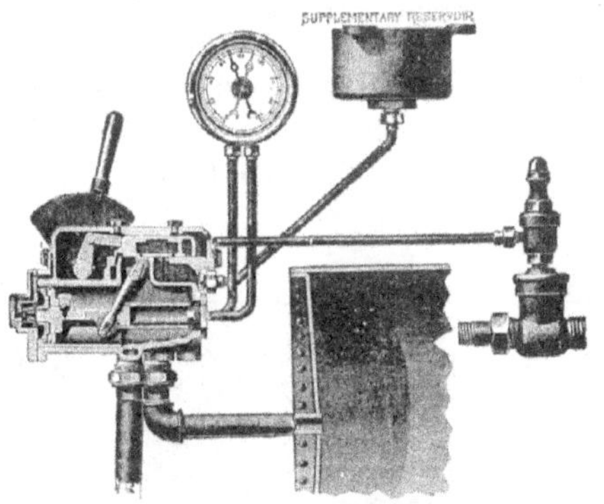

VOLUME 3

CONTEMPORARY AIR BRAKE PRACTICE

Edited by Mark Bussler

The Kinneys

Classic Trains and Railroad Engineering: Volume 3
Contemporary Air Brake Practice

Restored and Edited by Mark Bussler
Cover Design by Mark Bussler
Copyright © 2023 Inecom, LLC.
All Rights Reserved

No parts of this book may be reproduced or broadcast in any
way without written permission from Inecom, LLC.

Volume 3 of 8

www.CGRpublishing.com

The American Railway: The Trains, Railroads, and People Who Ran the Rails

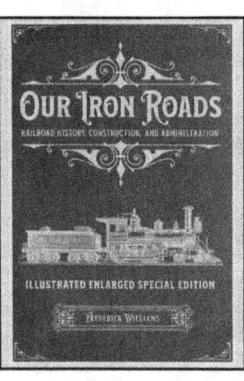

Our Iron Roads: Railroad History, Construction, and Administration

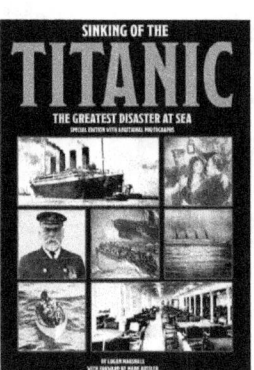

Sinking of the Titanic: The Greatest Disaster at Sea

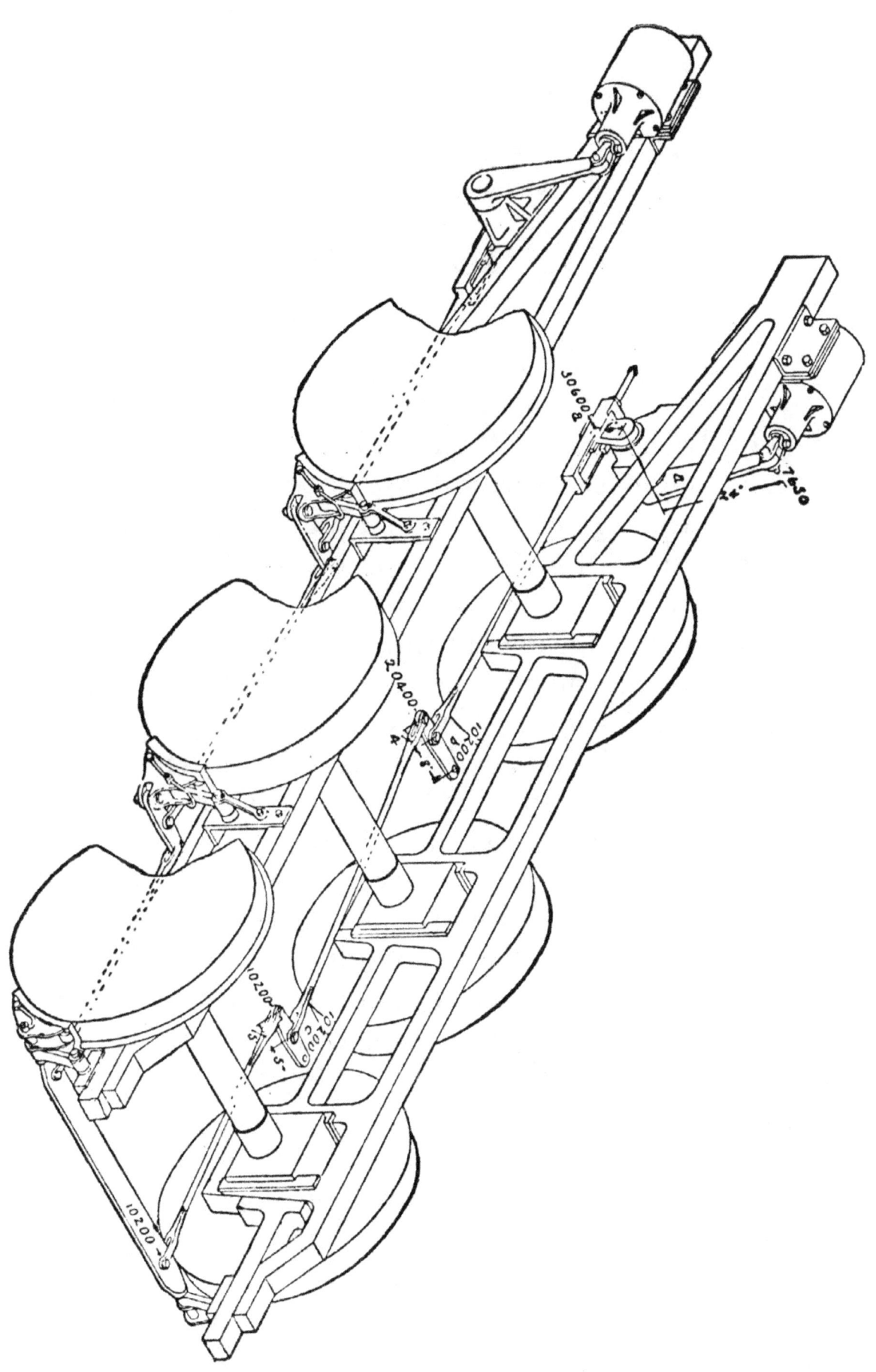

INTRODUCTION

MODERN Railroad Air-Brake Practice is an important subject in the Art of Railroading or Technique of Modern Transportation.

The increased weight of locomotives, the heavier tonnage of trains, and high speeds have made a study of Air-brake practice of supreme importance.

Changed conditions in railway operation have brought about improvements in the construction and operation of Air-brakes.

The latest improvements are fully described and illustrated, followed with questions and answers. The Westinghouse, New York, and Dukesmith systems are covered, making this volume the most comprehensive treatise and instruction on the subject extant.

It is written in a style which railroad men will readily understand and the subjects are divided into sections, with a series of questions and answers following each section.

It is believed this volume of the Prior Self-Educational Railway Series will be of great benefit to beginners as well as to advanced students and men experienced in the subject.

Table of Contents

009 - Chapter 1: Why the Air Brake is Misunderstood

021 - Chapter 2: Westinghouse Air Brake Equipment

203 - Chapter 3: Westinghouse Air Brake Defects

226 - Chapter 4: New York Air Brake Equipment

319 - Chapter 5: The Dukesmith Air Brake Control System

327 - Chapter 6: Operation, Handling, and Maintenance of the Dukesmith System

342 - Chapter 7: The Philosophy of Air Brake Handling, Rules & Tables for Computing Brake Power, Etc...

391 - Chapter 8: The Straight Air Brake as Used on Electric Traction Cars

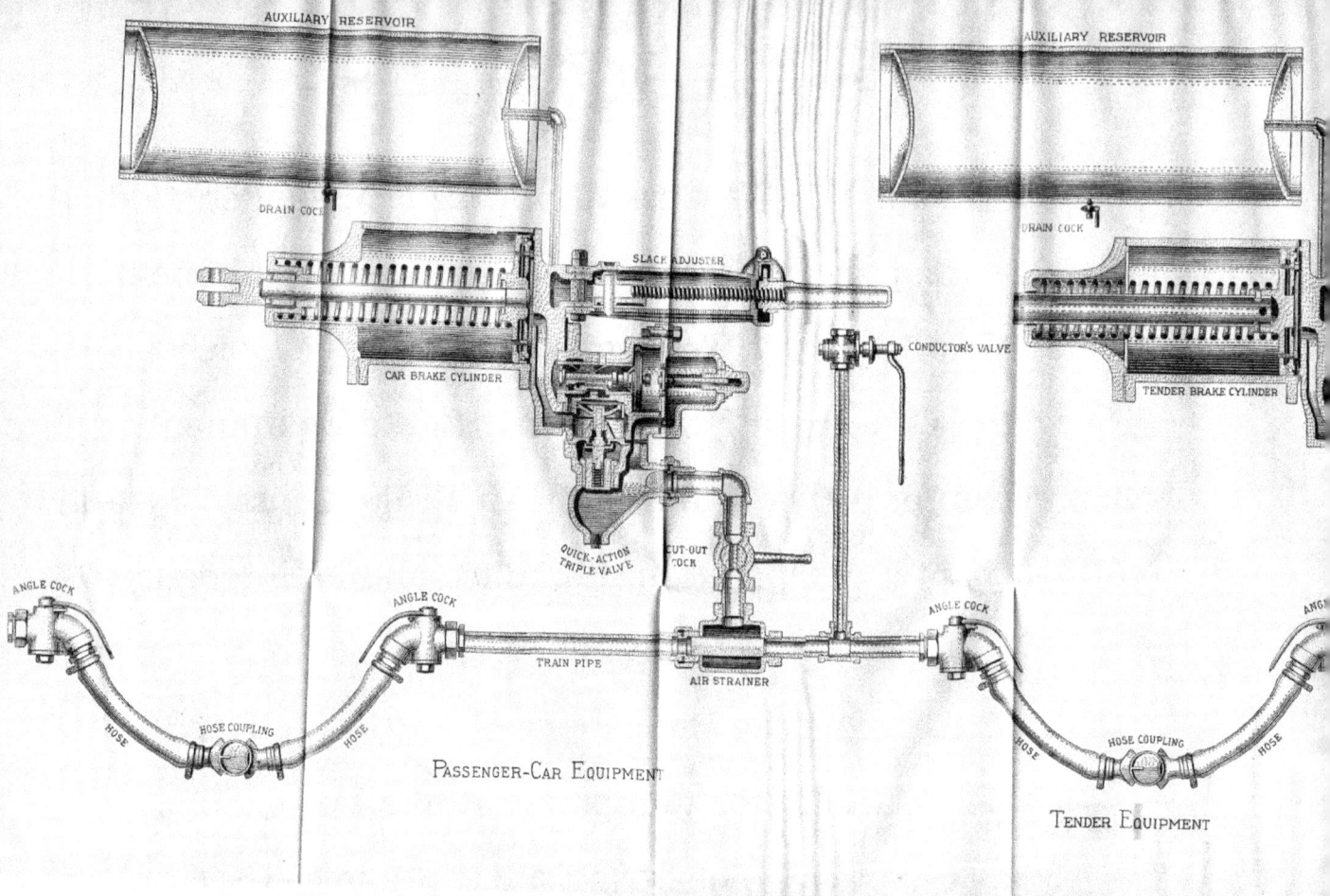

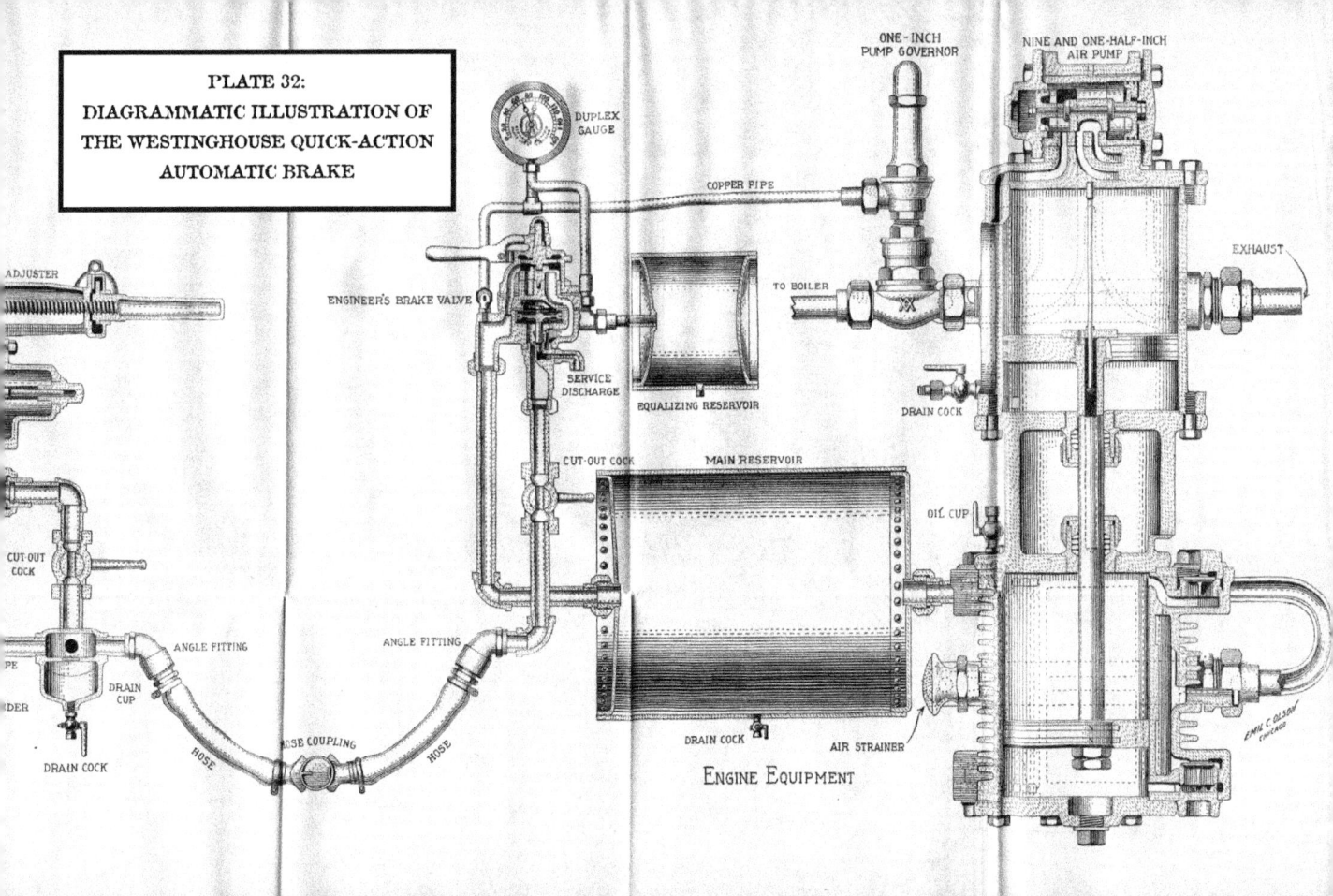

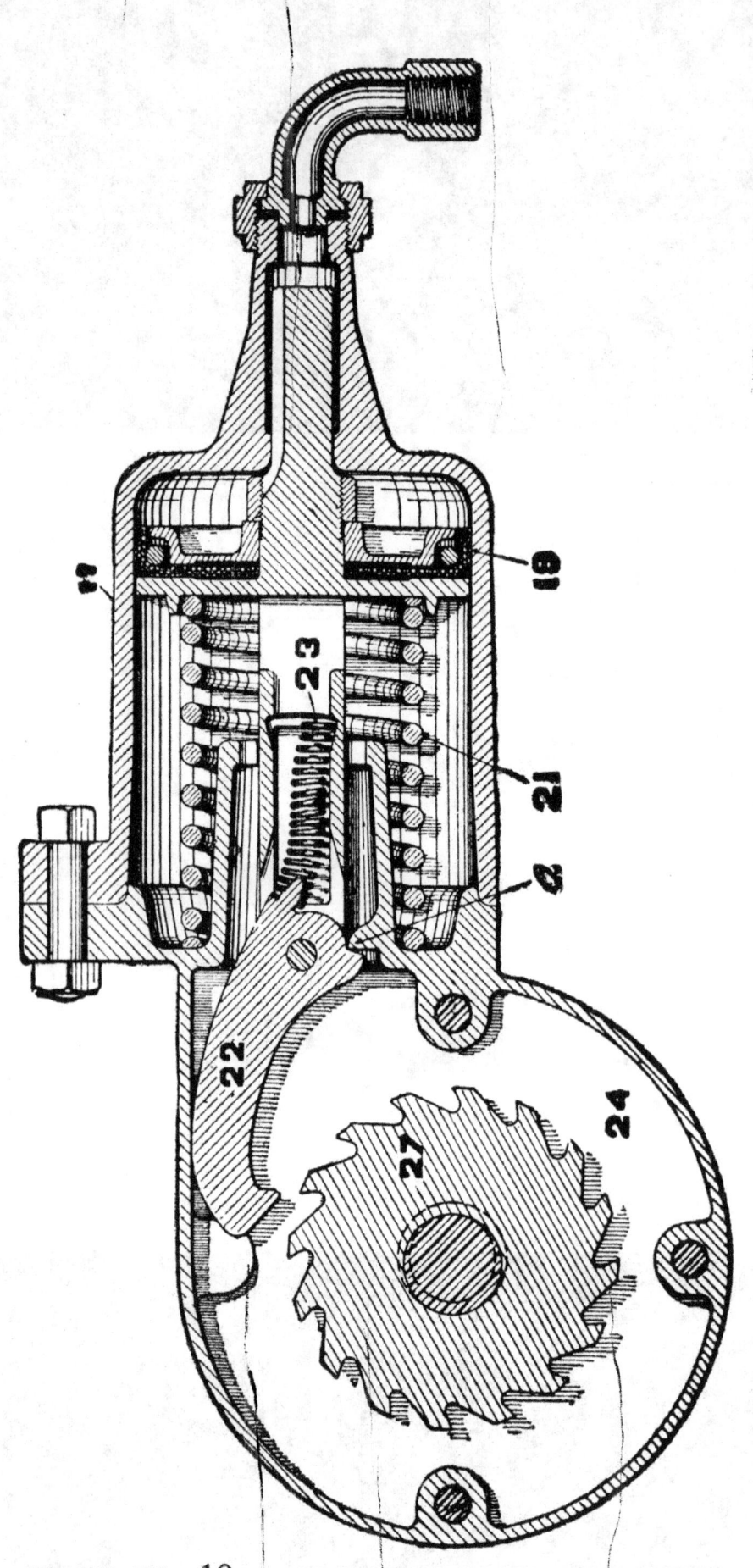

PLATE NO. 10.—AUTOMATIC SLACK ADJUSTER.

SECTION 1

CHAPTER I *

SOME REASONS WHY THE AIR BRAKE IS SO LITTLE UNDERSTOOD BY RAILROAD OFFICIALS AND EMPLOYES, AND WHY IT IS SO BADLY NEGLECTED

In writing a book of instruction an author has no right to presume upon his reader's previous knowledge of the subject, as there is no way of his knowing how far that knowledge may extend. Therefore, as this book is meant to contain full and complete instructions on modern air-brake practice, its use and abuse, I will take it for granted that you, my reader, are desirous that I leave nothing unsaid which may in any way throw light on the subject.

If the human memory could be depended upon to reproduce impressions made upon it after the manner of a phonograph, it would be the easiest thing in the world to acquire an education on any given subject, but as this is not the case, we must, first of all, bring ourselves to realize that

---

*Be sure and read the Introduction first.

MODERN AIR-BRAKE PRACTICE

if the knowledge we acquire is to be of any real value to us we must conform to the natural mental laws in our method of acquiring it, if we are to have any assurance that our memory will reproduce that knowledge at a time when it is most needed.

The mental laws by which the action of the mind is mainly controlled are those of Logic and Association. This may sound like Greek to you, but as it is very essential that you should know, at least in a general way, what is meant by the laws of Logic and Association, I will explain by saying that the truth or falsity of every statement is determined by Logic, and by the law of Association you are enabled to remember and trace one circumstance to another.

To make this plain, if I should say to you that the principle on which the automatic air brake operates is that any material reduction in the trainpipe pressure will cause the brake to set, then, if you should see the brake set on a car to which no engine was attached, you would logically say there must be a leak in the trainpipe somewhere, or the pressure could not have been reduced.

That you may understand how the law of Association enables you to remember things, I

ITS USE AND ABUSE

will just ask you to think for a moment of your home, and immediately there comes to your mind a mental picture of familiar faces, scenes and objects that a few minutes ago were buried in the depths of your memory. Now, supposing you wish to recall some bit of knowledge that has apparently slipped your memory, if you can take up the thread at any given point, the law of Association will carry your thoughts along, step by step, until you finally perceive the point you had forgotten, and which will cause you to suddenly exclaim, "O, pshaw, I remember now, it's so and so!" Now haven't you often gone through just this sort of experience?

Well, then, when you study any subject in a systematic way you will find that after you have once mastered it you can take it up at almost any given point, and by the laws of Association and Logic, recall and prove up your previously acquired knowledge.

The air brake is, comparatively speaking, a simple piece of mechanism, and as all machinery must conform to the laws of Logic in order to perform its functions correctly, it will be an easy matter for you to master a knowledge of the air brake provided you will keep firmly in mind the fact that the action of any one part of the apparatus always depends on the action of some other part in order to produce a certain result. For instance, if you had your train fully charged, and the gauge on the engine showed a pressure of 70 and 90 pounds, and the angle-cock was closed between the tender and the head car, you might even throw the handle of the brake valve to the emergency position, and still the brakes wouldn't set. Why? Simply because the action of the triple valve depends on the changing of the pressure in the trainpipe, and with the angle-cock closed on the head car the brakes couldn't set, even if you should knock the engineer's brake valve clear off the engine.

Therefore, when you have mastered a perfect knowledge of the air-brake system, the law of Association will force you to remember the functions of the different parts of the equipment, and by the law of Logic you will be enabled to tell exactly when the apparatus is working properly.

It may sound strange to make such a statement, but it is a fact, nevertheless, that the main reason why the air brake is so badly neglected is because it is automatic, or self-acting.

The average man, whether he be an official or employé, seems to feel perfectly safe on any kind

ITS USE AND ABUSE

of a train so long as he knows it has air brakes on it, and if any one were to ask him if he thought there was any danger of the brakes failing to stop the train, would laugh and say, "O, no, not at all, as all the engineer has to do is to make an emergency application, and the train will stop all right."

This would be a perfectly true statement if the air-brake equipment was always kept in its proper condition, but there are so very many things that can and do go wrong to prevent the brakes from doing their duty that the question of keeping them in good order is a very serious problem, indeed, and one that is arousing a deep interest in the minds of all railroad men. One of the strongest evidences of this fact is shown in the enactment of the national law spoken of elsewhere in this chapter.

It does not require much of a mechanical mind to grasp the fact that an air brake on a car would be worse than useless if the packing leather in the brake cylinder was dry, and allowed the air to escape, for no matter how good the engineer might be at handling his brake valve, the brake on that car could not be made to hold.

This is only one of a score of things which

might prevent the brakes from doing their duty, but because the brake will "work itself" the majority of men fail to see why the brake should not also 'take care of itself." But, like all other mechanism, it requires proper attention.

Another reason why the average man is lulled into the belief that the air brake needs but very little attention, is because a very few good air brakes on a train will produce results simply wonderful when compared with the old hand brake.

Such over-confidence in the power of the brakes to always stop the train is very much like the Irishman who bought two currycombs for his horse, because the dealer said to him if he would "buy one of his new patent currycombs, he could keep his horse on half the usual feed," whereupon Pat replied, "Faith, thin, I'll just take two, and I won't nade to buy any feed at all, at all."

Another reason why the air brake is so little understood and so badly neglected is because of its extreme simplicity, for with just ordinary attention it will continue to do its work, with more or less efficiency, for a considerable length of time, and as a consequence it is neglected until the brake-cylinder leather becomes dry and worthless; or the piston travel becomes too

ITS USE AND ABUSE

great; or the triple valve becomes gummed and dirty, and causes the brake to stick; or the strainers in the cross-over pipe becomes clogged; or the seats of some of the valves become worn and leaky, when the brake is "cut out," and the weight of that car is left to be stopped by the next car on which there happens to be a good air brake. But the average man fails to realize either the danger or expense of having the brakes "cut out," simply because so long as he knows a car to have a "self-acting brake" on it he feels safe, when as a matter of fact a hand-braked car is much safer for the railroad company than one with the air brakes cut out. For if it were not equipped with air the car would be carried with the non-air cars, and the train crew would have to look after it accordingly.

When short trains and slow speed were the order of the day it was perfectly safe to handle trains with only one-third of the cars air-braked, but in this twentieth century when long trains of heavy cars are shot over the country, up and down hill, like a Kansas cyclone or a scared wolf, the question of stopping power is of the highest importance, which means that *every* car in the train should not only have a "quick-action" brake on it, but that the brake *must* be in perfect order, and the "piston-travel" right up to where it belongs, and the enginemen and trainmen possessed of the proper knowledge of how best to manipulate and control the brakes in order *to prevent accidents* — as the great variety of accidents which may happen from bad handling, or not having a sufficient number of air brakes, is too numerous to mention.

It is safe to state that there is not a single railroad of any importance that does not pay out annually three times as much money on account of bad brakes and bad handling of brakes as they pay for a general manager, but because of the many different channels through which the expenditures are made they are not charged up directly to the brakes.

For instance, an engineer in coming into a station with a passenger train is making the stop with "one application" (the old way) and, after his brake cylinders and auxiliaries have equalized their pressure, and he is drifting along, depending upon the weight of the train to stop him at his usual place (because after equalization the automatic brakes cannot be applied any harder), a woman or child in crossing the track is killed. The amount of money the company has to pay out as a result of this "bad handling" of the

ITS USE AND ABUSE

brakes will run up into thousands of dollars, and yet the engineer excuses himself by simply saying: "The brakes failed to work."

Again, railroad companies are out thousands of dollars annually on account of damaged merchandise, caused by the brakes being "thrown into the emergency" when there was no real danger ahead to require the emergency application to be used, or because of a defective triple valve.

Uneven piston-travel causes more trains to be parted while running along, draw-heads pulled out, wheels flattened, etc., than any other one cause; hence it is evident that the brakes should not only be kept in perfect working order at all times, but the men who handle them should understand thoroughly how to properly manipulate and keep them in order.

The American Congress, realizing the vast importance of having the air-brake equipment kept up to somewhere like it should be, recently enacted a law, which became effective in September, 1903, requiring all railroads to have at least fifty per cent of the cars in all trains equipped with air brakes in good condition. And as the law would be a dead letter if the "good condition" clause was not lived up to, it is easy to see that railroads are forced to look after the instruction of their men as much as possible, and in order to do so many roads which are not already so provided, are putting on regular air-brake instructors as rapidly as conditions will permit, and are voluntarily increasing the number of air-braked cars in freight trains.

Some idea may be formed of the average man's knowledge of the "equalization of pressure" by the following true story: A certain engineer on a mountain road was going down a pretty stiff grade, and after making a great number of "reductions" from his trainpipe, and not feeling the train slow up as he expected, turned to the head brakeman, who happened to be riding on the engine, and said: "Hey, Bub, you'd better be gittin' back, 'cause I ain't got but a few more squirts left in this thing." And still he was considered a good runner by his employers.

In order to insure safety in the handling of trains it is absolutely essential that every one whose duties in any way connect him with the air brake, should not only know what all the parts are that constitute the air-brake equipment, but must also understand the philosophy of handling the brakes under any and all cir-

ITS USE AND ABUSE

cumstances, as the requirements of his position may demand.

In addition to all of the many reasons previously mentioned as to why the air brake is so little understood by the average engineman and trainman alike, a very common one is because of the unsystematic manner in which the study of the air brake is usually begun. The engineman, if he gives the subject any study at all, usually begins by trying to master the mysteries of the brake valve, or the pump, and the trainman usually thinks there is nothing for him to learn except how to "cut it in, or cut it out," and gives as his excuse that "the engineer handles the brake, and, besides, it is automatic, and works itself."

The experience of late years has abundantly proven that if an air-braked train is to be handled with safety it is absolutely necessary that every man on the train thoroughly understands at least the principle on which the brake operates, and must be able for a certainty to tell when the brakes are in perfect working order by making a careful test before starting, or when any change is made in the train.

A serious accident happened recently by the simple act of a brakeman turning up the handle

MODERN AIR-BRAKE PRACTICE

of a pressure-retaining valve. He heard the air escaping at the "retainer," and thinking he would "stop the leak," turned up the retainer handle, and as a consequence the brake on that car could not be released from the engine, which allowed the wheels to become overheated, causing them to burst, which ditched the train and killed three men. This would never have happened if that brakeman had only understood the mere principle on which the brake operates.

SECTION 2

CHAPTER II

THE WESTINGHOUSE AIR-BRAKE EQUIPMENT—THE PARTS AND THEIR DUTIES

The full and complete equipment of a modern quick-action automatic air brake is composed of twelve essential parts, as follows:

First: The steam-driven air pump which supplies the compressed air.

Second: The main reservoir in which the compression air is stored.

Third: The engineer's brake valve by which is regulated the flow of air from the main reservoir into the trainpipe for charging and releasing the brakes, and from the trainpipe to the amostphere for applying the brakes.

Fourth: The duplex air gauge, which shows simultaneously the pressure on the trainpipe (black hand), and in the main reservoir (red hand).

Fifth: The pump governor, which regulates the supply of steam to the pump, causing it to automatically stop when the desired maximum of pressure has been accumulated in the air-brake apparatus.

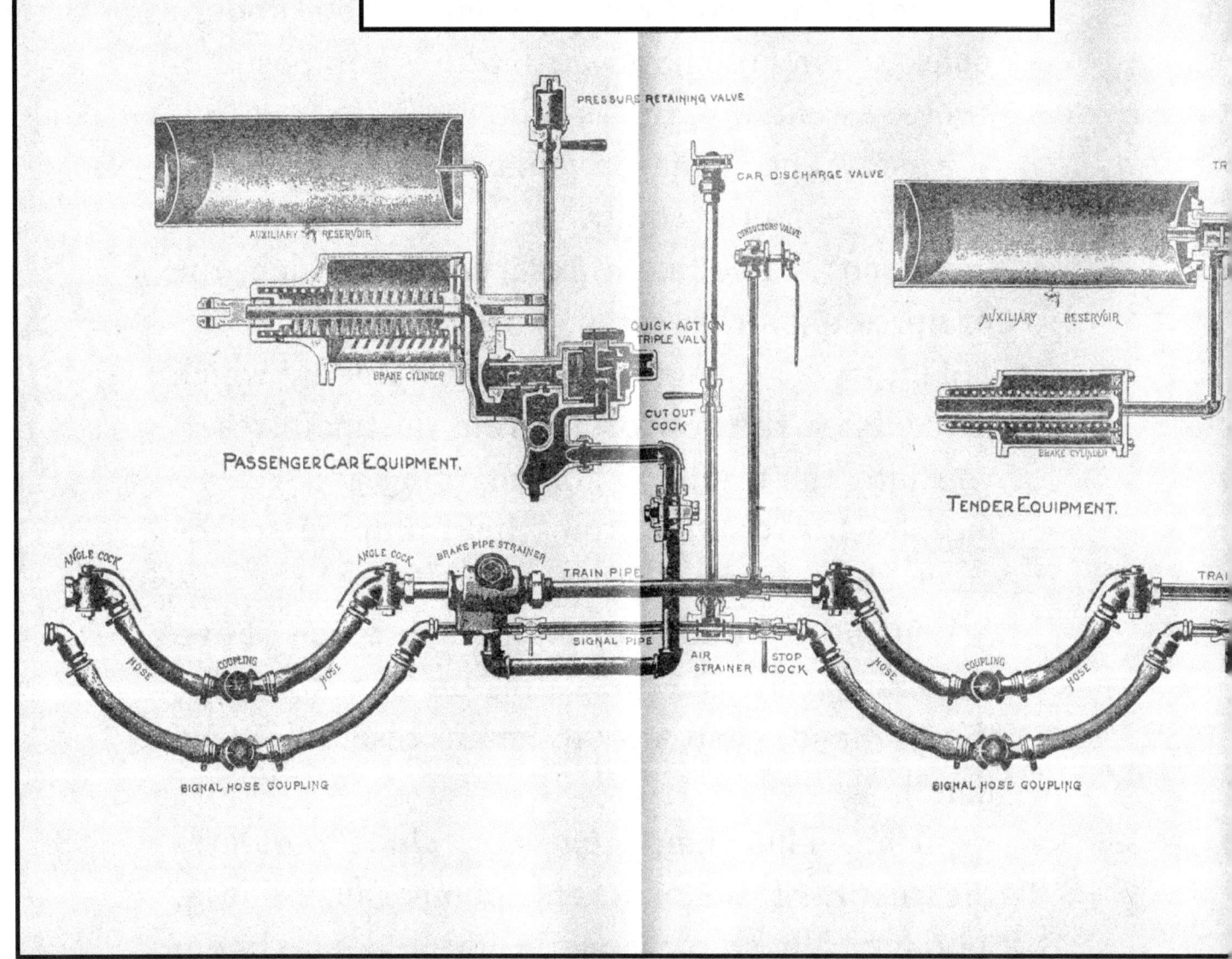

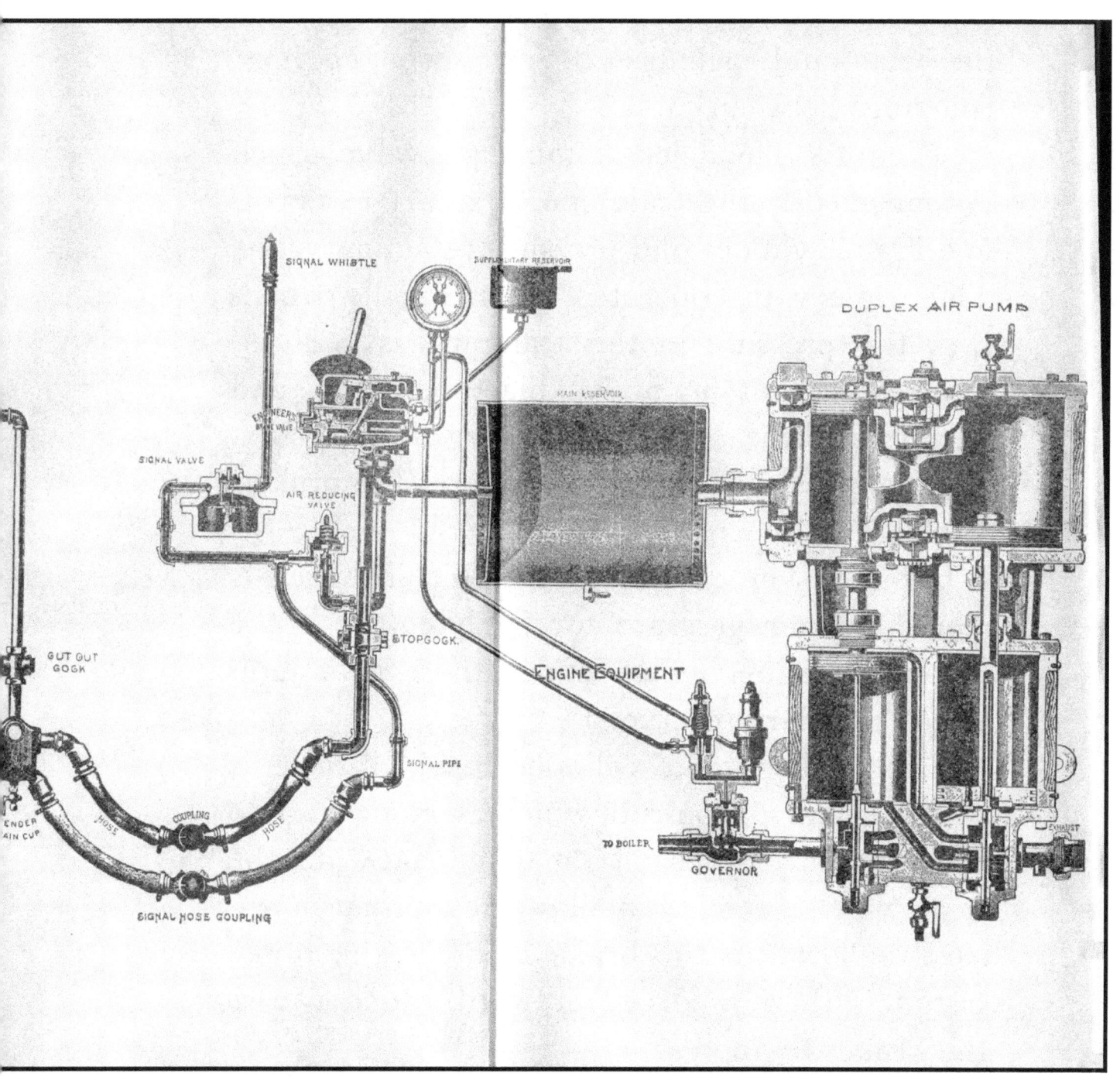

MODERN AIR-BRAKE PRACTICE

Sixth: The trainpipe, which connects the engineer's brake valve and each triple valve in the train, and includes the air hose and hose couplings between cars.

Seventh: The quick-action triple valve, which is connected to the trainpipe, auxiliary reservoir and brake cylinder and pressure-retaining valve. The triple valve operates automatically whenever the pressure in the trainpipe is reduced lower than that in the auxiliary reservoir, and performs three functions: charges the auxiliary, applies the brakes and releases the brakes, as will be fully explained hereafter.

Eighth: The auxiliary reservoir, in which is stored the air pressure for applying the brake (on each car, engine, or tender, there is an individual auxiliary reservoir).

Ninth: The brake cylinder, in which there is a piston and piston-rod, which is connected to the brake levers in such a manner that when the triple valve is moved to allow the auxiliary pressure to flow into the brake cylinder, the brake piston is thereby forced outward, which causes the brakes to apply.

Tenth: The pressure-retaining valve, which is connected to the triple exhaust by a small pipe. On freight cars the retaining valve is

ITS USE AND ABUSE

located on the end of the car near the top, just below the staff of the hand brake, and is for the purpose of enabling the brakeman to retain a pressure of 15 or 50 pounds in the brake cylinder while the engineer is recharging the auxiliary reservoir. While the handle of the retaining valve is turned up the brake cannot be released from the engine, neither can it be "bled off" by the bleed cock of the auxiliary, for the reason that the cylinder must discharge its air through the triple exhaust, and when the retaining valve is closed it means that the triple exhaust is also closed. It is very important that brakemen thoroughly understand the operation of the pressure-retaining valve, as many accidents are due to ignorance or negligence in the working of this device.

Eleventh: The automatic slack-adjuster automatically maintains the travel of the brake cylinder piston at a given distance. For instance, if the piston-travel is set for eight inches it will automatically keep it there. The slack-adjuster is piped direct to the brake cylinder, so that every time the brake is applied the adjuster is operated automatically.

Twelfth: On passenger cars there is commonly in use a valve known as the conductor's valve, which is connected directly to the trainpipe and by means of which the conductor can apply the brakes from the car in case of danger. This valve is now being superseded by one known as the Dukesmith Car Control Valve, which is a combined Retaining Valve, Conductor's Valve and Release Valve. This new valve is located in the same place as was the old conductor's valve, and in making an emergency application the conductor pulls the cord the same as he did before, and after having applied the brakes he resets the valve, the same as he did with the old conductor's valve, and when it is desired to retain the pressure in the brake cylinder he moves the handle of the Dukesmith Control Valve to retaining position after the same manner as any ordinary retainer, but a feature of this valve which is of great importance is the fact that by its use the conductor or trainman can release the brake on a car when the triple valve fails to go to release position, thereby avoiding the great danger of having to stop the train in order to release a stuck brake. A still further attachment to the Dukesmith Control Valve is the Automatic Release Signal, which is for the purpose of automatically signaling the trainmen from the inside of the car what the brake under the car is doing, that is, it tells when the brake is set or released,

ITS USE AND ABUSE

what the piston travel is, whether the brake is leaking or releasing off, and should too much pressure be accumulated in the brake cylinder than the standard amount, the Release Signal automatically blows down whatever extra pressure is in the brake cylinder, thereby reducing the liability of sliding the wheels or stalling the train.

As the very heart of the automatic air-brake equipment is the triple valve, it is necessary that both enginemen and trainmen thoroughly master this feature first of all.

It is not necessary that trainmen should know all about the care of the pump, the ports in the brake valve nor how to handle the air as an engineer, but they should know and *understand* all about the triple valve, and be able to make an intelligent report of any defects that may be found in the car equipment, how to make a proper test, and why correct piston travel is positively essential to good brakes, by mastering a knowledge of the Laws of Leverage.

Enginemen should not only be thoroughly familiar with the points just outlined for trainmen to learn, but, in addition, should know all about the action of the pump and the pump governor; the different kinds of automatic and straight air brake-valves, their parts and their action; how to determine and maintain the proper braking power on engine and tender; the construction and operation of the whistle-signal apparatus; why different air pressures are necessary; the best manner of nandling different trains under any and all circumstances, and how to detect and report intelligently any trouble that may arise in any part of the equipment.

To the ordinary mind this may at first thought appear very difficult of accomplishment, but such, however, is not the case, provided the study of the air brake is taken up systematically, and one thing is mastered at a time, taking each part in its regular order.

This cannot be done in a minute or a month, but requires time and patience. There is nothing mysterious about the air brake, as it is simply a question of one pressure working against another at all times, and all there is to learn is how and when the several pressures are separated or joined together, and when and to what extent you wish to let the pressures flow together or be kept apart, in order to secure a given result.

All this is done by a system of very simple valves and pistons, reservoirs and cylinders, all connected by suitable pipes for the purpose of allowing the compressed air to pass from one part of the equipment to the other, or to the atmosphere, as the case may be.

The first part of the air-brake equipment we will consider will be

THE TRIPLE VALVE

Naturally the first question you will ask is 'Why must there be a triple valve?"

It is because the brake charges, sets and releases automatically, and as this requires three distinct services, it follows that a device capable of doing a triple service must be had, and as these three things are done by one part of the equipment it is called the triple valve (meaning three valves in one, or a valve that charges the auxiliary reservoir, a valve that sets the brakes and a valve that releases the brakes).

As there are several kinds of triple valves in use, but as the same principle operates them all, l will first describe the action of the "plain" triple in making a full service application of the brakes, releasing the brakes and recharging the auxiliary reservoir (taking it for granted that the auxiliary and trainpipe are charged to 70 pounds to begin with).

In order to clearly understand the duties and action of the triple you must always bear in mind that on each car there must be a trainpipe, an auxiliary reservoir, a brake cylinder and the triple valve.

MODERN AIR-BRAKE PRACTICE

The trainpipe is the channel through which the compressed air passes between the engineer's brake valve and the triple.

The auxiliary reservoir is where the air is stored under each car, ready for use.

The brake cylinder is where the air is applied in setting the brakes, and the triple valve performs the triple duty of charging the auxiliary, applying the air to the brake cylinder and releasing the air from the brake cylinder.

But before describing the air brake let us draw a comparison with something that will help to fix in our mind what action *must* take place in order to set the brake.

The best thing to compare the air brake with in order to exemplify the principle on which it operates, is a bottle of soda pop, for the reason that gas is mixed with the soda when it is bottled. A bottle of champagne would make a better comparison, owing to the higher pressure with which the wine is bottled, but as it is a little too expensive for the average railroad man to become very familiar with, I will just use the ordinary bottle of soda pop.

If you wanted to fill a glass with pop, the first thing you would have to do would be to break the wire that holds the cork, when the pressure in the bottle would force the cork out and let the soda flow into the glass.

Therefore, figuratively speaking, the brake cylinder represents the glass, the auxiliary the bottle, the compressed air in the auxiliary the soda, the triple valve the cork and the trainpipe pressure the wire, and when you take the trainpipe pressure away from the triple (or break the wire that holds the cork), the pressure that is in the auxiliary forces the triple out and lets the air pass from the auxiliary into the brake cylinder and sets the brake, by forcing the cylinder piston out against the levers, which in turn forces the shoes up against the wheels.

By this you will understand that in order to set the brakes the pressure in the trainpipe *must be reduced lower than that in the auxiliary*, otherwise the triple would not move and open the port between the auxiliary and brake cylinder.

The Parts of the Plain Triple Valve consist of only six things, besides the casing which holds them all, and are shown in plate 1 (which shows the way the new plain triple now used for driver brakes would look if it was cut in half), and they are designated as follows: 23 is called the triple piston; 24 is the slide valve; 25 is the graduating valve; 26 is the graduating stem, and 27 is the graduating spring; 32 is the U spring over the slide valve.

The casing is so shaped that one part of it forms a cylinder for the triple piston to move in, and is marked B, and adjoining it is a chamber having a flat side (called the slide valve seat), for the slide valve to slide on, and is marked C.

The flat side of this chamber, which forms the seat on which the slide valve rests, has two ports cut through it; the one marked f leads to the brake cylinder, and the other, marked h, leads to the atmosphere. (See plate 1.)

In the slide valve there are also two ports; one passes clear through the valve, as shown by the letters l, p-p, and the other is a groove cut in the bottom of the valve, and marked g, and when the valve is moved toward the left end of chamber C (in other words, moves down), the port through the valve marked p connects with the port in the seat marked f, so that the air in the auxiliary can pass through the valve and valve seat and on through pipe connection X directly into the brake cylinder; and when the slide valve is in the opposite end of chamber C the groove g in the bottom of the slide valve connects the two ports f and h together, so that

ITS USE AND ABUSE

one end of the groove rests directly over the port leading to the brake cylinder, and the other end rests over the port leading to the atmosphere, thus forming a direct opening between the brake cylinder and the atmosphere; therefore, as the triple is so connected to the auxiliary by pipe connection Y that the auxiliary pressure is always in direct communication with chamber C, in which the slide valve moves, and as the port in the seat marked f is the only way for the air to get in or out of the brake cylinder, with this kind of a triple, it is very evident that when the slide valve is moved along on its seat until the port in the valve marked p-p comes opposite the port in the seat marked f, the air in the auxiliary is free to pass into the brake cylinder, and set the brake. And when the slide valve is forced back again to its original position, as shown in plate 1, the air in the brake cylinder is free to pass out to the atmosphere through ports f, g, h and exhaust port k, and thereby release the brakes. Therefore, as the flow of air from the auxiliary to the brake cylinder, and from the brake cylinder to the atmosphere is dependent upon the movement of the slide valve, it is necessary that you next understand how this movement is accomplished.

MODERN AIR-BRAKE PRACTICE

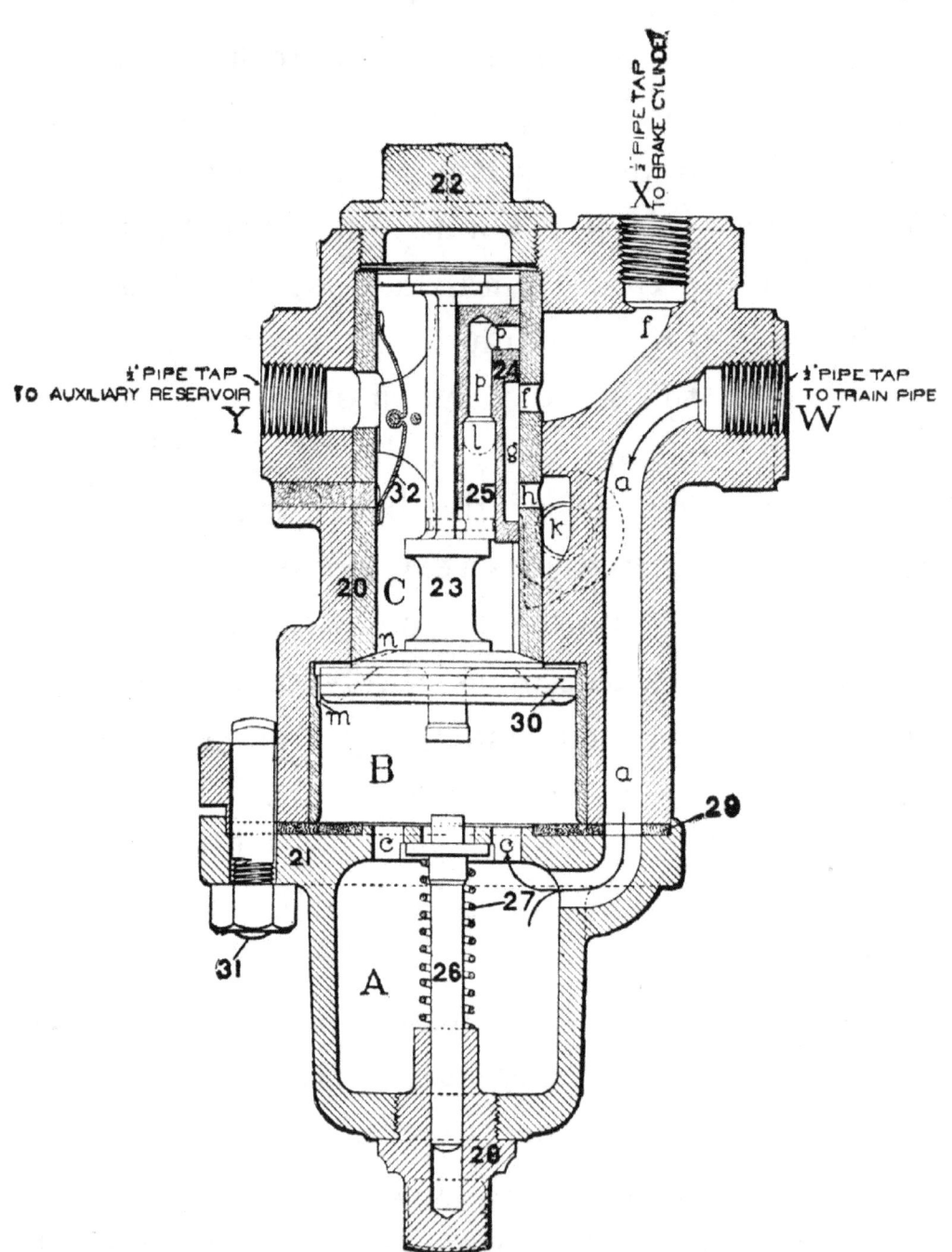

PLATE NO. 1—NEW STYLE PLAIN TRIPLE-VALVE.

ITS USE AND ABUSE

DESCRIPTION OF PLATE 1—NEW DRIVER BRAKE
PLAIN TRIPLE

W is the trainpipe connection.
X is the cylinder connection.
Y is the auxiliary connection.
23. Triple piston and stem.
24. Slide valve.
25. Graduating valve.
26. Graduating stem.
27. Graduating spring.
30. Triple piston packing ring.
32. U, or slide valve spring.

The air passages and ports are explained in the text.

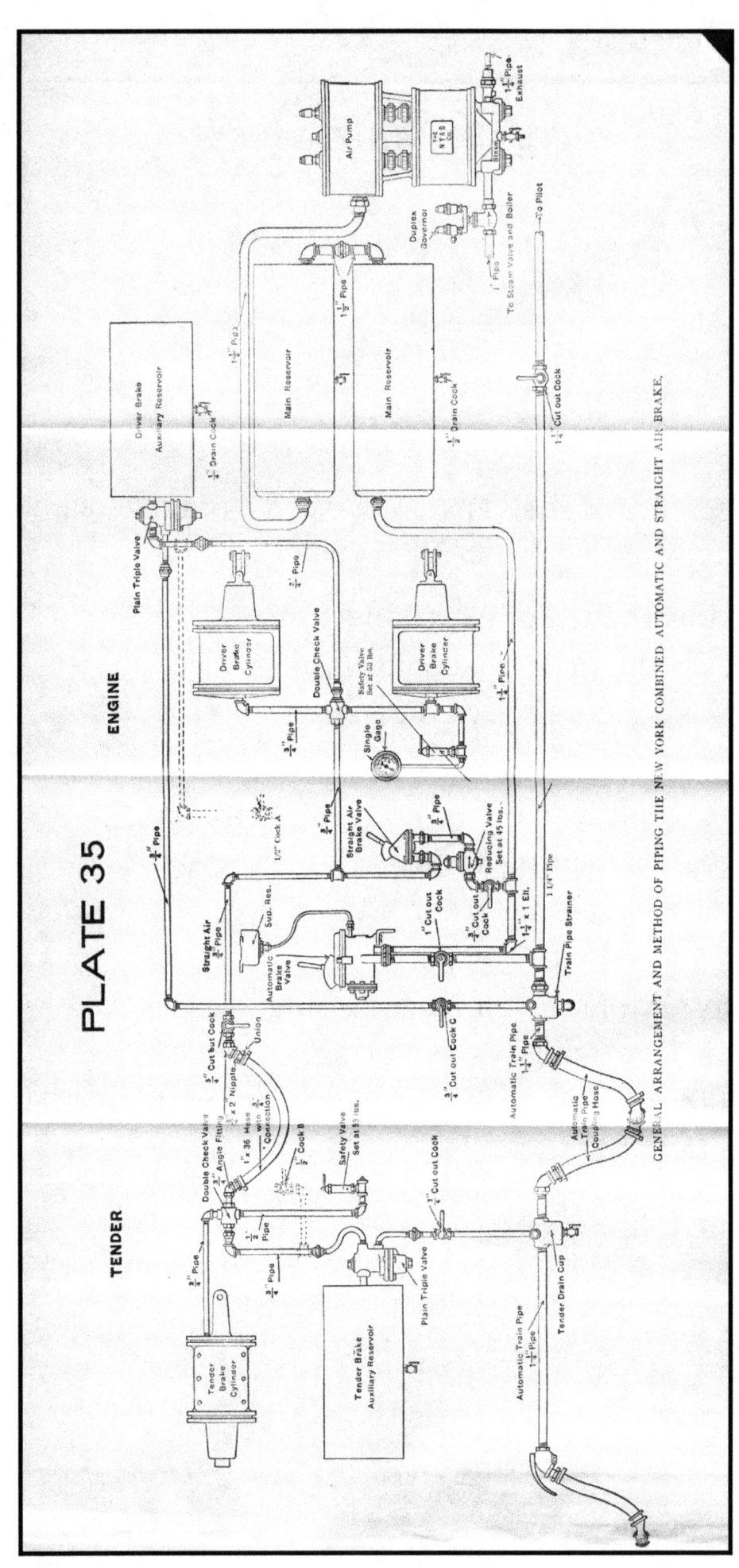

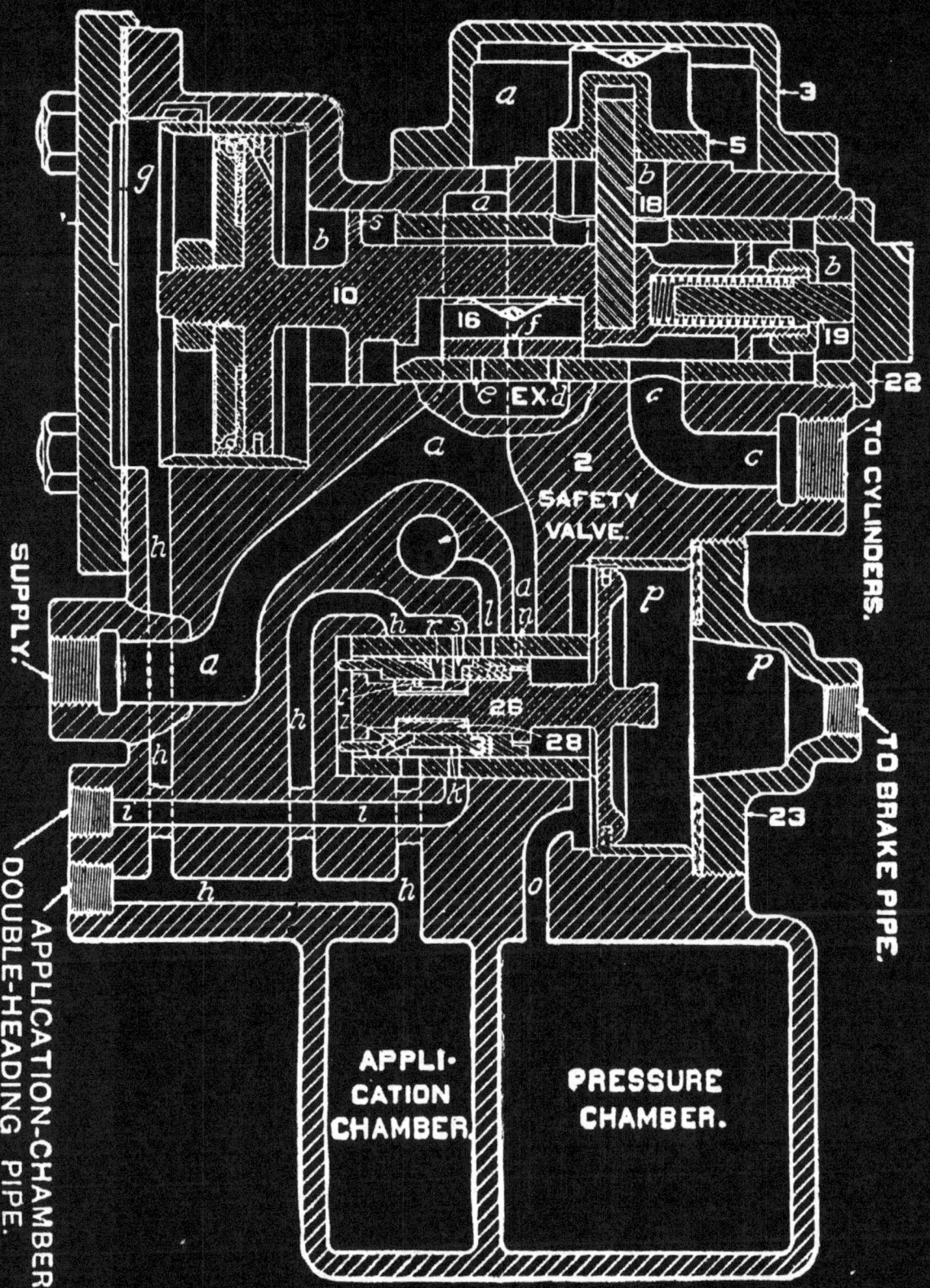

MODERN AIR-BRAKE PRACTICE

The stem of the triple piston extends into chamber C, in which the slide valve moves, and the valve is hung on this stem; there is a packing ring (30) around the triple piston, making a tight joint against the walls of cylinder B, and as one end of this cylinder is always open to chamber C (which always contains auxiliary pressure) and the other end of cylinder B is always open to the trainpipe, you will at once see that the triple piston stands between the auxiliary and trainpipe pressure at all times, and if these pressures are equal, and the piston is in full release position, as shown in plate 1, should the pressure on the trainpipe side of the piston become lower than that on the slide valve side, the piston would be moved by the auxiliary pressure, and of course draw the slide valve with it, causing the port in the valve marked p to come opposite the port in the seat marked f, and allow the air from the auxiliary to pass into the brake cylinder and set the brake.

Now that the air is in the brake cylinder, the next point to learn is how to release the brake.

To Release the Brake it is necessary to force the slide valve back to the position it occupied before the brake was set, as shown in plate 1.

To do this we use the pressure stored in the

main reservoir, on the engine, for when the engineer places his brake valve in full release position the main reservoir pressure quickly raises the pressure on the trainpipe side of the triple piston and forces it back to the position shown in plate No. 1, and, as the slide valve has to go back with it, the groove *g* in the bottom of the valve is placed so that one end of it rests over the port marked *f* in the valve seat, and the other end rests over the port marked *h* in the valve seat, consequently the air in the brake cylinder is free to pass out to the atmosphere through ports *f*, *g*, *h* and through a passage around the casing to the triple exhaust marked *k*. The air having thus escaped from the brake cylinder the heavy spring in the cylinder, marked 9, in plate 7, drives the brake piston back from the levers, which allows the shoes to drop away from the wheels, and the brake is released.

The whistling noise heard when the brakes are releasing on passenger cars is caused by the air escaping through the small ports in the triple (on freight cars the air exhausts through the pressure-retaining valve on top of the car), and if this whistling is weak, when releasing after a full application has been made, it indicates that either a portion of the air has already escaped from the cylinder through a bad packing leather around the brake piston, or there is too much piston travel, which allowed the air to expand in the cylinder more than it should have done; in other words, a high pressure will rush out quicker than a low pressure, for, as you know, the faster wind blows the louder it whistles.

Recharging the Auxiliary.—Having set the brakes and released them, it now becomes necessary to recharge the auxiliary reservoir, to be ready for the next application.

You must keep in mind that the brake cylinder gets its power from the auxiliary, and the latter must always be kept charged ready to meet all demands made upon it by the cylinder. If the auxiliary is only partly charged, the force with which the brakes set will be correspondingly weak.

Also remember that just as soon as the slide valve moves to let the air out of the brake cylinder that in doing so the feed grooves between the trainpipe and auxiliary are opened to admit air again into the auxiliary.

You will now look at plate 1, and trace the course of the air from the trainpipe through the triple to the auxiliary.

Begin at the point indicated by W, and follow

ITS USE AND ABUSE

the arrows; you will notice the air travels through a passage (*a-a*) in the casing, to a chamber indicated by A, and from this chamber there are two openings (*c, c,*), which allow the air to pass into the cylinder in which the triple piston moves, as indicated by B. As the air passes from chamber A it strikes the plain side of the triple piston and forces it to the extreme end of cylinder B, and as the piston is supposed to be a tight fit in cylinder B, the only chance the air has to get into chamber C is by passing through a small groove cut in the wall of cylinder B, as indicated by *m*. This is called the "feed groove." As this groove *m* is only as long as the head of the piston is thick, you will at once see that the piston must be all the way back before the air can enter this groove; you will also notice that the piston only forms a seat about half way from its center to its outer edge; in other words, there is a shoulder on the slide valve side of the piston, and this necessitates another groove to be cut in this shoulder, which is shown by the letter *n*. The air can now pass from cylinder B by way of the feed grooves, *m* and *n*, into chamber C, and over the top of the slide valve through the pipe connection Y into the auxiliary.

In order, therefore, to make it plain to you

how the auxiliary is charged to its proper pressure of seventy pounds to the square inch, we will just suppose that the pump on the engine (which, when modern brake valves are used, is controlled by the main reservoir pressure) will only pump up to seventy pounds pressure, and no more; in other words, the pump will keep working until all the space into which the compressed air from the pump is allowed to flow is filled to seventy pounds before it stops.

If the space to be filled by the pump is merely the main reservoir, the pump will stop when the main reservoir is charged to seventy pounds, provided the governor is set at seventy; but if the engineer places the handle of his brake valve in position so that the air in the main reservoir can flow direct into the trainpipe, it means that there is just that much more space to be filled before the pump will stop; then if the auxiliary is cut into the trainpipe, by opening the cut-out cock on the cross-over pipe, it means that there is still more space for the air to flow into, and as the pump will not stop until there is seventy pounds in the main reservoir, and as the main reservoir cannot get its seventy pounds until the trainpipe has its seventy pounds, and as the trainpipe cannot get its seventy pounds

ITS USE AND ABUSE

until the auxiliary gets its seventy pounds, it follows that the pump will continue to work until the auxiliary, trainpipe and main reservoir are all equally charged up to seventy pounds, for the reason that air, like water, will continue to flow until it finds its level, and when we speak of the pressure being "equalized" we mean that they have come to a level with each other.

Owing to the smallness of the feed groove in the triple through which the air passes to get into the auxiliary, the trainpipe will naturally fill quicker than the auxiliary, and cause the pump to stop temporarily, but as soon as the trainpipe pressure is again lowered by the air passing through the feed grooves into the auxiliary, the pump will again start, and continue to compress air until every bit of space is filled to seventy pounds.

If the main reservoir, trainpipe or auxiliary reservoir leaks, while the brake valve is in the position we are now speaking of, the pump will not stop at all, and a great many leaks will very soon wear a pump out.

Right here I will mention a few important things to remember when charging up a train: first, leaks of any kind will prevent getting the required pressure in the time it should be gotten, and bad leaks will prevent it entirely.

Second, the strainer and feed grooves in the triple must be kept clean to allow the air to pass freely.

Third, the packing ring around the triple piston must be a good fit to prevent the auxiliary charging too rapidly, and to insure against charging too quickly is the reason for having a shoulder on the slide valve side of the piston; for if any air leaks around the packing ring it cannot enter the auxiliary except through the second feed groove, as shown by n in plate 1, unless the shoulder on the piston has a bad seat.

A still greater reason for having the packing ring (30) tight, is to insure the brake against "sticking," as it will if the trainpipe pressure equalizes with the auxiliary without moving the slide valve.

The reason for having the feed grooves so small in the triples is to allow all the auxiliaries on the train to charge as nearly together as possible, and also to assist in making the triple sensitive to the slightest reduction of trainpipe pressure, for, if the feed groove was large, when the air was drawn from the trainpipe a considerable amount of air from the auxiliary would

ITS USE AND ABUSE

flow back into the trainpipe before the piston moved; but, as it is, the feed groove is so small and so short that it requires less than a two pound reduction to cause the triple piston to move and shut off communication between the auxiliary and trainpipe.

For the same reason (sensitiveness) the piston packing ring must have a good fit, or else the auxiliary and trainpipe pressures will equalize, and thereby fail to move the piston when desired in setting or releasing the brakes. This is especially true on long trains.

If everything was tight, and all the parts working as they should, and trainpipe pressure was kept constantly at seventy pounds, you could charge a one hundred car train as quickly as you could one car, as under such perfect condition the air will pass through the feed grooves at the rate of one pound a second, but as this is never the case in actual practice, it will take about five minutes to charge up a short train of ten cars, and about twelve to fifteen minutes for a train of thirty or forty cars, with comparatively no trainpipe leaks, and where there are leaks it naturally takes much longer.

Always fight trainpipe leaks like you would a rattlesnake, as this trouble does more to prevent the proper action of the brakes than any other one thing.

So far I have only had occasion to speak of but one kind of an application of the brakes, and that is "full service application"; but there are three kinds of applications, which will be fully explained in their proper place: one is called "full service application," one is called "partial service application," and the third is called "emergency application."

As yet I have only described the duties of the triple piston and the slide valve, but there are four other parts to the plain triple that must now be explained to you, which you will see by again referring to plate 1, and designated as follows:

The graduating valve, which works in the slide valve, is marked 25; the graduating stem is marked 26, and the graduating spring which surrounds it and holds it to its seat is marked 27; the U spring is marked 32. Now let us see why we need these parts.

The graduating valve is what enables us to make a partial service application, for without it the pressure in the auxiliary reservoir would be reduced much below that in the trainpipe, after a ten pound reduction, before the triple would

ITS USE AND ABUSE

lap itself, as there would be nothing to stop the flow of air from the auxiliary into the brake cylinder, until the auxiliary pressure becomes low enough for the trainpipe pressure to overcome the friction on the seat of the slide valve; but with the graduating valve in good condition, when a reduction of say ten pounds is made on the trainpipe, the triple will automatically lap itself as soon as a fraction over ten pounds has left the auxiliary.

This is done as follows: when the trainpipe pressure is reduced below that in the auxiliary the triple piston moves and carries with it the graduating valve, for, as you will see by looking at plate 1, the graduating valve is connected directly to the stem of the triple piston by a small pin, as shown by the dotted lines, and, when the piston moves, the graduating valve is carried from its seat in the slide valve and opens port p, so that when the slide valve is in service position the auxiliary air can pass through the slide valve by way of ports l and p, then through port f in the seat of the slide valve and on through pipe connection X direct into the brake cylinder; as only ten pounds was drawn from the trainpipe, just as soon as a fraction over ten pounds flows from the auxiliary, the trainpipe pressure being now the strongest forces the triple piston towards the auxiliary end of its cylinder, but it can only force it a very short distance, for the reason that the distance between the end of the slide valve and the shoulder on the stem of the piston is only three-sixteenths of an inch, and when the piston has moved this distance it is stopped by the slide valve, because the auxiliary pressure, aided by the U spring, is firmly holding the slide valve, on account of the friction being greater on the slide valve seat than it is around the edge of the triple piston, and when the piston is thus stopped by the slide valve, the graduating valve is now back on its seat, and no more air can flow from the auxiliary into the brake cylinder, until the trainpipe pressure is again reduced and the graduating valve again unseated by the movement of the triple piston.

The slide valve does not move when the second reduction is made, but stands in the same position it assumed on the first reduction. Consequently, as soon as the graduating valve is unseated the air will again flow into the brake cylinder; but when the air in the brake cylinder finally becomes as strong as it is in the auxiliary (or equalizes) the pressure in the auxiliary no

ITS USE AND ABUSE

longer falls below that in the trainpipe, and therefore the graduating valve remains off its seat, because the triple piston does not now move back as it did when the first reduction was made, as the pressure in the trainpipe is now as low or lower than it is in the auxiliary, and the brakes are now fully applied.

Hence we can make a full service application without the graduating valve, but we must have this valve in making a "partial service application."

If the engineer simply wants to slow his train up, but does not want to come to a full stop, he can draw off any amount of air from the trainpipe he desires, and when he laps his brake valve, the triple valve will, by means of the graduating valve, let a corresponding amount of air from the auxiliary into the brake cylinder and automatically lap ports *l-p-p* in the slide valve, but if the engineer should draw his trainpipe pressure down below the point at which the auxiliary and brake cylinder equalize, he would not only be wasting the trainpipe pressure, but would have trouble when it came time for him to release his brakes as will be explained later on.

We now understand what the graduating valve

MODERN AIR-BRAKE PRACTICE

is for; now let us see what the graduating stem and spring has to do with it.

As I have already mentioned, the third kind of an application is called the "emergency." When this kind of application is made it is only in case of danger, and therefore it is desired that the air in the auxiliary should be passed into the brake cylinder as quickly as possible, and in order to do this it is necessary to have the entire slide valve clear the port in the seat through which the air has to pass.

In making ordinary stops this very quick action is not required, and in order to prevent the slide valve making the full stroke, there is a projection on the trainpipe side of the triple piston which strikes against the graduating stem (26), and as this stem is held to its seat by the graduating spring (27), the strength of this spring combined with the pressure in the trainpipe causes the triple piston to stop, and in doing so the slide valve is held in such a position that port p is in register with port f, and of course the brakes are applied gradually.

But if the pressure in the trainpipe is reduced suddenly, the auxiliary pressure causes the triple piston to strike the graduating stem a hammer blow and overcomes the tension of the spring

ITS USE AND ABUSE

so that the slide valve entirely clears the port in the seat, and the auxiliary pressure immediately equalizes with the brake cylinder. (This refers to the plain triple. The emergency action of the quick-action triple will be described later on.)

The U spring (32) is placed over the slide valve for the reason that if the brake is applied and all the air is let out of the trainpipe, and the car cut off from the engine, the brake could not be "bled" off by the release valve on the auxiliary if the slide valve could not be lifted off its seat by the brake cylinder pressure, but as there is a slight lift to the slide valve for this purpose, the U spring is required to reseat the valve, so that when the auxiliary is again recharged no air can get under the slide valve and pass out to the atmosphere through port *h* in the valve seat.

If there is a great deal of oil on the slide valve seat it will prevent the slide valve from being forced up by brake cylinder pressure, when a single car is being "bled off," and the brake cannot be released at all until the air finally leaks out around the packing leather in the cylinder. In such a case the release signal is very handy.

MODERN AIR-BRAKE PRACTICE

THE WESTINGHOUSE QUICK-ACTION TRIPLE VALVE

So far I have only spoken of the plain triple, but as all cars are now supposed to be equipped with the "quick-action triple" we will next ascertain what is the difference between the two kinds of triples, and what the advantage is in having the quick-action triple.

When an engineer applies the brakes he has to draw the trainpipe pressure down by letting it escape to the atmosphere through a port in the brake valve, and as the triple pistons will not move until the trainpipe pressure is reduced below that in the auxiliary reservoirs, it naturally follows that on a train, of say thirty cars, equipped with plain triples, the brakes on the head end will set before the ones on the rear end, for the reason that the air in the front end of the trainpipe has to get out of the way before the air in the rear end can escape, and whenever the pressure on the trainpipe side of any triple is reduced lower than the auxiliary side, that triple will move and set the brake at once, and the main difference between the plain and quick-action triple is that the trainpipe pressure can be reduced faster with a "quick-action" triple than it can with a plain one, and consequently

ITS USE AND ABUSE

the brakes on a long train can be applied more rapidly with "quick-action" triples.

To make this plain to you, suppose you were in a crowded opera house and a cry of "fire" was heard, as it recently happened in Chicago, everybody would make a rush for the front door at once, but as the door would only let so many out at a time, those in the rear would have to wait until those in front got out first, and if it was a bad fire the result would be a horrible catastrophe. This refers to the plain triple.

Now suppose that the opera house was so built that in addition to the regular front entrance there was another big door in the side of the building which opened into a large hall, then when the cry of fire was heard a portion of the audience would escape through the regular front entrance and the others would get out through the side door, thus emptying the burning building so quickly that everybody is saved. This refers to the quick-action triple, for with the plain triple there is but one way of getting the trainpipe pressure away from the triple piston, and that is through the brake valve (the front door), but with the quick-action triple there is an extra outlet through which the trainpipe pressure can escape when an emergency appli-

MODERN AIR-BRAKE PRACTICE

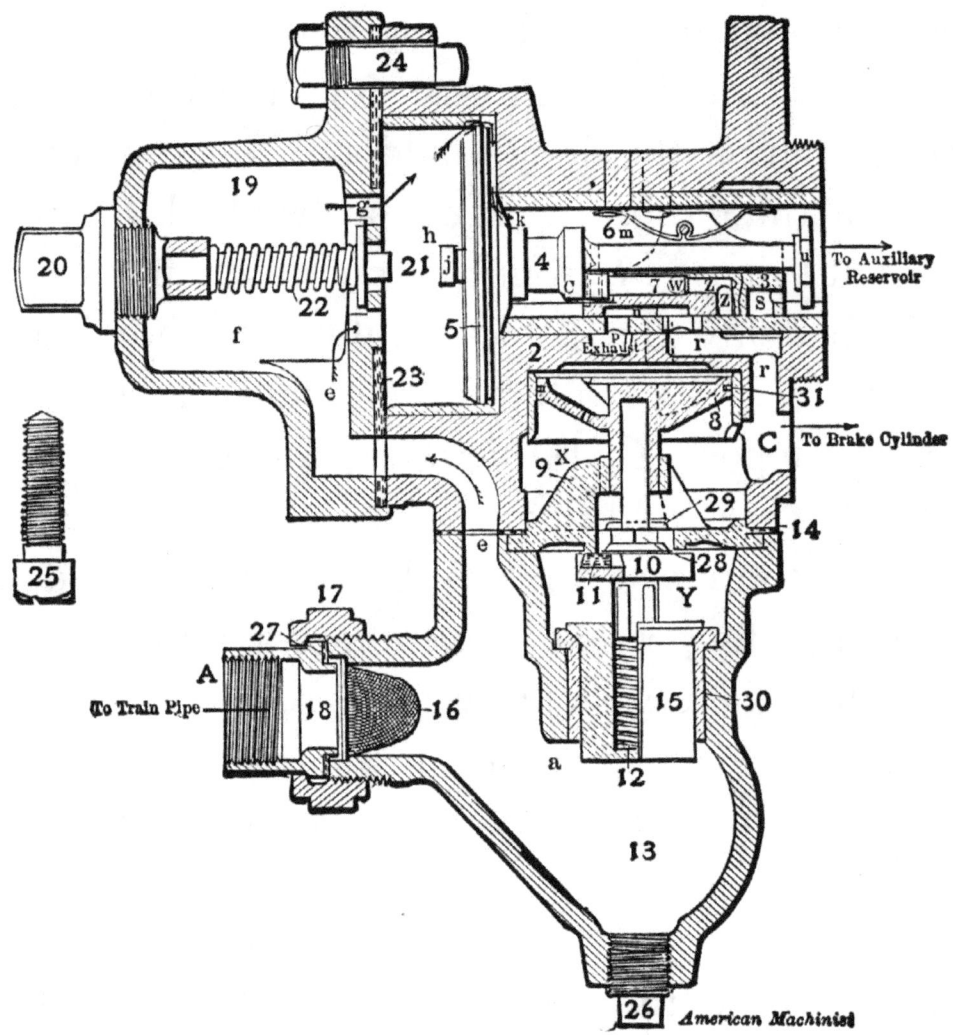

PLATE NO. 2—QUICK-ACTION TRIPLE IN RELEASE AND CHARGING POSITION.

ITS USE AND ABUSE

DESCRIPTION OF PLATE **2** — QUICK-ACTION TRIPLE VALVE, RELEASE AND CHARGING POSITION

A. Trainpipe connection.
B. Auxiliary reservoir connection.
C. Cylinder connection.
3. Slide valve.
4. Triple piston and stem.
5. Triple piston packing ring.
6. U or slide valve spring.
7. Graduating valve.
8. Emergency valve piston.
9. Emergency valve seat and guide.
10. Rubber seated emergency valve.
12. Check valve spring.
14. Check gasket.
15. Check valve.
21. Graduating stem.
22. Graduating spring.
23. Triple gasket.

The air passages and ports are described in the text. The feed groove, i, is now open.

MODERN AIR-BRAKE PRACTICE

cation is made, and thus cause the brakes on the entire train to be applied in about two seconds. This extra outlet is called the "emergency valve," and will be explained when I describe plate 5.

The parts contained in the quick-action triple which are not in the plain one, are shown in plates 2, 3, 4 and 5, and are indicated as follows: The emergency piston is marked 8; the guide for this piston, which also forms a seat for the emergency valve, is marked 9; the emergency valve is 10; the check-valve spring is 12; the check valve is 15, and the gasket which separates chamber X from chamber Y is marked 14. This gasket, you will notice, extends clear across the triple, but a portion of it is cut away just over the emergency valve, so that when that valve is unseated, as it is in an emergency application, the air in chamber Y can pass into chamber X and the brake cylinder, and another hole is cut in this same gasket at e, so that the train-pipe pressure, which enters the triple at A, can pass freely into chambers f and h.

PLATE 2 — QUICK-ACTION TRIPLE IN RELEASE AND CHARGING POSITION

The quick-action triple has five positions: release, charging, service, lap and emergency.

ITS USE AND ABUSE

Release and charging positions are really one and the same, and are shown in plate 2. While the air is being released from the brake cylinder by way of the ports in the slide valve seat, etc., as previously described in plate 1, the auxiliary is being charged by way of the feed grooves described in plate 1 as *m* and *n*, but in plate 2 they are marked *i* and *k*.

In plate 2 you will observe that a different set of figures and letters are used from those employed in plate 1, to point out the different ports, etc., but this need not worry you, for, as the poet says, "a rose by any other name would smell just as sweet," and whether you call it cylinder *h*, as in plate 2, or B, as in plate 1, you know that it is the cylinder in which the triple piston moves. So, to clear you up on this point, notice that in plate 2 the trainpipe connection to the triple is marked A, while in plate 1 it is W. Now look at the arrows in plate 2 and you will see that after the air enters the triple at A it passes through a passage in the casing, the same as in plate 1, to a chamber having two openings into the cylinder containing the triple piston, just like plate 1, and from this cylinder the air passes through the same two feed grooves that in plate 1 are marked *m* and *n*, but

MODERN AIR-BRAKE PRACTICE

in plate 2 are marked *i* and *k*, on into the slide-valve chamber, and instead of entering the auxiliary at the pipe connection Y, as in plate 1, it passes right on through the slide-valve chamber into the auxiliary, so you see whether it is a plain or quick-action triple the auxiliary pressure is always on the slide-valve side of the triple piston, and trainpipe pressure is on the opposite side.

Having familiarized yourself with the parts of the triple as described in plate 1, you will see by plate 2 that the same parts are contained in the quick-action triple, and perform the same duties, so that the only difference between the two kinds of triples is the emergency attachment (which I have explained by reference to Figs. 8, 9, 10, 12, 14 and 15), and in charging an auxiliary or releasing a brake the air has to travel the same routes whether a plain or quick-action triple is used, but in setting the brakes in emergency is where the difference comes in between the two kinds of triples. (See plate 5.)

PLATE 3—SERVICE POSITION OF QUICK-ACTION TRIPLE VALVE

In this position you will notice that the triple piston has moved in its cylinder until the projec-

ITS USE AND ABUSE

tion *j* strikes against the graduating stem, which stops it, and in making this movement the stem of the piston has drawn the slide valve to a position which places the port marked *w*, *z-z* in register with the port in the seat marked *r*, thus allowing the auxiliary pressure to pass into the brake cylinder through pipe connection C and set the brake. (If this is not perfectly clear to you, read again what I said about setting the brakes in my description of plate 1.)

PLATE 4 — LAP POSITION OF QUICK-ACTION TRIPLE

Lap position means that all ports are closed, and the reason why the triple automatically laps itself is due to the fact that when the slide valve is moved to service position the graduating valve is held off its seat at *w* by the triple piston and when the pressure in the auxiliary becomes a little less than trainpipe pressure the piston is forced back by the trainpipe pressure until the graduating valve strikes its seat in the slide valve, and stops the flow of the auxiliary air into the brake cylinder.

The reason the slide valve is not moved when the graduating valve moves is because the auxiliary and trainpipe pressures are so nearly equal that the friction of the slide-valve seat

MODERN AIR-BRAKE PRACTICE

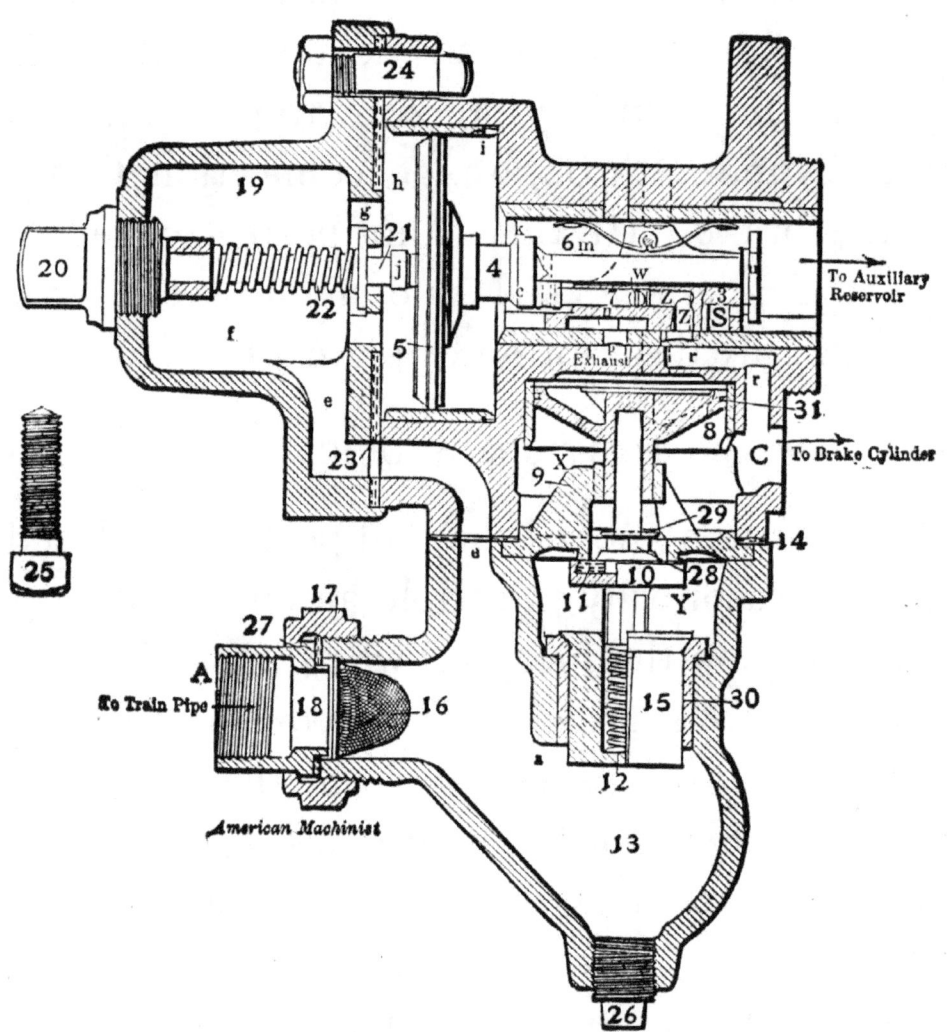

PLATE NO. 3—QUICK-ACTION TRIPLE IN SERVICE POSITION.

ITS USE AND ABUSE

DESCRIPTION OF PLATE 3—QUICK-ACTION TRIPLE VALVE, SERVICE POSITION

A. Trainpipe connection.
B. Auxiliary reservoir connection.
C. Cylinder connection.
3. Slide valve.
4. Triple piston and stem.
5. Triple piston packing ring.
6. U or slide valve spring.
7. Graduating valve.
8. Emergency valve piston.
9. Emergency valve seat and guide.
10. Rubber seated emergency valve.
12. Check valve spring.
14. Check gasket.
15. Check valve.
21. Graduating stem.
22. Graduating spring.
23. Triple gasket.

The feed port, i, is now closed.

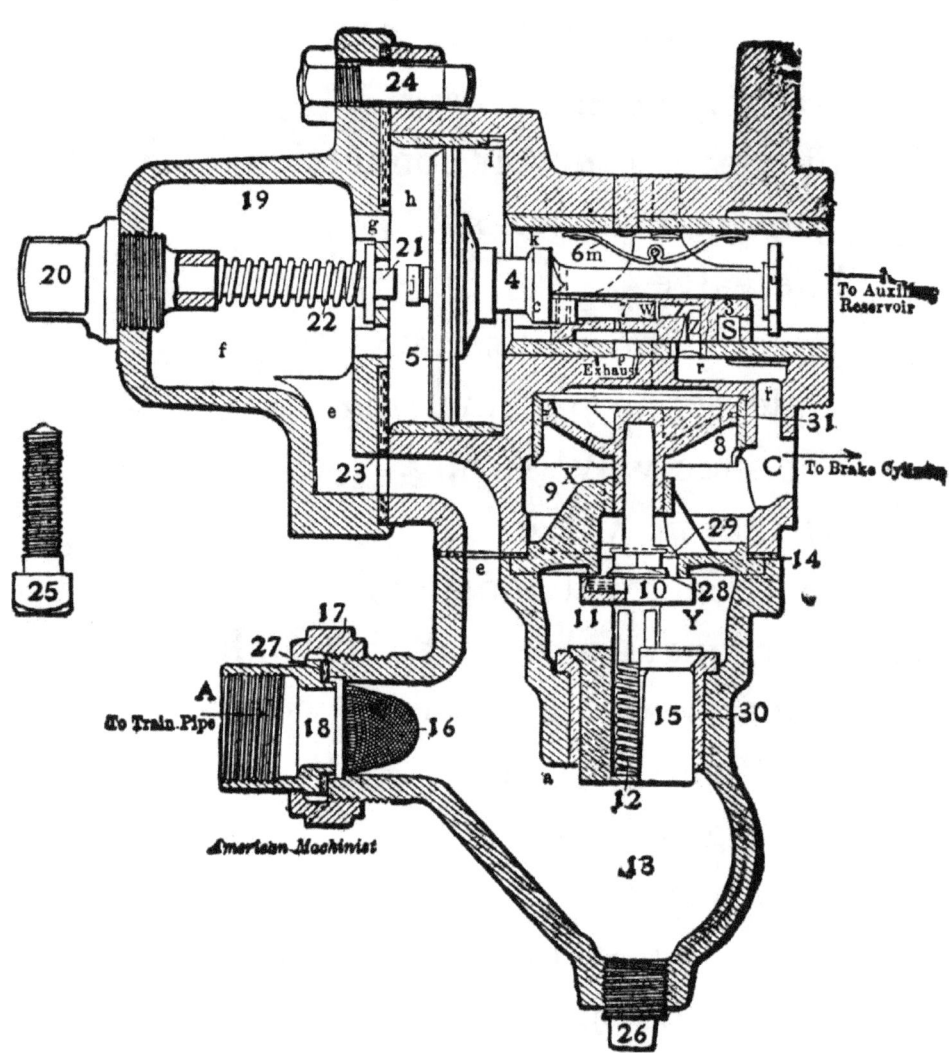

PLATE NO. 4—QUICK-ACTION TRIPLE IN LAP POSITION.

ITS USE AND ABUSE

DESCRIPTION OF PLATE 4—QUICK-ACTION TRIPLE VALVE, LAP POSITION

A. Trainpipe connection.
B. Auxiliary reservoir connection.
C. Cylinder connection.
3. Slide valve.
4. Triple piston and stem.
5. Triple piston packing ring.
6. U or slide valve spring.
7. Graduating valve.
8. Emergency valve piston.
9. Emergency valve seat and guide.
10. Rubber seated emergency valve.
12. Check valve spring.
14. Check gasket.
15. Check valve.
21. Graduating stem.
22. Graduating spring.
23. Triple gasket.

All ports are now closed.

combined with the tension of the slide valve spring (marked 6, in plate 5), prevents it, and as this keeps the exhaust port closed, and the position of the triple piston keeps feed groove i closed, all ports are now closed and the valve is said to be on lap.

Remember, the triple will not lap itself unless the auxiliary pressure has a chance to get lower than trainpipe pressure, which means that if an engineer reduces his trainpipe pressure below the point at which the auxiliary and brake-cylinder pressures equalize, the only means of holding the air in the brake cylinder (aside from the packing leather around the brake piston and the closing of the triple exhaust) is the check valve (15), or the packing ring (30) of the triple piston, for while it is true that the piston would seat against gasket 23, still this gasket so soon becomes hard that it cannot be relied upon to stop the auxiliary pressure from flowing back into the trainpipe.

The reason the check valve has to be depended upon to keep the brake-cylinder pressure from flowing back into the trainpipe, after an extra heavy reduction has been made, is because nine times out of ten the air in chamber Y will reduce as fast as the trainpipe pressure is reduced, on

ITS USE AND ABUSE

account of the volume in Y being so small that the slightest possible leak in the seat of the check valve will let it out, and after the trainpipe pressure has been drawn down sufficient to allow the auxiliary and brake cylinder to equalize, the leak from chamber Y is supplied by the brake cylinder, for whenever the pressure in Y becomes less than that in the brake cylinder the emergency valve (10) is forced off its seat by the brake-cylinder pressure until it equalizes again with chamber Y, when the spring (12) reseats valve 10, which is done very quickly, consequently if the trainpipe pressure was entirely exhausted and the check valve leaked very bad the brake cylinder would very quickly be robbed of its pressure, and let the brake off. It is, therefore, very bad practice to ever reduce the trainpipe pressure below the point at which the auxiliary and brake cylinder equalizes, except in an emergency.

In making an emergency application the check valve is raised off its seat 120 times a second.

PLATE 5—EMERGENCY POSITION OF QUICK-ACTION TRIPLE VALVE

A sudden reduction of trainpipe pressure is necessary to cause the triple to assume emergency position.

MODERN AIR-BRAKE PRACTICE

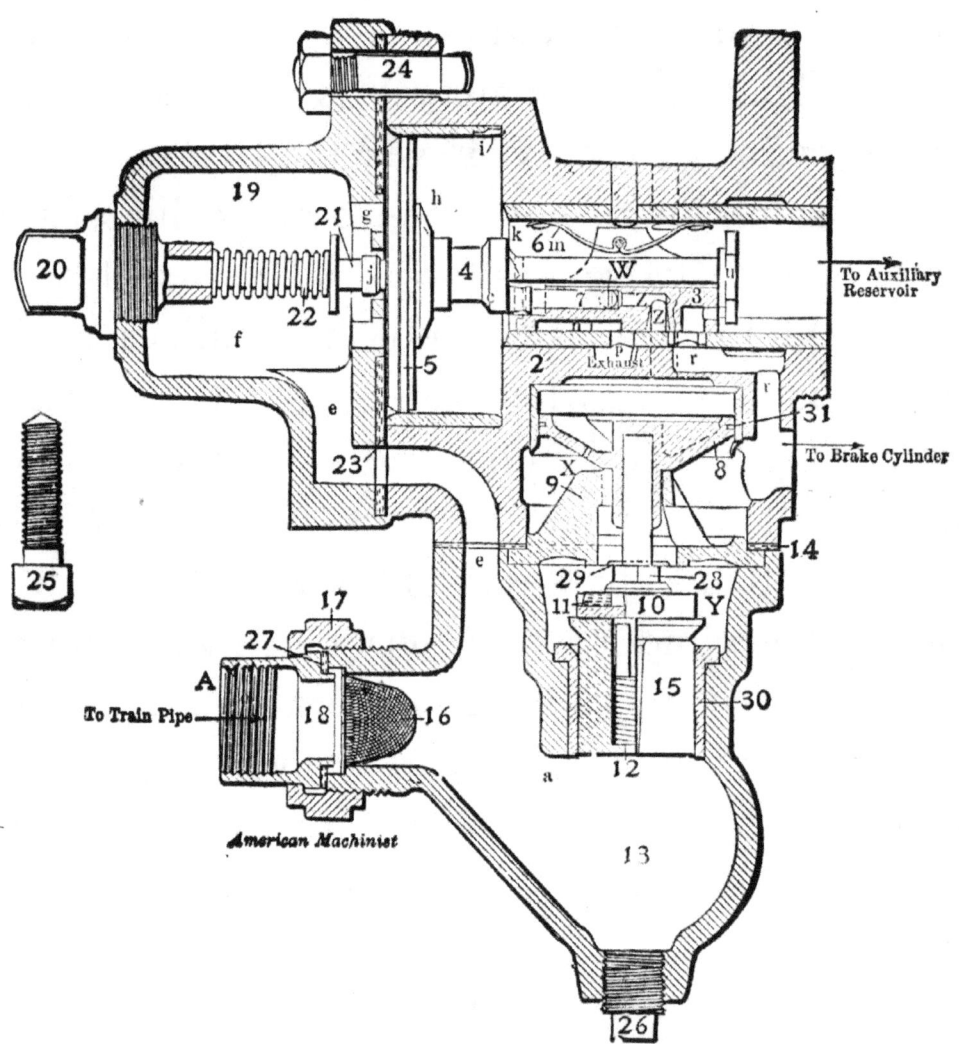

PLATE NO 5—QUICK-ACTION TRIPLE IN EMERGENCY POSITION.

ITS USE AND ABUSE

DESCRIPTION OF PLATE 5 — QUICK-ACTION TRIPLE VALVE, EMERGENCY POSITION

- A. Trainpipe connection.
- B. Auxiliary reservoir connection.
- C. Cylinder connection.
- 3. Slide valve.
- 4. Triple piston and stem.
- 5. Triple piston packing ring.
- 6. U or slide valve spring.
- 7. Graduating valve.
- 8. Emergency valve piston.
- 9. Emergency valve seat and guide.
- 10. Rubber seated emergency valve.
- 12. Check valve spring.
- 14. Check gasket.
- 15. Check valve.
- 21. Graduating stem.
- 22. Graduating spring.
- 23. Triple gasket.

MODERN AIR-BRAKE PRACTICE

When a sudden reduction is made it causes the triple piston (4) to strike the graduating stem (21) such a hammer blow that the graduating spring (22) is unable to stop it from making its full stroke, and as it has now traveled further than it did in service position, the slide valve has also been moved a correspondingly greater distance on its seat, which brings a big slot, or in some triples, a removed corner (not shown) in the slide valve over a port in the seat (indicated by dotted lines behind port Z), and allows the auxiliary pressure to fall on the emergency piston (8), which strikes the stem of valve 10 and forces it from its seat (which is kept closed by spring 12 and the trainpipe pressure in Y), and valve 10 being thus unseated, the air from Y rushes into the brake cylinder.

As all this is done so very quickly that the trainpipe pressure has as yet reduced but very little, the remaining trainpipe pressure forces the check valve up and also rushes into the brake cylinder until it equalizes with what is left in the trainpipe, when spring 12 reseats the check valve, preventing the air in the brake cylinder from flowing back into the trainpipe.

At the same time that the big slot in the back of the slide valve reached its position over the

ITS USE AND ABUSE

port in the seat leading to the emergency piston, another small port in the slide valve, marked S in plate 4, is placed in register with port r in the valve seat, taking the place of port Z, which allows the auxiliary pressure to flow into the brake cylinder on top of what went in from the trainpipe.

The opening around the emergency valve is so much larger than port S in the slide valve that virtually no air enters the brake cylinder from the auxiliary until the check valve closes on the charge received from the trainpipe.

It is this air from the trainpipe that gives the added percentage of brake power after an emergency application; for the air which enters the brake cylinder from the trainpipe has the same effect as shortening the piston travel, because it forces the auxiliary pressure to equalize just that much higher than it would if the brake cylinder was empty when the auxiliary pressure started to flow into it.

On account of the trainpipe pressure having two outlets (one by way of the brake valve, and the other by way of valve 10) when an emergency application is made, it is reduced so suddenly that the next triple is thrown into quick action, because the pressure that was holding

MODERN AIR-BRAKE PRACTICE

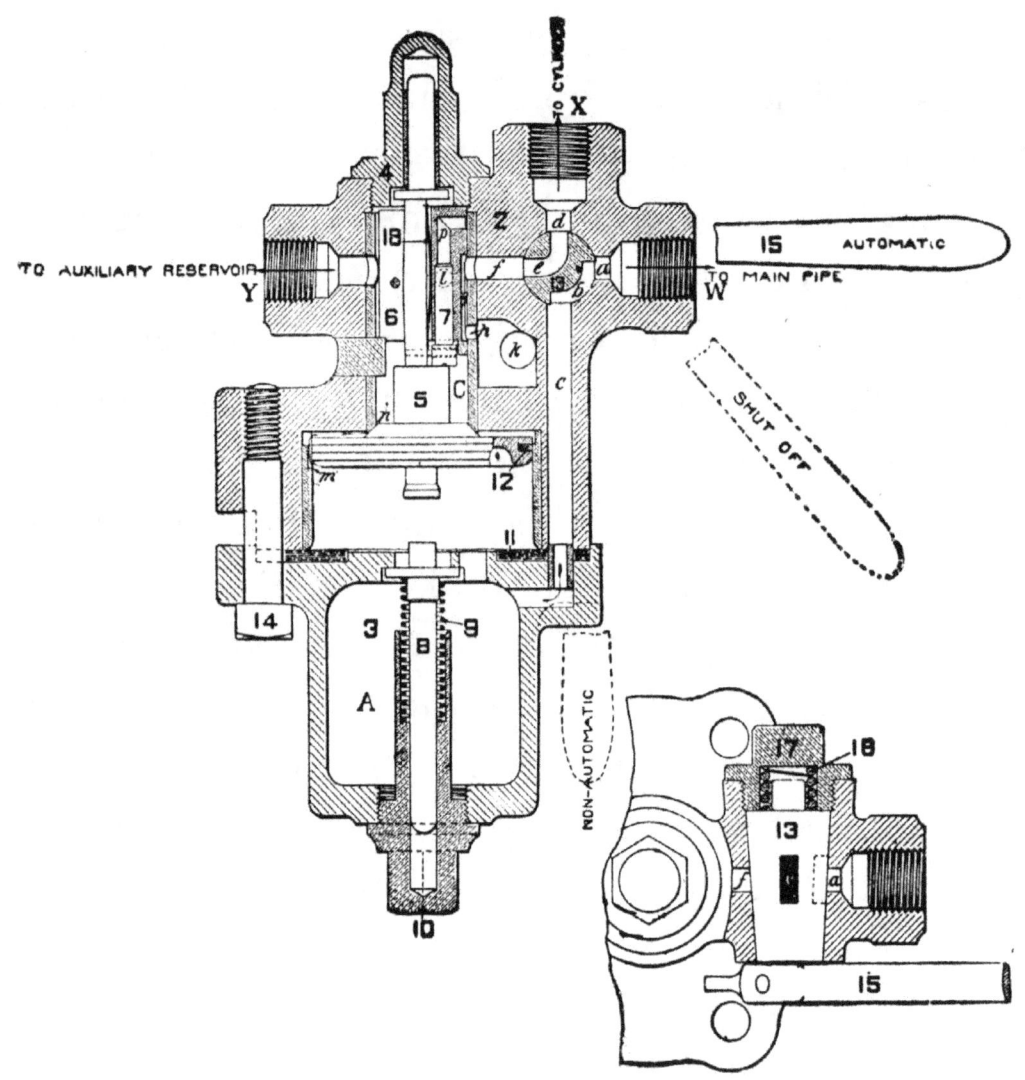

PLATE NO. 6—PLAIN TRIFLE VALVE. (OLD STYLE.)

ITS USE AND ABUSE

DESCRIPTION OF PLATE 6—PLAIN TRIPLE VALVE, RELEASE POSITION (OLD STYLE)

W. Trainpipe connection.

X. Cylinder connection.

Y. Auxiliary reservoir connection.

15. Handle of cut-out plug 13.

18. Slide valve.

5. Triple piston.

7. Graduating valve.

8. Graduating stem.

9. **Graduating spring.**

that triple to release position immediately rushes back into the empty space just created in the trainpipe by the first reduction, and as it can't be in both places at the same time, the triple is left without sufficient trainpipe pressure to hold it, when the pressure on the auxiliary side of that triple piston drives it to emergency position, which in turn creates a vacancy in the trainpipe on that car which the next car tries to fill, and so on, till all the brakes on the entire train are set in emergency, and it all happens so quick that the triples on a train of fifty cars can be thrown into quick action in about two seconds.

PLATE 6—PLAIN TRIPLE, OLD STYLE

This plate illustrates the common form of plain triple, and before the advent of the quick-action triple, it was the standard for passenger cars. It is now sometimes used on driver and tender brakes having cylinders of ten inches, or less; but with larger cylinders the new plain triple, as shown in plate 1, is used.

The principal difference between these two kinds of plain triples is the arrangement of the cut-out cock. In plate 6 you will notice that the cut-out cock is attached right to the triple, and by turning the handle, which controls plug 13, you can make the triple work "automatic" by placing it horizontal, and to cut it out place it at an angle of forty-five degrees; to make it work "straight air" place the handle perpendicular, for then plug 13 is turned so that the end of the passage which is shown to be in register with port d, would then be in register with port a, and the other end of e would register with d, which would allow trainpipe pressure to flow direct into the brake cylinder through ports a, e and d; in other words, the triple valve proper and auxiliary reservoir would not be used when the handle was turned for "straight air." This is so seldom done nowadays that there is a lug cast on the handles of all such plain triples to prevent cutting them in straight air.

When it becomes necessary to bleed off a brake that is set with a plain triple of this kind, drain the auxiliary before closing the cut-out cock, for, when cut out, the position of the passage is changed so that the air in the brake cylinder cannot escape through the triple exhaust.

With the new plain triple, plate 1, the cut-out cock is on the pipe leading from the triple to the brake cylinder. The ports in this triple are made large to accommodate a large volume of air.

In the new plain triple the ports are necessarily larger, on account of handling a greater volume of air.

PLATE 7—TRIPLE VALVE, AUXILIARY RESERVOIR AND BRAKE CYLINDER COMBINED

This plate illustrates a freight equipment. The brake cylinder (2) is bolted directly to the auxiliary reservoir (10), and while the supply pipe (*b*) runs through the auxiliary and into the cylinder, still the air in the auxiliary cannot get into the cylinder except by way of the ports in the triple, as previously described, for the left end of pipe *b* is connected with the triple at C, as shown in plate 4.

The gasket between the auxiliary and brake cylinder is not for the purpose of separating them, but is to make the cylinder air-tight at that end, and when the brakes are set the other end of the cylinder is made air-tight by the packing leather (7) around the piston head (3) which is held to its place by the expansion ring (8) and follower (6). Spring 6 is to force the pis-

ITS USE AND ABUSE

ton back when the air is let out of the cylinder.

To prevent the brakes from setting on account of trainpipe leaks, there is a small leakage groove (*a*) cut in the wall of the cylinder for about three inches from the extreme left end, or pressure head, so that any small amount of air that might be let into the cylinder through the triple, from any cause, would escape to the atmosphere, instead of pushing the piston out, by passing through the leakage groove by the piston head, and out around piston 3.

The Release Valve (17) or "bleeder," is for the purpose of drawing the air from the auxiliary, and when a car is set out, and *especially* when a brake is cut out, the release valve should be held open until all the air in the auxiliary has escaped, for if any air is left in it the brake will again set whenever the trainpipe pressure is reduced lower than that in the auxiliary. Whenever a brake cannot be released from the engine, but has to be "bled off," either at the auxiliary, or by the release signal valve inside of the car, always cut that brake out at the first opportunity and *drain* the auxiliary.

MODERN AIR-BRAKE PRACTICE

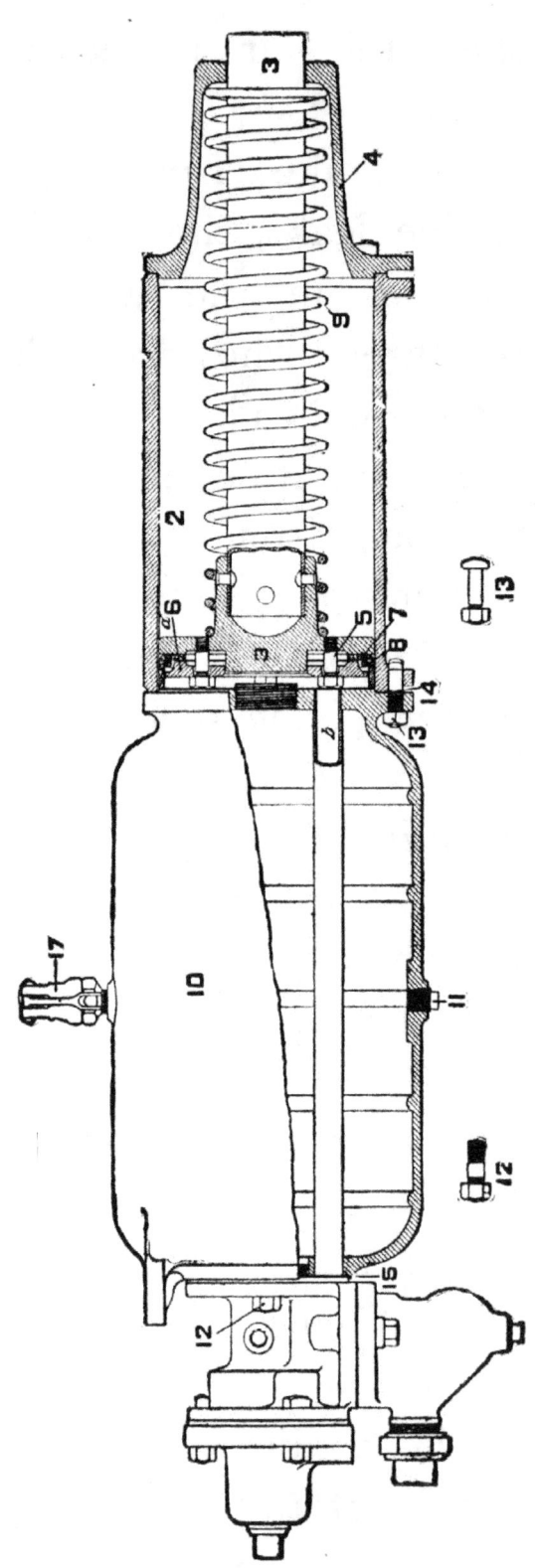

PLATE 7—TRIPLE VALVE, AUXILIARY RESERVOIR AND BRAKE CYLINDER COMBINED

ITS USE AND ABUSE

DESCRIPTION OF PLATE 7—TRIPLE VALVE, AUXILIARY RESERVOIR AND BRAKE CYLINDER COMBINED

2. Brake cylinder.
3. Brake piston.
4. Non-pressure head of brake cylinder.
6. Follower.
7. Packing leather.
8. Expansion ring.
9. Release spring.
10. Auxiliary reservoir.
11. Drain plug.
17. Release valve (or bleed cock).
a. Leakage groove.
b. Supply pipe, between triple and cylinder.

MODERN AIR-BRAKE PRACTICE

More money is paid out annually by railroad companies on account of freight wrecks caused by bleeding off "stuck" brakes than would pay for a tolerably good railroad, for the reason that where a car is not equipped with a release signal on top, and a brakeman bleeds the auxiliary as the train is pulling out and then climbs aboard, that brake is almost sure to stick again, while the train is moving too fast to allow the brakeman to get at the auxiliary bleed cock, and as a consequence the wheels are either flattened, draw-heads pulled out, train stalled, or the wheels become overheated, so that they burst and wreck the train. Where cars are equipped with the new release signal the brakeman can keep a brake off that is inclined to stick, until the train is in a safe position to allow him to get down and cut the brake out.

If the auxiliary release valve leaks and it cannot be stopped by one or two quick jerks, to dislodge the dirt that is causing it to leak, cut the brake out, as no air can accumulate in the auxiliary, thus making that brake worthless, but the leak is drawing air from the trainpipe, which affects the rest of the brakes. Should the release valve become clogged so that no air could be drawn through it, you can remove the

ITS USE AND ABUSE

drain plug (11) in the under side of the auxiliary. This plug will not have to be removed, of course, where a car is equipped with the release signal as the brake cylinder can be emptied independent of the action of the triple, by simply pressing down on the valve of the release signal.

PLATE 8—PRESSURE-RETAINING VALVE

Many enginemen and trainmen utterly fail to realize the importance of this little device, and in view of the wonderful aid it is to handling trains down heavy grades, it is surprising that, by the average man, it is less understood than almost any part of the equipment.

A true story is told of an engineer who had just made a stop at the foot of a heavy grade on which the brakeman had turned up a few "retainers," and just as he was about to pull out, the brakeman asked if he should turn the retainers down, when the engineer "hollered" back, "No, you needn't mind, I can kick 'em off with my brake valve."

Now let us see if he could kick them off. In the first place a retaining valve, as the name implies, is for the purpose of retaining a certain amount of pressure in the brake cylinder after the triple valve has been moved to release position. It is simply a cork for the triple exhaust, and when you look at plate 8 you will readily understand this.

Into the triple exhaust a small pipe is attached and extends from the triple to the top of the car at the end where the hand-brake staff is, and onto this pipe is attached the retaining valve at the connection marked X. The handle (5) controls a plug (6) similar to the cut-out plug (13) in the plain triple. When the handle is turned as you see it in plate 8, port c through the plug is in register with port b-b, and the air which comes from the triple exhaust is forced against the seat of the valve 4, which raises and allows the pressure to escape to the atmosphere through port d. As port d is controlled by valve 4, the air will exhaust only while this valve is up, and as the weight of the valve, combined with the size of the ports, requires a pressure of fifteen pounds to keep it up, just as soon as the pressure in the brake cylinder has been reduced to a fraction less than fifteen pounds to the square inch, the valve will seat and retain the remaining pressure in the brake cylinder until the handle is turned down. When the handle is turned down it brings port a in register with the lower part of b, and port c is turned to

ITS USE AND ABUSE

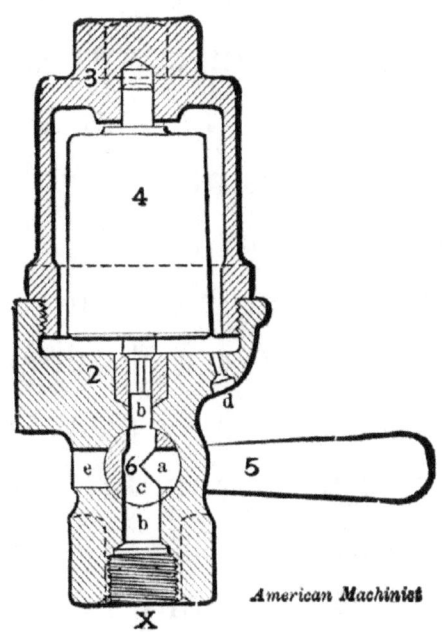

PLATE NO. 8—PRESSURE-RETAINING VALVE.

DESCRIPTION OF PLATE 8 — PRESSURE-RETAINING VALVE, IN RETAINING POSITION

X. Triple exhaust connection.
4. Retaining valve weight.
5. Handle.
6. Cut-out plug.

register with port *e*, and thereby allows all the air in the brake cylinder to escape to the atmosphere.

Therefore if the handle of the retainer is kept turned down the engineer can release the brakes from the engine, but if the handle is turned up (unless the brake leaks off) it will stay set until the handle is turned down.

Retainers were formerly made to hold only ten pounds in the brake cylinder, but are now made to hold fifteen or fifty pounds.

With the retainer handle turned up, the second application of the brakes will give a much higher brake-cylinder pressure, if the auxiliary has been allowed time enough to recharge, because the pressure that is already in the cylinder will force the auxiliary to equalize much higher than it would if the cylinder was empty to start with (in the same manner that the emergency application causes an added pressure on account of the trainpipe pressure entering the cylinder before the auxiliary pressure has a chance to get in). For this reason it is best to apply the brakes and recharge the auxiliaries as soon as possible after passing the summit of a mountain grade, and besides it gives an increased reserve of brake power.

ITS USE AND ABUSE

THE AUTOMATIC SLACK ADJUSTER

The question of correct piston-travel is of the highest importance, and the automatic slack adjuster is for the purpose of keeping it as nearly uniform as possible, which should be eight inches when running.

PLATE 9—SLACK ADJUSTER COMPLETE

Plate 9 shows how the adjuster is attached to the pressure head of the brake cylinder. One end of cylinder lever (5) is bolted to a cross head, which moves in a guide (4) that is bolted to the pressure head of the cylinder. The cross head is held to its place by a threaded rod (1), which has a ratchet nut where its opposite end extends through the adjuster body (3), and when it is desired to reduce the piston travel, it is done by moving the cross head *away* from the cylinder head a distance equal to the amount of slack to be taken up; and to increase the travel move the cross head *toward* the cylinder.

When no air is in the cylinder the threaded rod can be turned either way with a wrench, and four turns of the rod will equal one inch of piston travel.

In running along, whenever the piston travel exceeds eight inches the adjuster automatically

MODERN AIR-BRAKE PRACTICE

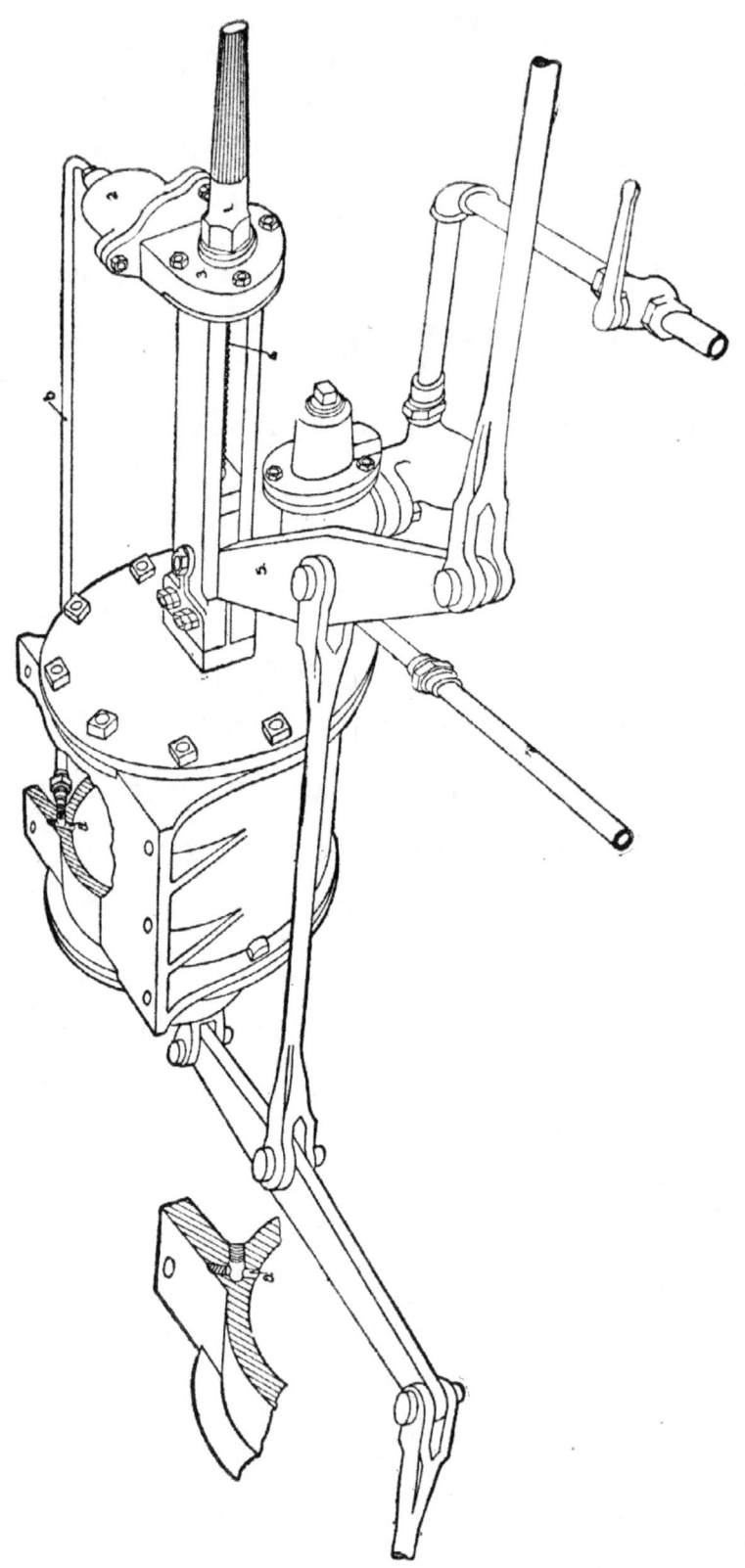

PLATE NO. 9—AUTOMATIC SLACK ADJUSTER, COMPLETE

ITS USE AND ABUSE

DESCRIPTION OF PLATE 9

5. Cylinder lever.
1. Threaded rod.
3. Ratchet-nut wheel casing.
2. Adjuster cylinder.

a and *b*. Pipe connection between brake cylinder and adjuster cylinder.

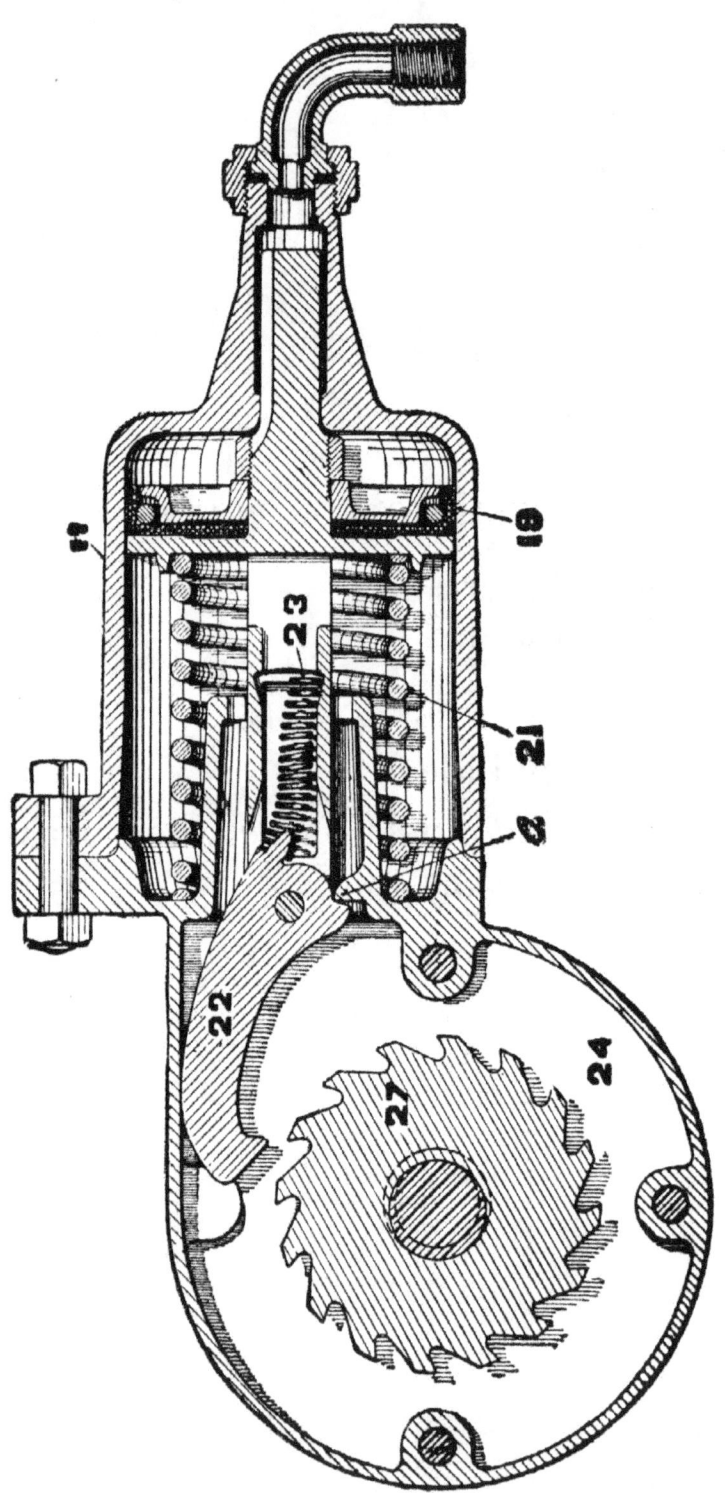

PLATE NO. 10.—AUTOMATIC SLACK ADJUSTER.

ITS USE AND ABUSE

DESCRIPTION OF PLATE 10

27. Ratchet-nut wheel.
22. Pawl.
 a. Projection for lifting pawl.
23. Piston.
21. Release spring.

takes up one thirty-second of an inch every time the brake is released, and therefore whenever new shoes are put on (which necessitates letting the adjuster well back), the brake should be fully applied and whatever travel the piston shows *over* 6½ inches should be taken up by turning the ratchet nut, as the running piston travel is from one to two inches greater than it is when the car is standing still.

Don't try to turn the ratchet nut while the brake is set, and never alter the dead levers or bottom rods unless, with all adjuster slack out, the piston-travel is less than 5½ inches, or when the adjuster has been taken up to its limit and the travel is too long, and not then in the latter case if any brake shoes need renewing.

Plate 10 illustrates the adjuster in cross section. 27 is the ratchet nut which is attached to the threaded rod; 22 is the pawl which moves the ratchet nut; 23 is the piston, to which the pawl is attached, and 21 is the spring which drives the piston back after the cylinder pressure has escaped from in front of it, and as the adjuster cylinder is connected to the brake cylinder by a small pipe, whenever the air in the brake cylinder forces the brake piston out eight inches, brake-cylinder pressure is admitted

ITS USE AND ABUSE

against piston 23, which forces the pawl back so that it engages the ratchet-nut wheel, and when the air is released from the brake cylinder the air in the adjuster cylinder (11) escapes through the non-pressure end of the brake cylinder, and spring 21 pushes the piston and pawl forward, thus turning the ratchet-nut wheel the distance of two teeth, which takes up one thirty-second of an inch of piston-travel. The pawl is released by striking a projection (a), which keeps it up.

Plate 11 illustrates the degree of angularity at which the port in the brake cylinder should be tapped according to the size of the cylinder. As this port is only one-eighth of an inch, it may easily become clogged, so that if the adjuster fails to work you should at once ascertain if the air passages are open between the brake and adjuster cylinders by loosening the union swivel on the adjuster cylinder connection.

Whenever the adjuster has operated to the limit of the screw and the pawl fails to release, so that the ratchet-nut cannot be started back with a wrench, if it be the old style adjuster, remove the ratchet nut cover and carefully pry the piston outward until the pawl can be raised, then slack back the nut about a turn, which will let the piston return to the end of its cylinder

MODERN AIR-BRAKE PRACTICE

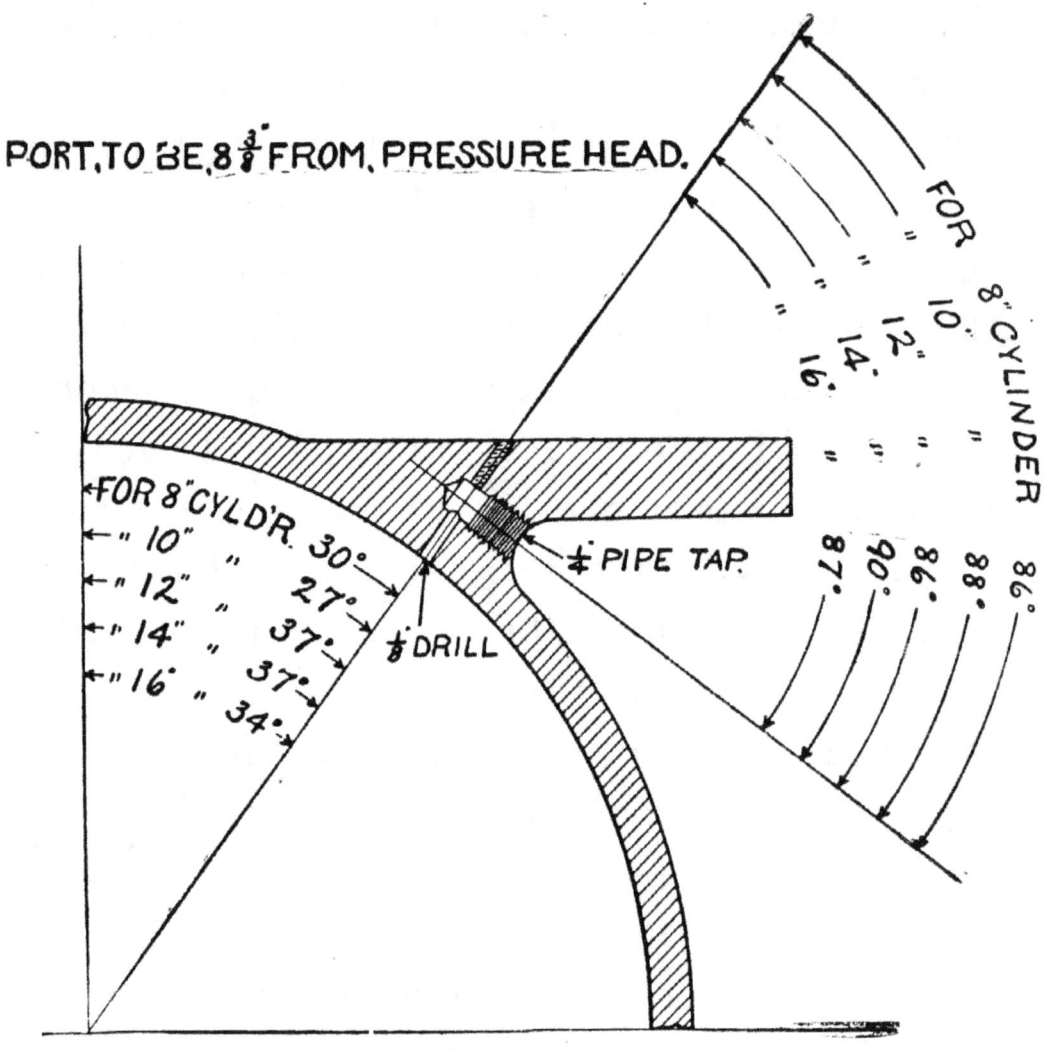

PLATE NO. 11.—AUTOMATIC SLACK ADJUSTER—SIZE OF CYLINDER PORT.

DESCRIPTION OF PLATE 11

The illustration shows the angularity at which the brake-cylinder port should be drilled for the different sized cylinders.

ITS USE AND ABUSE

and keep the pawl free from the ratchet nut as before.

An improvement has lately been added by inserting a stop screw next to the ratchet-nut casing, which holds the threaded rod a short distance from its extreme travel, so that in case the pawl sticks it is only necessary to back out the stop screw, when the pawl will release itself automatically. The adjuster cylinder should be cleaned and oiled every time the brake cylinder is oiled.

PLATES 12 AND 13—THE CAR CONTROL VALVE AND RELEASE SIGNAL

As the high-speed brake and the automatic slack adjuster have both become necessary as the result of changed conditions in the operating of railroads, in like manner, owing to the great increase of railroad traffic and other conditions, the Car Control Valve and Release Signal have also become a necessity, when the saving of time, property and human life are taken into account. Railroads today are not only running very heavy trains, but are running many of them, and, on some roads, they are vastly too close together for the comfort of the train crews who have to operate them. As long as railroads are compelled to employ new men and as long as many of the older employés continue to be indifferent to the proper care, maintenance and operation of the air-brake equipment, just so long will we continue to have dangerous delays to trains, damage to merchandise and rolling stock, and frequent taking of human life, as the direct result of bad handling or bad condition of the air brakes.

The question of keeping the brakes in good order is a continually growing trouble, and to know for a certainty that they are in good condition is of vital importance.

The new device which is meant to largely overcome these distressing conditions is known as the Car Control Valve and Automatic Release Signal, shown in plates No. 12 and No. 13.

As the automatic reducing valve of the high speed brake is attached to the brake cylinder and as the automatic slack adjuster is also operated by brake-cylinder pressure, in like manner the release signal is also connected to the brake cylinder, so that whenever there is air in the brake cylinder that fact is instantly made known by the release signal target.

The release signal is composed of a cylinder in which is contained a piston, piston-rod and return spring, being in reality a miniature of the

ITS USE AND ABUSE

brake cylinder, with this difference: the cylinder of Style A Signal has a square metal signal fastened to it, so that whenever the brake-cylinder pressure causes the brake to apply, the same pressure forces the release signal cylinder up, so that the metal signal is brought to full view of the trainmen, thereby signaling them that the brake on that car is set. When the engineer releases the brake and the brake cylinder is empty, there is then no pressure under the release signal piston, and the return spring forces the metal signal down out of sight. Should the engineer be testing brakes, and, while the brakes were being held on, if the release signal should be seen to gradually drop, it would mean that the brake on that car was in bad order, for the signal would not go down unless the air was escaping from the brake cylinder. Should a signal be seen to drop and no air was found to be coming out through the retaining valve, or triple exhaust, it would mean that the brake was "leaking" off, but if the signal dropped and at the same time air was heard to be escaping through the retainer, or triple exhaust, it would mean that the brake was "releasing" through the triple. This point is very important to remember, for if a

MODERN AIR-BRAKE PRACTICE

brake "releases" when it should stay set, the triple valve, or auxiliary gasket, or auxiliary release valve needs attention. But, if a brake "leaks" off, the packing leather in the brake cylinder needs oiling. Therefore, if you report a defective brake as having leaked off, the repairman will know at once what to do and thereby save considerable time.

Should the signal remain up after the engineer has released the rest of the brakes, it means that the triple on that car is unfit for service, and should this happen while the train is in motion serious trouble is likely to follow. But, as the piston-rod of the release signal is made of a piece of pipe, and as there is a valve on the signal, the brakeman can, by simply opening up the valve, drain the brake cylinder independently of the action of the triple without stopping the train and without getting off.

Should a brake have too much piston-travel the release signal will indicate it, for with the proper piston-travel and an auxiliary pressure of seventy pounds, a twenty-pound trainpipe reduction will cause the signal to go to its full stroke, but the same reduction with too much piston-travel will cause the signal to stop at half-mast. This stopping at half-mast is caused by reason

ITS USE AND ABUSE

of the tension of the return spring, combined with the lowered pressure in the brake cylinder.

The release signal is located in the most convenient place on the inside end of passenger coaches, usually on the wall of the toilet room. There is a metallic casing into which the signal drops when no air is in the brake cylinder, but every time the brake sets the signal is raised above the top of the case, so that it can be seen from either end of the car. For instance, when the brakes are being tested at terminal points, or when any change has been made in the make-up of the train by setting out or picking up cars, the action of the release signal will notify every one alike just what condition the brakes are in.

The release signal needs no more attention than is ordinarily given to the retaining valve, as it is so simple in construction that it is almost impossible for it to get out of order. It needs but little oiling, as the oil from the brake cylinder is usually sufficient to care for the one-inch piston in the release signal cylinder.

The Dukesmith Car Control Valve performs three functions, namely, it applies the brake, retains the brake or releases the brake when the triple fails to go to release position.

The Car Control Valve is located in the same

position as the old style conductor's valve, and performs the same service, in addition to the other two just mentioned.

There are three pipe connections to the Car Control Valve as follows: one from the trainpipe, one from the triple exhaust and one from the brake cylinder direct. There are four positions on the valve which are: Normal, Lap, Release and Emergency Application. When the handle is in normal position all ports except the triple exhaust are closed; when the handle is in lap position all ports are closed, thereby retaining the pressure in the brake cylinder; when the handle is in full release it exhausts the air from the brake cylinder direct without regard to the triple valve, and when the handle is in emergency application position all exhaust ports are closed excepting the trainpipe exhaust, which causes the brakes to apply. With this valve the brakes can be applied gradually or in emergency, according to the quickness with which the handle is moved.

There is nothing about this valve to get out of order as it consists merely of a brass plug working in a cast iron case.

On the brake cylinder pipe leading from the Car Control Valve there is a connection made to the Automatic Release Signal, as shown in plate 12.

ITS USE AND ABUSE

When a passenger car is equipped with this device the brake on that car is absolutely under the control

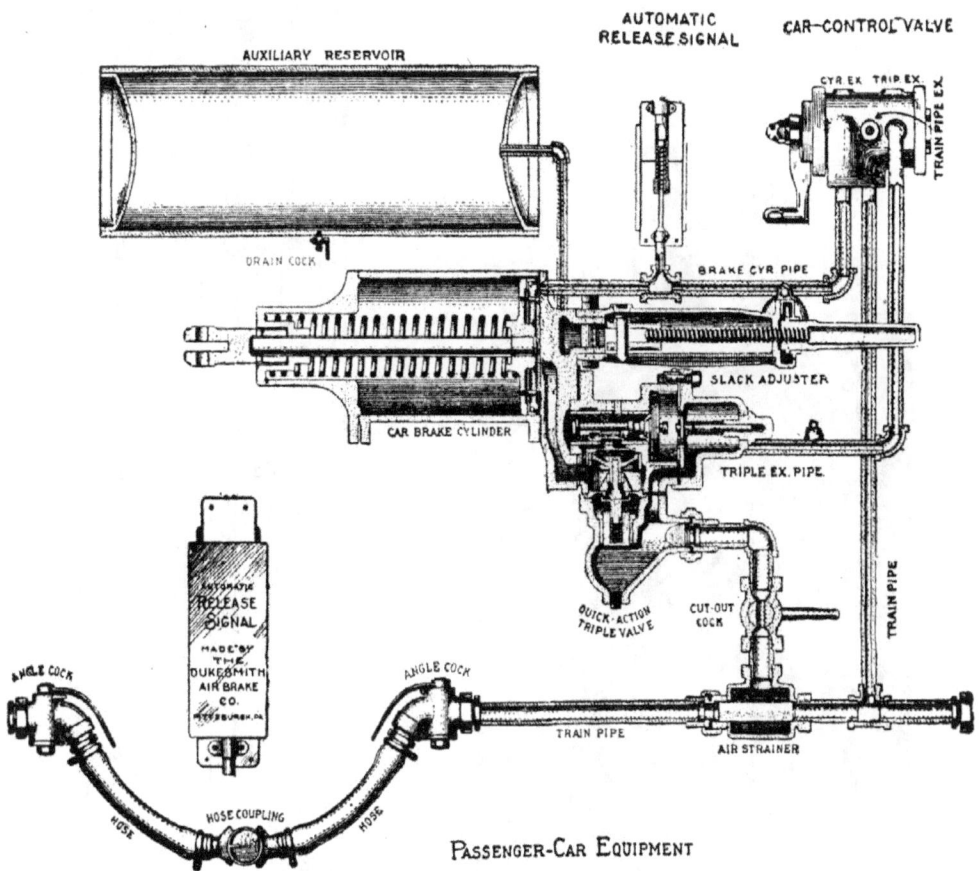

PLATE 12—THE DUKESMITH CAR CONTROL VALVE, SHOWING HOW THE VALVE IS CONNECTED TO THE TRAINPIPE, TRIPLE EXHAUST AND BRAKE CYLINDER; ALSO SHOWING HOW THE RELEASE SIGNAL IS CONNECTED TO THE BRAKE CYLINDER.

of the trainman, for the reason that he can either apply, retain or release the brake as the case may be.

DESCRIPTION OF PLATE 13

Style A Release Signal is shown as it would appear when the brake is partially applied. The front cover can be removed by simply sliding it up.

Style B Release Signal is for use on locomo-

MODERN AIR-BRAKE PRACTICE

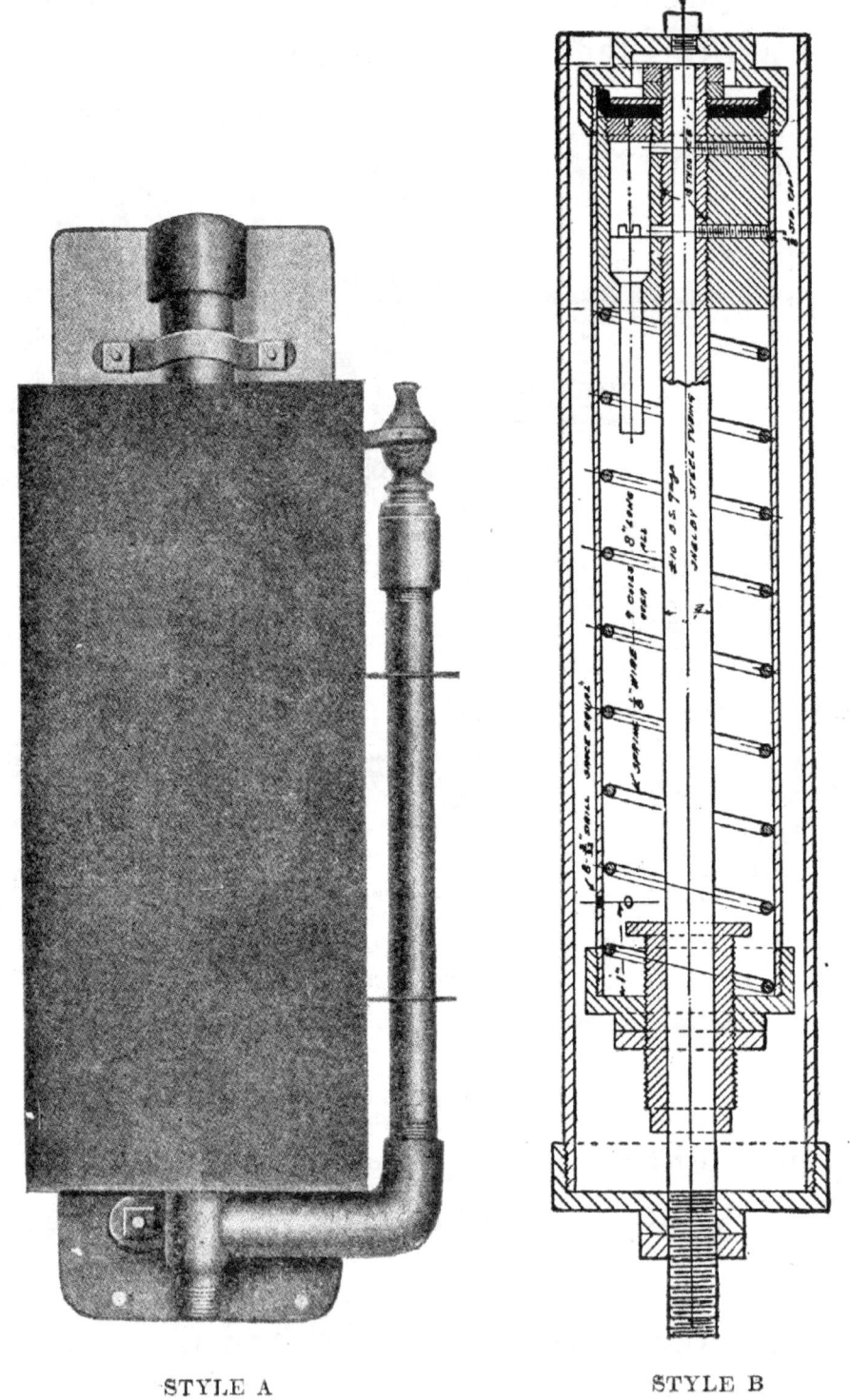

STYLE A STYLE B

PLATE 13—THE DUKESMITH AUTOMATIC RELEASE SIGNAL. STYLE A AS USED FOR PASSENGER CARS, AND STYLE B AS USED FOR ENGINES.

ITS USE AND ABUSE

tives principally, but may be used equally as well on passenger cars. The casing of Style B is circular, in order to occupy as small a space as possible in the cab of the engine. The automatic release feature of Styles A and B consists of a graduating sleeve which strikes the stem of a valve in the signal piston whenever the pressure in the brake cylinder becomes greater than is desired. In the top of the signal cylinder there is an oil plug to permit of oiling the cylinder when required, although it rarely needs to be oiled by hand, for the reason that the oil in the brake cylinder takes care of the signal cylinder.

The Release Signal is located at the end of passenger coaches, usually on the wall of the toilet room in plain view of everyone. There should also be a release signal on each outside end of the coach on opposite sides, enabling inspectors and trainmen to quickly test the air brakes.

Style A Release Signal is provided with a small exhaust valve in order to enable the trainmen to release by hand a brake which has failed to release in the usual way.

THE CONDUCTOR'S VALVE

What is known as the conductor's valve is merely an additional stop cock attached to the trainpipe of passenger coaches.

MODERN AIR-BRAKE PRACTICE

There is a branch pipe running from the trainpipe up through the body of the coach, usually in the toilet room, and on this branch pipe is a stop cock, or valve, so that in case the conductor is unable to signal the engineer, or an emergency arises making it necessary to stop the train as quick as possible, the conductor can let the air out of the trainpipe by simply opening this valve.

If he wishes to make a gradual stop he has only to open the valve gradually, but if he wishes to stop quick, he must open the valve quick, and also must hold it open until the train is stopped, for if the engineer should fail to lap his brake valve, as soon as the conductor's valve was closed the brakes would release, on account of the main reservoir pressure driving the triples to release position. As previously explained the old style conductor's valve is now being replaced by the Car Control Valve which combines in one valve the conductor's valve, the retaining valve and the cylinder release valve.

Having explained the construction and action of the plain and quick-action triple valves, the pressure-retaining valve, the slack adjuster, the release signal, the car control valve and the conductor's valve and having shown how these parts, together with the auxiliary and brake cylinder,

ITS USE AND ABUSE

are combined to form the car equipment, now let us see by what means the air is compressed, where it is stored and how it is manipulated from the engine. This brings us to

PLATE 14—THE EIGHT-INCH PUMP

The eight-inch pump is so called on account of the bore of the cylinders being eight inches. It has two cylinders, the one on top (3) is the steam cylinder, and the one below (5) is the air cylinder. They are joined together by a neck (4), and in the top of the air cylinder and bottom of the steam cylinder there are stuffing boxes (56) through which passes a piston-rod, on each end of which there are piston heads (12 and 13). The piston-rod (10) is hollow for a sufficient depth to admit the stem (17) of the reversing valve (16). The reversing plate (18) is bolted on top of steam piston (10) so that it strikes the button on the stem 17 as the piston approaches the end of its down stroke, and strikes the shoulder of the stem 17 as it makes the up stroke, for the purpose of changing the position of the reversing valve (16), which reverses the stroke of the pump.

The valves through which the air is received and discharged are all in the lower, or air end of the pump.

The Action of the Steam End of the Pump is as

follows: Steam from the boiler enters the pump at the union swivel 54, and besides filling the chamber which contains the main valve (7), passes through a port in the wall of this chamber and through a passage (not shown in plate 14) to the chamber in which the reversing valve works, thereby constituting the main valve chamber and the reversing valve chamber as the two steam chests of the pump.

From the reversing valve chamber the steam passes through a small port into the space occupied by the reversing piston (23), as shown in plate 14, and as the combined area of piston 23 and small piston 9 is greater than the area of the large piston 8, the main valve (7) is forced down until the small piston strikes the stop pin (50) and thus uncovers the port in bushing 26, which admits steam to the underside of main piston 10, forcing it up.

DESCRIPTION OF PLATE 14—EIGHT-INCH PUMP, ON THE UP-STROKE

54. Boiler connection. 7. Main valve. 7-8 and 7-9. Large and small piston of main valve. 25 and 26. Main valve bushings. 50. Stop pin. 23. Reversing piston. 16. Reversing valve. 17. Reversing valve stem. 18. Reversing plate. 10 and 11. Main steam and air pistons 3. Steam cylinder. 4. Neck. 5. Air cylinder. 57. Main steam exhaust. 41. Drain cock. 30 and 32. Discharge valves. 31 and 33. Receiving valves. 53. Main reservoir connection.

ITS USE AND ABUSE

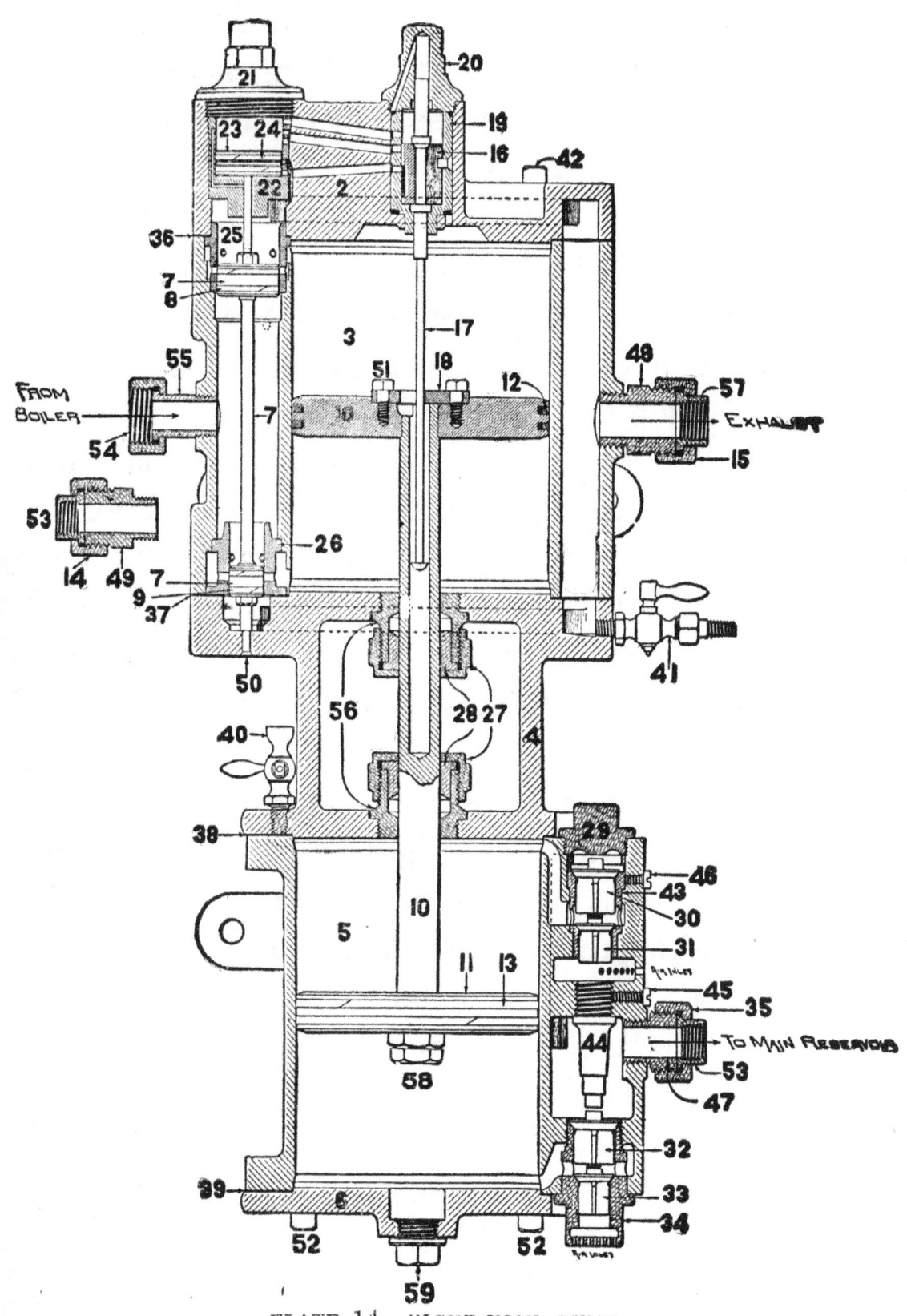

PLATE 14—EIGHT-INCH PUMP.

As the main piston moves up, plate 18 strikes the shoulder of stem 17 and thus changes the position of the reversing valve, so that the top port in its chamber is closed to piston 23, and the two lower ports are connected by the cavity in the reversing valve, which allows the steam to flow from off the top of piston 23, and pass under it into the exhaust passage across the head, as shown by dotted lines, to the main exhaust. When the pressure is thus shut off from piston 23, the main valve raises and causes the small piston to close the steam port to the underside of the main piston, and opens the exhaust port leading into the passage in the bottom of the cylinder, shown by dotted lines, and out at the main exhaust, at the same time piston 8 of the main valve closes the top exhaust port in bushing 25 and opens the supply port through the bushing, and thus admits steam on top of the main piston, which drives it down.

In making the down-stroke, plate 18 engages the button on stem 17 and again changes the position of the reversing valve, which again admits steam on top of the reversing piston, which causes the main valve to move down as before, and piston 8 uncovers a port in the bushing 25 which exhausts the steam from off the top of the main piston, and at the same time piston 9 opens the supply

ITS USE AND ABUSE

port in bushing 26, which admits steam to the underside of the main piston, and at the same time closes the lower exhaust. The pump has now made a complete double stroke.

Drain cock 41 must always be opened before the pump is started, and left open until the pump is warmed up, or until there is about thirty pounds pressure in the main reservoir, and great care must be taken to start the pump slow, to avoid pounding and jarring, as the condensation cannot be compressed, and there must be an air cushion for the piston head to strike against in the lower cylinder.

The Action of the Air End of the Pump is as follows: There are four air valves, two are called receiving valves (31 and 33), and two are called discharge valves (30 and 32). There are two valve cages (34 and 43), and as the discharge valves have a greater area than the receiving valves, in the eight-inch pump, the flow of air past the valves is determined by the lift each valve has; the receiving valves have a lift of one-eighth of an inch, while the discharge valves have a lift of three thirty-seconds of an inch, or one-thirty-second less than the receiving valves.

These standards must never be changed, as too much lift of any of the valves will cause the pump to pound, and not enough lift will cause it to run hot.

MODERN AIR-BRAKE PRACTICE

The way in which the pump receives and discharges air is as follows: When piston 11 is drawn up by steam piston 10 there is a partial vacuum formed in the air cylinder beneath piston 11, and as the atmospheric pressure is about fifteen pounds to the square inch, the receiving valve 33 is forced off its seat by the air rushing in to fill up the space created by the partial vacuum, and if the piston was to stop when it reached the top, the valve would be seated by its own weight when the pressure inside and out of the cylinder equalized; but as the piston reverses just as it reaches the top, the valve is forced to its seat and held there by the compression of the air on top of it, and if the valve has too much lift the pound heard when the valve is seated is great in proportion.

When the piston starts on the down-stroke it compresses the air higher and higher as it nears the bottom, and when the pressure in the pump becomes greater than that in the main reservoir, the lower discharge valve (32) is forced up and the air from the pump rushes into the main reservoir, until the valve is seated by the main reservoir pressure becoming greater than that in the pump.

The action of the top receiving and discharge valves is the same as the lower ones, except on the opposite stroke.

ITS USE AND ABUSE

NINE AND ONE-HALF INCH AIR PUMP.

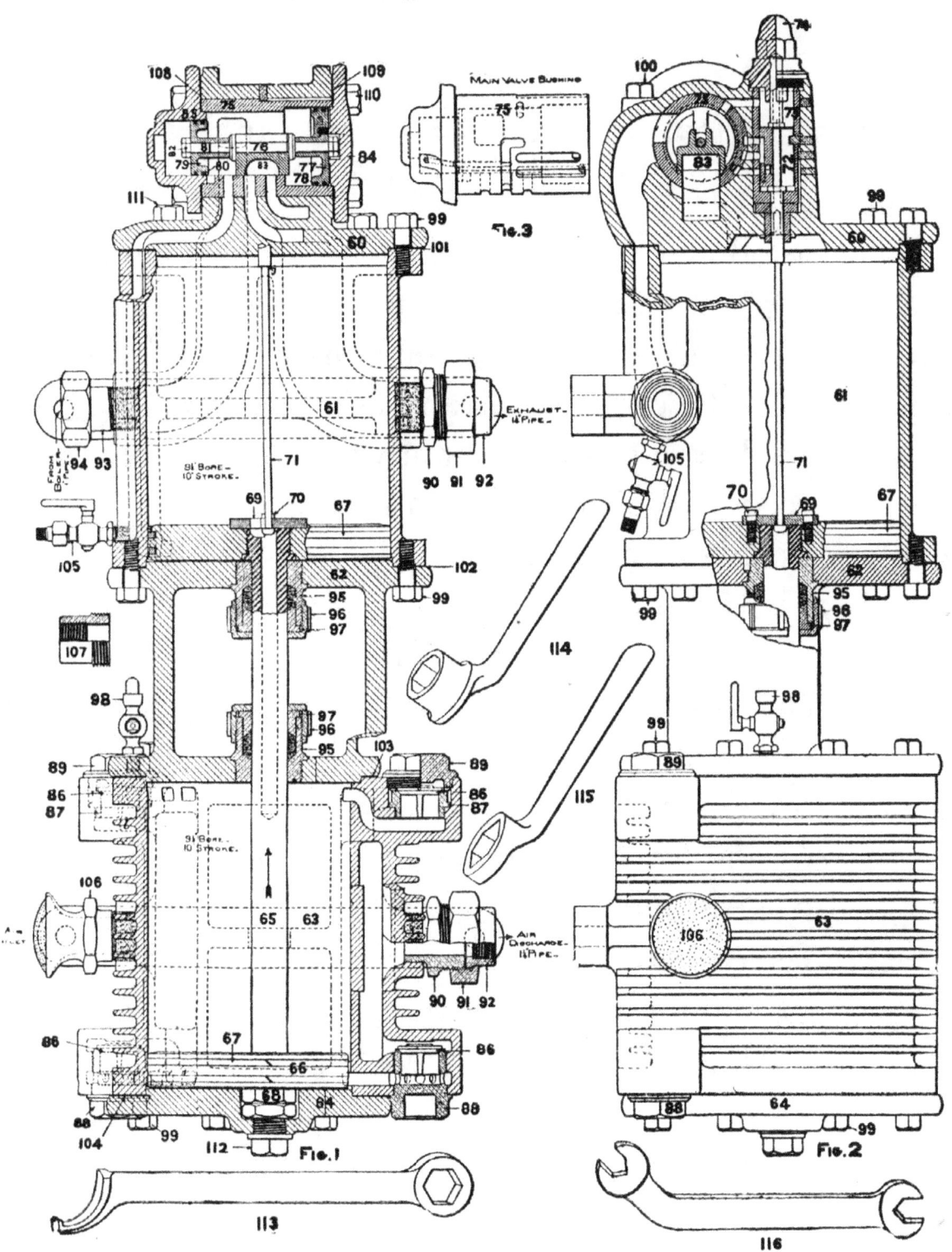

PLATE NO. 15—NINE AND ONE-HALF INCH PUMP.

MODERN AIR-BRAKE PRACTICE

DESCRIPTION OF PLATE 15—NINE AND ONE-HALF INCH PUMP

- 94. Boiler connection, showing by dotted lines how steam passes to main-valve chamber A. Main steam exhaust is indicated by dotted lines and figures 61-92.
- 77. Large piston of main valve.
- 79. Small piston of main valve.
- 83. Slide valve.
- 105. Drain cock.
- 71. Reversing-valve stem.
- 69. Reversing plate.
- 97. Stuffing boxes.
- 98. Oil cup.
- 65 and 67. Main steam and air pistons.
- 106. Air inlet.
- 86. Air valves.
- 92. To main reservoir.
- 75. Fig. 3. Main valve bushing.
- 72. Fig. 2. Reversing valve.

ITS USE AND ABUSE

PLATE 15—THE NINE AND ONE-HALF INCH PUMP

The 9½-inch pump differs from the 8-inch pump in several ways. In the first place it is larger by 1½ inches in the bore; second, the valve motion of the steam end is all contained in the top head, except the reversing valve stem, which is the same as in the 8-inch pump; third, the air valves are all the same size, and all have the same lift of three thirty-seconds of an inch, and the valves are placed so that the discharge valves are both on one side, and the receiving valves on the opposite side of the air cylinder; fourth, there is but one air inlet for the receiving valves, making it possible to strain all the air through one strainer, as indicated by 106, Fig. 1. The main piston is the same in construction as in the 8-inch pump; there are two heads (67) on one piston rod (65), and this rod is hollow to admit the stem (71) of the reversing valve (72), and the reversing valve stem is driven up or pulled down by the reversing plate (69) striking the shoulder (*j*) or the button (70), just as it does in the 8-inch pump.

As the reversing valve was the channel through which the steam had to pass to and from the top of the reversing piston in the 8-inch pump, in like manner the reversing valve in the

9½-inch pump controls the flow of steam to and from the plain side of piston 77 of the main valve, which in connection with the slide valve (83) controls the supply and exhaust ports in the steam cylinder.

To explain this it is necessary to use two sectional views of the pump, as shown in plate 15. In Fig. 1 the pipe connection 93 shows by dotted lines how the steam from the boiler is carried through a passage in the back of the pump to the main-valve chamber.

The main valve is composed of two pistons of unequal diameters, fastened to a suitable rod (76), and on this rod there are two shoulders between which a common D slide valve (83) is held. Fig. 3 represents the bushing in which the main valve and slide valve works.

The slide-valve seat has three openings: the one on the left, in Fig. 1, leads to and from the underside of the main piston; the one on the right leads to and from the top side of the main piston, and the one in the middle leads to the main exhaust, 92. Consequently when steam enters the main-valve chamber the piston 77, having the largest area, is forced to the extreme right, as in Fig. 1, against the head 84, which causes the slide valve to uncover a port

ITS USE AND ABUSE

in the seat so that the steam can pass from the main-valve chamber down through a passage in the side of the cylinder to the underside of the main piston, which forces it up, and the reversing plate strikes the shoulder, *j*, on the reversing-valve stem, which drives the reversing valve up and allows the steam in the reversing-valve chamber to pass through the lower horizontal port in the main-valve bushing (see Fig. 3) into the chamber between the head 84 and piston 77. As this balances the pressure on both sides of the large piston 77, the small piston 79 now pulls the slide valve to the opposite end of the chamber, which uncovers the supply port to the top of the main piston and allows the steam to force it down, and at the same time the steam from the underside is being exhausted by way of the cavity in the slide valve, which now has the lower supply port and the main exhaust connected.

The reason the small piston pulls the large piston over, after the pressure is balanced on both sides of piston 77, is because there is a small port between the plain side of piston 79 and the head 85, which is always open to the main exhaust, so that no back pressure can remain in the chamber indicated by 82, and no partial vacuum can be formed on that side of the small piston.

The main-valve chamber is always in communication with the reversing-valve chamber by a small port in the bushing (75), as shown in Fig. 2; cap nut 74 has a small port in it which allows live steam to always reach the top of the reversing-valve stem, for the purpose of keeping the pressure balanced on both ends of it.

As the main piston is now making its downstroke the reversing plate (69) engages the button on the end of the reversing-valve stem and draws the reversing valve down to the position shown in Fig. 2, which connects the second horizontal port in the bushing with the port which in Fig. 3 appears to be vertical and having a short extension to the right, and as this port is always open to the main exhaust, the steam between piston 77 and the head 84 is exhausted, which allows the steam in the main-valve chamber to again force piston 77 to the position shown in Fig. 1, which places the slide valve in position to allow the steam to exhaust from the top of the main piston, and at the same time connects the main-valve chamber with the underside of the main piston, causing it to be forced up, as before.

ITS USE AND ABUSE

Like the eight-inch pump, the stuffing boxes (95) must be kept well packed, and the gland nuts (96) just tight enough to stop leaks, but not tight enough to cause groaning. With metallic packing the nuts can be tightened more than they could if a fiber packing is used, for if you screw down too tight on a fiber packing it will ruin it.

The drain cock (105) must be handled in the same way as the one on the eight-inch pump, but in addition to this one there is one in the main exhaust (not shown in Fig. 1), and it also must be opened when starting the pump.

THE ELEVEN-INCH PUMP

The Westinghouse Air-Brake Company are now making an eleven-inch pump after the same pattern as the 9½-inch one.

As the 9½-inch pump can compress about a third more air in a given time than the 8-inch pump, in like manner the 11-inch pump can compress a third more air than the 9½-inch pump can within the same length of time.

Right and Left Hand Pumps are pumps having two sets of plugs on either side of the steam cylinder, so that the pump can be located on either side of the engine as desired. All 9½-

inch and 11-inch pumps are now made right and left.

To change a pump from right to left, or vice versa, remove the steam port fittings and opposite plug and exchange them, remove the exhaust port fitting and its opposite plug and exchange them.

In oiling either the 8, 9½ or 11-inch pump the steam end is oiled by a lubricator, and when first starting the pump, the oil should be allowed to flow at the rate of about fifteen drops a minute, but as soon as the pump is nicely warmed up, or say about thirty pounds pressure in the main reservoir, then the oil should be cut down to about *one drop a minute*, if that will keep the pump lubricated so that it won't groan. Some pumps require more oil than others, according to the work they have to do. There is now being supplied on all pumps, when so specified, an automatic oil cup for the air end of the pump on both the Westinghouse and New York Air Pumps. An automatic cup is very essential as too much or too little oil in either end of the pump is ruinous.

The air cylinder should be oiled regularly with good valve oil, as the old practice of oiling it only when the pump groans is now found to be bad practice. The old practice of having a good fat swab on the piston rod of the pump is a very good

ITS USE AND ABUSE

one, although it is a bad practice to expect sufficient oil to pass into the cylinder to lubricate it from this source, for the simple reason that the piston rod packing is supposed to be air tight.

Under no circumstances must oil be sucked in through the air inlet, as it will surely ruin the pump. Whenever the air cylinder is to be oiled, the pump should be throttled down to a very slow speed, and after first filling the oil cup, watch the stroke of the piston, and, when it is going down, quickly open the oil cup and allow the oil to be sucked in *before the piston starts up*. This causes the oil to be sprayed around the cylinder. If oil was poured in while the pump was cold, just as soon as it was started up the oil would be forced into the main reservoir, and eventually find its way to the brake valve, and gum up the rotary, feed valve and pump governor.

Some engineers say they can't oil a pump on the down-stroke for the reason that the oil blows back in their face; this is true only when the piston packing rings are leaky, and if the oil does blow back on the down-stroke, it tells you very plainly that new packing rings are needed, and needed bad, as one of the most common causes for the pump running hot is leaky packing rings. A leaky discharge valve might cause a back blow, but if the pump is completely stopped and you hold your finger slightly above the open oil cup you can tell if the trouble is there.

Never Use Anything but Good Valve Oil for either end of the pump, as the heat generated by the compression of air is so great that it requires oil of a high flashing point to withstand it. On a warm summer's day the air in a pump working against a ninety-pound pressure in the main reservoir is about 550 degrees, and on a cold winter's day, when the thermometer is thirty degrees below freezing, the pump generates a heat of 300 degrees against a ninety-pound main reservoir pressure. And if you run your pump faster than sixty or seventy full strokes a minute, or have leaky packing rings or leaky discharge valves, the heat is raised considerably higher.

The Westinghouse Air Brake Company are now manufacturing a compound air compresser, the main feature of which is that the steam cylinder is much smaller than the air cylinders,

ITS USE AND ABUSE

thereby effecting a considerable saving in fuel. The first compression of air raises the pressure to about forty pounds, and as this pressure operates against the air piston at the same time that the steam is operating against the steam piston, it is readily seen that the combined force of compressed air and steam enables the pump to be operated at a very material saving of fuel. As this new compound pump is not as yet in general use, a detailed description of it at this time is not absolutely necessary, but full information will be sent to anyone desiring it by addressing the Railway Publications Society, Chicago, Ill.

The air valves in the $9\frac{1}{2}$-inch pump operate same as in the 8-inch. But the lift of the air valves in the $9\frac{1}{2}$-inch pump are all the same, whereas they differ in the 8-inch pump, as previously explained.

PLATE 16—PUMP GOVERNOR

When an engine is equipped with a brake valve on which there is a feed valve attachment the pump governor controls the main-reservoir pressure.

But when the D-8 brake valve is used, the governor controls the trainpipe pressure.

MODERN AIR-BRAKE PRACTICE

THE THREE POSITIONS OF THE COMBINED RELEASE AND RETAINING VALVE ARE: FULL RELEASE, NORMAL, AND LAP.

THIS ILLUSTRATION SHOWS LOCATION OF COMBINED RELEASE AND RETAINING VALVE JUST ABOVE THE ENGINEER'S AUTOMATIC BRAKE VALVE.

ITS USE AND ABUSE

INTERIOR VIEW OF A MODERN AIR BRAKE INSTRUCTION CAR

While many railroads are now operating their own instruction cars, in order to benefit their employes, it is, nevertheless, a fact that such method of instruction, if unaccompanied by careful study on the part of the employe, is of little value, for the reason that the continual movement of the car only affords each man a few hours instruction in a whole year, and therefore it is of vital importance that each employe should possess a thorough text book on the subject in order to be prepared for his examination when the car returns.

MODERN AIR-BRAKE PRACTICE

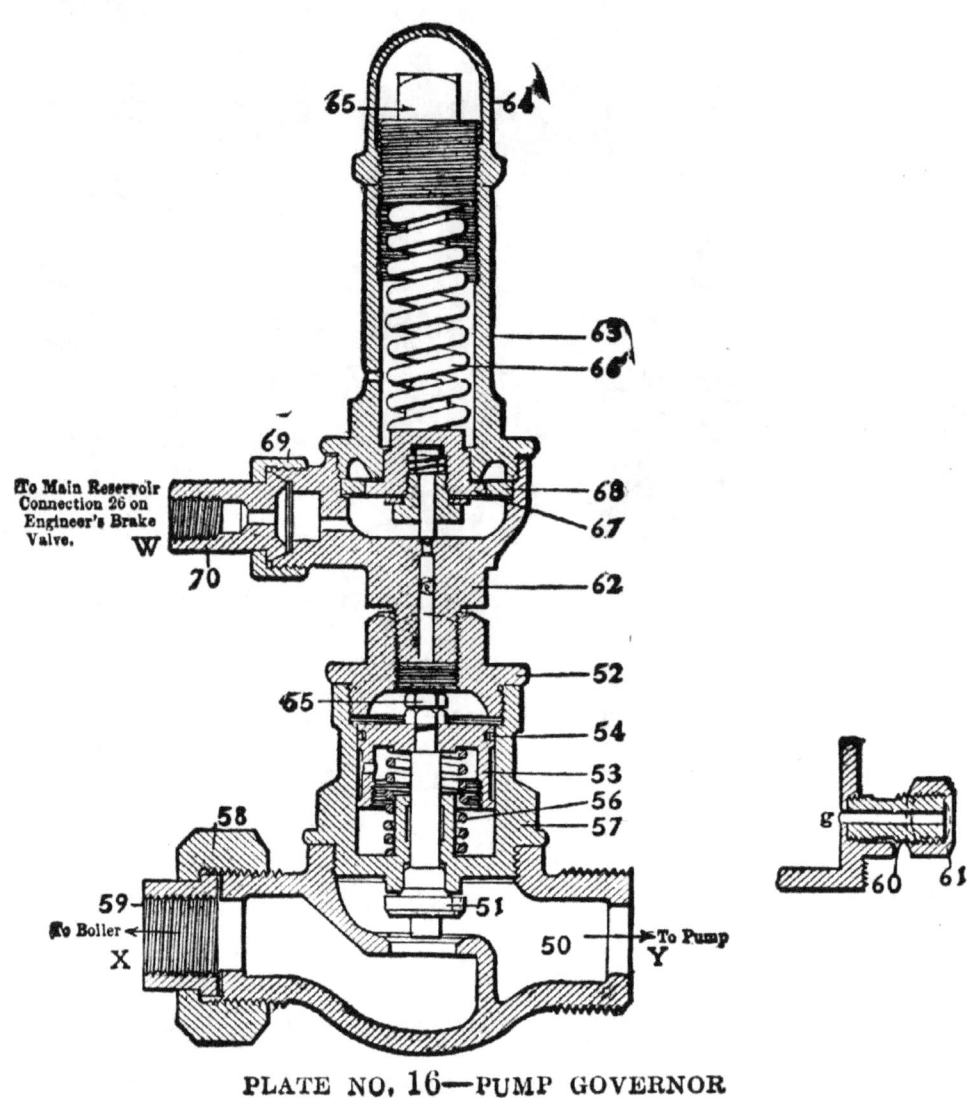

PLATE NO. 16—PUMP GOVERNOR

ITS USE AND ABUSE

DESCRIPTION OF PLATE 16—PUMP GOVERNOR

X. Boiler connection.
Y. To pump.
51. Steam valve.
53. Air valve.
56. Air-valve spring.
62. Vent port.
67. Diaphram and valve.
68. Diaphram ring.
66. Regulating spring.
65. Regulating nut.
W. Main reservoir.
61. Waste-pipe stud.

MODERN AIR-BRAKE PRACTICE

While the new style governor is very similar to the old style, the new one is much more reliable, as it is more positive in its action.

The governor is located on the steam pipe leading to the pump, as its purpose is to shut off the steam whenever the pump has compressed the required amount of air, and whenever the air pressure falls below standard the governor automatically reopens the valve in the steam pipe and keeps it open until the air pressure is again restored, when it again shuts off the steam.

This action is very simple. As the steam enters the governor at x, it passes under the steam valve (51) and through Y into the pump, and as long as the steam valve is unseated the pump will continue to work and compress air right up to boiler pressure; but as ninety pounds is all that is wanted in the main reservoir with the regular quick-action equipment, the tension spring of the governor must be set so that the steam valve will seat when ninety pounds is reached.

This is done as follows: you will notice that piston 53 rests on the stem of the steam valve, and that the area of piston 53 is several times greater than the area of the steam valve, which

ITS USE AND ABUSE

means that if the relative areas were as three is to one that when a fraction over fifty pounds of air got on top of piston 53 it would drive the steam valve to its seat against a steam pressure of 150 pounds.

The manner in which the air is admitted to the top of piston 53 to stop the pump, or kept from it to allow the pump to run, is as follows: A small pipe leading from the main-reservoir return pipe is connected to the governor at W, which allows main-reservoir pressure to always fill the chamber under diaphram 67, and as this diaphram is held down by a tension spring (66) and as there is a small pin valve attached to the center of the diaphram which closes the port leading to the top of piston 53, whenever the air pressure becomes greater under the diaphram than the tension of the spring, it will cause it to raise and unseat the pin valve, and allow the air to reach the top of piston 53, causing it to seat the steam valve and stop the pump. If the tension spring 66 is properly set the pump will stop when there is ninety pounds in the main reservoir. Whenever the main-reservoir pressure gets lower than the tension of the spring, the diaphram valve drops back to its seat and the air escapes from the top of piston 53 through a

MODERN AIR-BRAKE PRACTICE

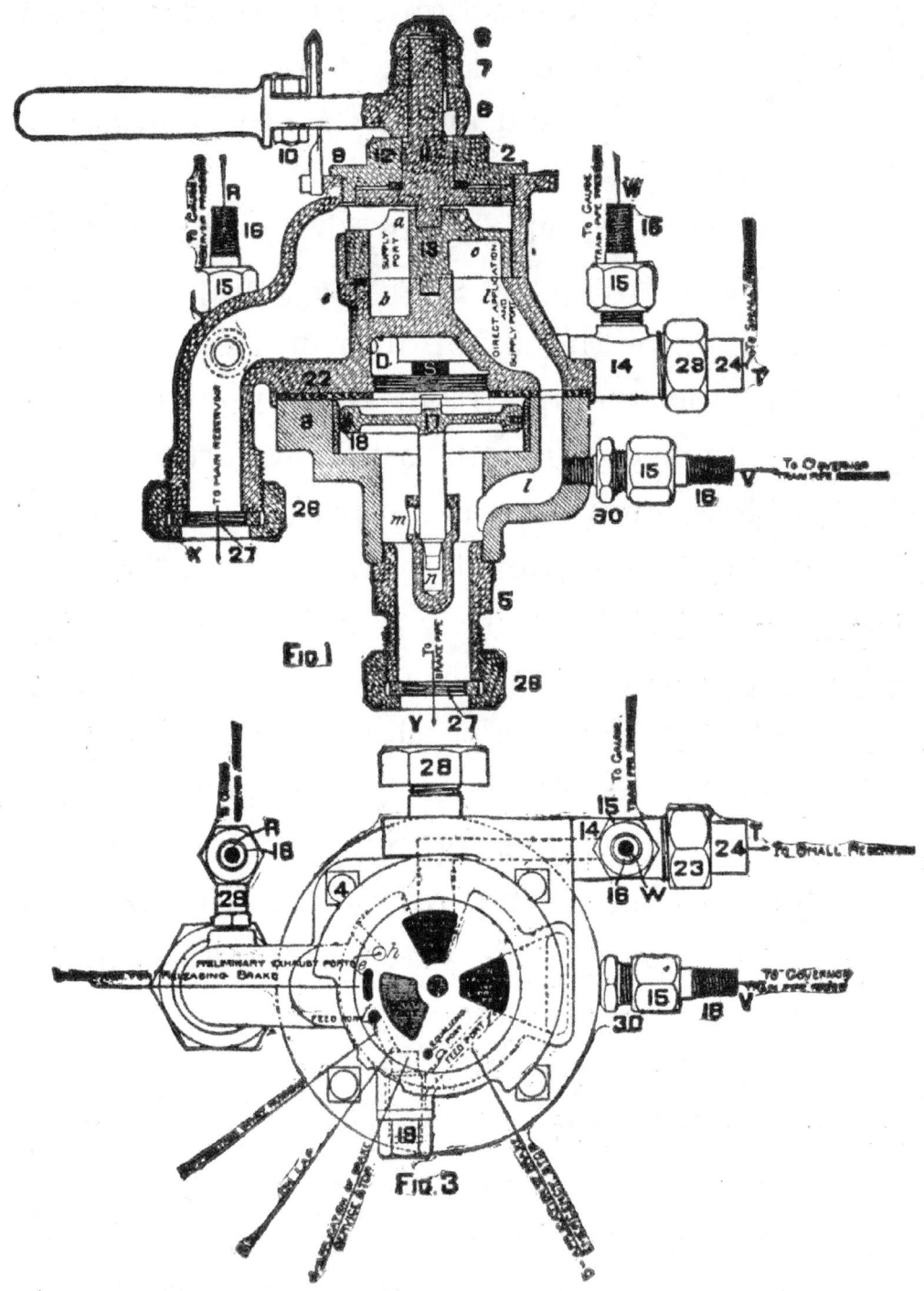

PLATE NO. 17—D-8 BRAKE VALVE AND ROTARY SEAT.

ITS USE AND ABUSE

DESCRIPTION OF PLATE 17—D-8 BRAKE VALVE

X. Main reservoir connection.

R. Gauge connection for red hand.

W. Gauge connection for black hand (or trainpipe).

T. To the little drum.

V. To the pump governor.

Y. Trainpipe.

17. Equalizing discharge valve.

18. Rotary valve.

22. Body gasket. Fig. 3 shows rotary seat and preliminary exhaust port h, and equalizing port g, both lead into cavity D.

small vent port (62) which allows spring 56 to aid the steam in lifting the steam valve from its seat.

If the vent port 62 is not kept open the pump will be slow in starting, for the air could only get off the top of piston 53 by passing down around packing ring 54 and out at the waste-pipe connection (*g*); stud 60 is tapped in the back of the governor under piston 53, to carry off any steam that might leak by the stem of valve 51, or any air that might leak around packing ring 54, consequently should both the vent port and the waste pipe become clogged the governor would not shut off the pump, and the main-reservoir pressure would run up to boiler pressure.

PLATES 17 AND 18—D-8 ENGINEER'S BRAKE VALVE

In applying the brakes with the quick-action triple, it is not only necessary to reduce the train-pipe pressure lower than that in the auxiliary, but it is absolutely necessary that the reduction be made *gradually* to prevent the emergency action.

The old style brake valve, or three-way cock, had only three positions, application, lap and release, and while some men seem to think the new brake valve has only two positions, "on" and

ITS USE AND ABUSE

"off," there are, however, five positions, as follows: full release, running position, lap, service application, and emergency.

There are two kinds of brake valves, one has no feed-valve attachment, and is known as the D-8, and depends upon the pump governor to regulate the trainpipe pressure. The other kind has a feed-valve attachment for controlling the trainpipe pressure, which leaves the pump governor to control the main-reservoir pressure, and is known as the F-6 and G-6 brake valve, according to the kind of feed valve there is on it. The F-6 has the old style feed valve, and the G-6 has the new slide valve feed valve, as shown in plates 21 and 22.

It is not necessary to go into details in describing the D-8 brake valve, as it is now practically superseded by the F-6 and G-6, therefore I will simply explain the differences between the two kinds of brake valves, and will fully explain the F-6 under plates 20 and 21.

The D-8 brake valve uses the pump governor to control the trainpipe pressure of seventy pounds, and the connection is made at V (plate 17); the "excess" is controlled by what is known as the excess pressure valve (19, Fig. 3, of plate 17).

MODERN AIR-BRAKE PRACTICE

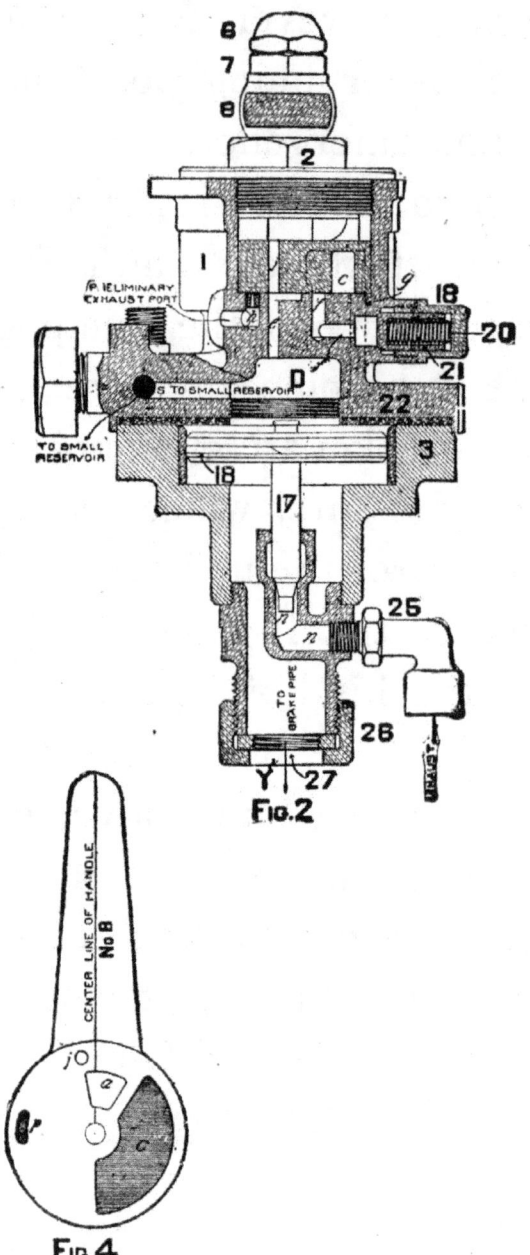

PLATE NO. 18—D-8 BRAKE-VALVE AND ROTARY.

DESCRIPTION OF PLATE 18—D-8 BRAKE VALVE

Fig. 2 shows trainpipe exhaust *n, n,* 25.

21. Excess pressure valve.

Fig. 4. The rotary valve and handle.

ITS USE AND ABUSE

When the handle of the D-8 brake valve is in full release position the pump will shut off at seventy pounds, and the pressure in the main reservoir and trainpipe would be the same, but if the handle is in running position the excess pressure valve will not open to admit air into the trainpipe until there is twenty pounds in the main reservoir, and as it requires twenty pounds to hold this valve open, the trainpipe will get a pressure of seventy pounds before the pump will shut off, thus leaving an excess pressure of twenty pounds in the main reservoir.

If the handle is placed on lap while the trainpipe pressure is below seventy pounds, the pump will run the main reservoir pressure up to boiler pressure, for the governor cannot shut the pump off unless there is seventy pounds in the trainpipe; on the other hand, if the handle is in running position no air can get into the trainpipe until there is twenty pounds of excess in the main reservoir, and as a consequence the many leaks that commonly occur in the main reservoir and trainpipe connections cause the brakes to creep on before the pressure can be restored to keep them off. It was mainly on this account that the F-6 brake valve was invented, for with this valve the pump governor is controlled by the main reservoir pressure, and will stop the pump at ninety pounds in the main reservoir, no matter in what position the handle is, and, as the trainpipe pressure is controlled by the feed valve, whenever that pressure falls below the standard of seventy pounds, if the handle is in running position the feed valve will open and let the main reservoir pressure in, and thus keep the brakes from dragging.

Another difference between the two kinds of brake valves is that with the D-8 valve, when making a service application, the air from cavity D over the equalizing discharge valve (17) is exhausted to the atmosphere through a separate little port in the casing, marked h in Fig. 2 of plate 18, whereas the preliminary exhaust h, in the F-6 valve, is connected with the main or emergency exhaust, marked k in Fig. 2 of plate 20, thus making one port less through the casing of the F-6 brake valve.

Therefore there are the following differences between the D-8 and the F-6 brake valves: 1st, with the D-8 valve the excess pressure is gotten *before* the trainpipe begins to charge, if the handle is in running position; 2nd, with the D-8 valve the trainpipe pressure is controlled by the pump governor, instead of the feed valve attach-

ITS USE AND ABUSE

ment, as it is with the F-6; 3rd, with the D-8 valve, if the handle is left in either lap, service or emergency position, the pump will run the main reservoir pressure up to boiler pressure, or will shut off when there is only seventy pounds in the main reservoir if the handle is left in full release from the starting of the pump, whereas with the F-6 valve, the pump will be shut off by the governor, if properly set, when the main reservoir reaches ninety pounds, no matter what position the handle of the valve is in; 4th, with the F-6 valve the excess pressure is gotten *after* the trainpipe pressure is pumped up; 5th, with the D-8 valve, if the excess pressure valve should happen to be in bad order, and it usually is, if the handle was left on lap for any considerable length of time after making a service application, the main reservoir pressure would be raised so high that, with a short train, when the handle was thrown to release position the auxiliaries would be overcharged, and the wheels slid on the next application, unless the engineer was very careful, whereas with the F-6 valve the most that could get in the auxiliaries, if the governor was correct, would be ninety pounds; 6th, when an emergency application is made with the D-8 valve, the black hand on the gauge will rise instead of fall, because in this position the equalizing port to cavity D is open to the main reservoir pressure. The construction of the D-8 valve, with these differences, is the same as the F-6 or G-6, except that the D-8 has an excess pressure valve while the F-6 or G-6 has a feed valve attachment, which will be explained in regular order.

PLATES 19 AND 20—THE F-6 (1892 MODEL) ENGINEER'S BRAKE VALVE

The engineer's brake valve is the device on the engine by means of which the engineer is enabled to charge up, and keep charged, the trainpipe and auxiliaries; apply the brakes, and keep them applied, release the brakes, and keep them released, and to do these several things he has either to place the main reservoir in communication with the trainpipe, or open the trainpipe to the atmosphere, or shut off all communication, as the case may be, according to whether he is applying or releasing the brakes, keeping them set, or running along.

There are just four things that constitute the essential parts to a modern Westinghouse brake valve, viz: the rotary valve, the handle that controls the rotary, the equalizing discharge valve

ITS USE AND ABUSE

and the feed valve attachment, or trainpipe governor. Of course there are gaskets, springs, packing rings, the equalizing reservoir, etc., but they are matters of detail.

There are five positions in which the handle of the brake valve can be placed.

The first, or extreme left position is "full release," and is the position the handle should always be in when releasing brakes, or when it becomes necessary to charge up quickly, for in this position the air from the main reservoir flows through the largest ports in the rotary direct into the trainpipe.

The second position is called "running position," because the handle should be carried in this position while running along, for the reason that in this position the rotary valve is placed so that all the air that passes from the main reservoir into the trainpipe must go through the feed valve attachment, and as this attachment will only allow seventy pounds of air to get into the trainpipe (if set correctly, and unless the high-speed apparatus is being used), it enables the pump to maintain an excess pressure in the main reservoir, for if the pump governor is set at ninety pounds, and the feed valve set at seventy, there will naturally be twenty pounds

MODERN AIR-BRAKE PRACTICE

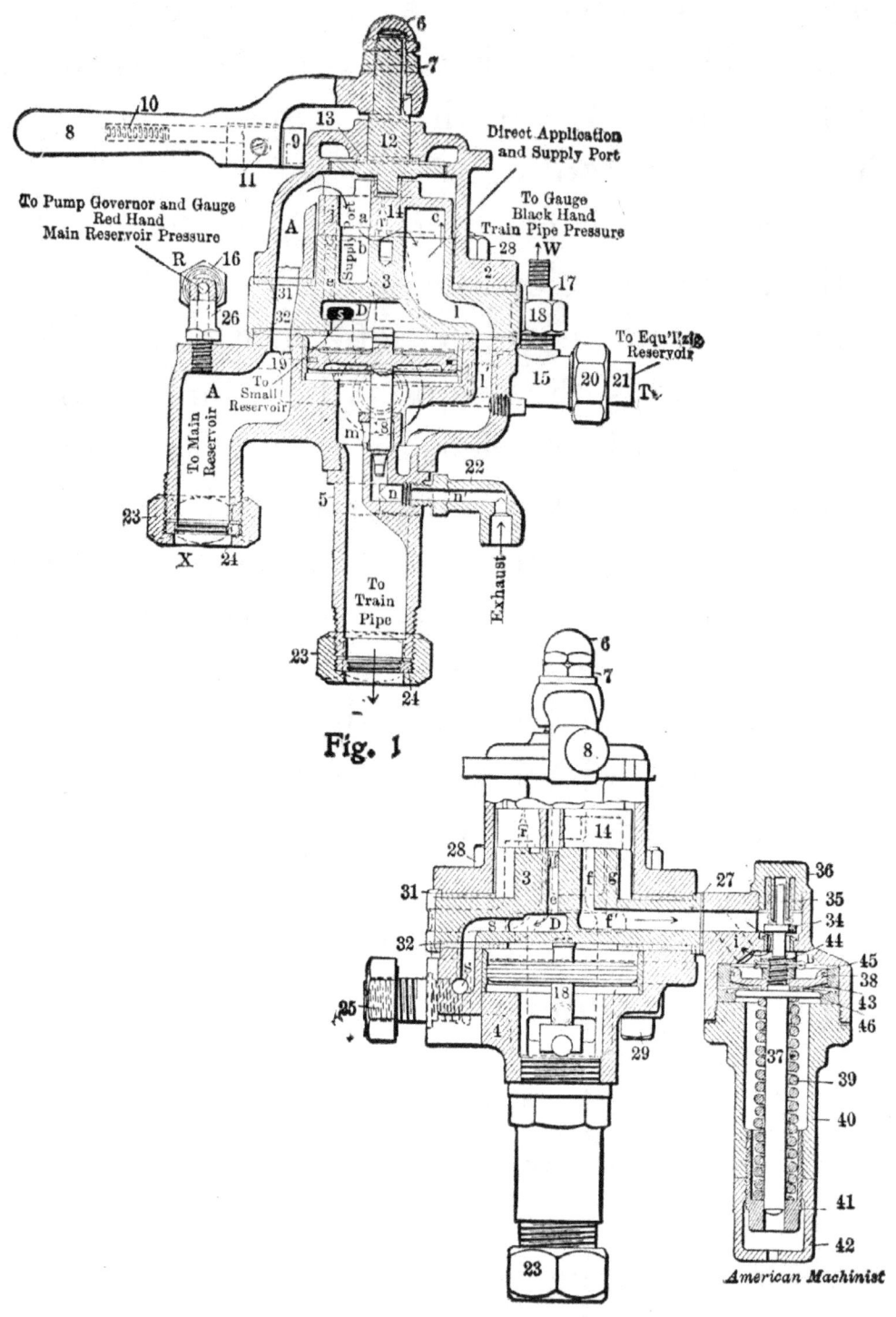

PLATE NO. 19—F-6 BRAKE VALVE AND OLD STYLE FEED VALVE

ITS USE AND ABUSE

greater pressure in the main reservoir than in the trainpipe before the pump is stopped by the governor.

Another reason why the handle must always be carried in running position while the train is running along, is because whenever the pressure in the trainpipe leaks down below the standard of seventy pounds, the feed valve will open automatically and allow the main reservoir pressure to again flow into the trainpipe until that pressure is restored, when it will automatically close itself, and allow the pump to again create the "excess" in the main reservoir.

The third position on the brake valve is "lap," and when the handle is in this position all ports are closed, so that no air can pass either into the trainpipe or out of it. After applying the brakes, the handle should be brought to lap carefully, and held there until it is desired to further reduce the trainpipe pressure or release the brakes, as the case may be, and when releasing the brakes the handle must be placed on full release position for a few seconds, according to the length of train and the amount of excess carried, before it is allowed to rest on running position.

The fourth position is called "service application position," because in this position the air is allowed to escape gradually from the trainpipe. In this position the air on top of the equalizing discharge valve is allowed to escape through the small preliminary exhaust port in the seat of the rotary so gradually that a sudden reduction on the trainpipe is prevented, for as the pressure on top of the discharge valve is allowed to escape, the trainpipe pressure below gradually forces it from its seat and thereby opens the trainpipe exhaust. If the handle is left in service position until ten pounds is drawn from the top of the discharge valve and then placed on lap, the valve will not seat until a fraction over ten pounds has escaped from the trainpipe, when the pressure on top will then be the greatest and force the discharge valve back to its seat, and thereby close the trainpipe exhaust.

The fifth position is called "emergency application position," because when the handle is in this position the rotary connects the main trainpipe supply port with the main exhaust port, and the air is allowed to escape from the trainpipe direct to the atmosphere, regardless of the equalizing discharge valve, and this sudden reduction of trainpipe pressure allows the triples to be forced to their full stroke, as explained

ITS USE AND ABUSE

under plate 5, and thus causes the quick action, or emergency application. Emergency position should never be used except in case of danger. Owing to the rough manner in which some enginemen handle their brakes, this position is often called "criminal application position."

The parts of the F-6 brake valve are as follows: the handle, which controls the rotary, is marked 8, in Fig. 1; the lug (9) is forced out by a spring (10) so that the handle may be stopped in any desired position, and when placing the handle in any of the positions be sure that the bolt in the handle is right up against the lug on the brake valve, for the reason that the rotary valve is moved in exact accord with the handle. If the bolt or lug is worn the movement of the rotary will be correspondingly changed out of its proper alignment; 12 is the stem to one end of which the handle is fastened by nuts 6 and 7, and the other end is dove-tailed or keyed into the top of the rotary, so that whatever way the handle is turned the rotary has to turn with it; 13 is a small leather gasket for the purpose of preventing any air from leaking out around the stem, as main reservoir pressure is always on top of the rotary and under the shoulder of stem 12, forcing it up against the casing. This gasket

MODERN AIR-BRAKE PRACTICE

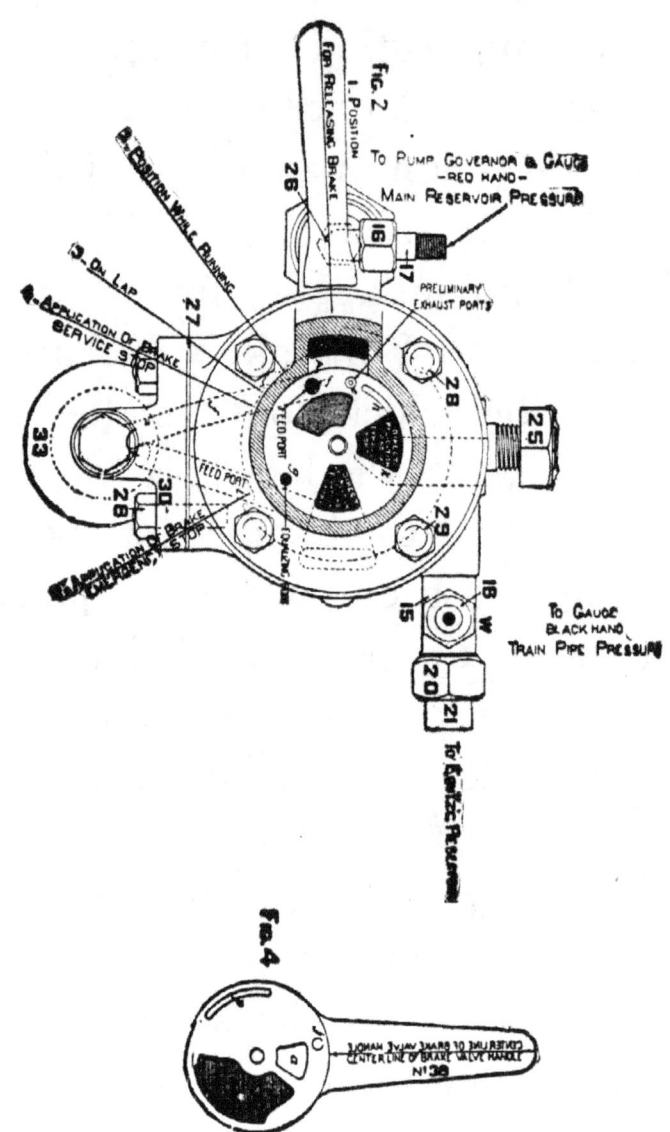

PLATE NO. 20—F-6 BRAKE VALVE—ROTARY AND SEAT.

DESCRIPTION OF PLATE 20—F-6 BRAKE VALVE

Fig. 2 shows rotary seat and five positions of the handle.

Fig. 4 is the rotary and handle.

ITS USE AND ABUSE

sometimes gets gummed up so badly that it causes the handle to move very hard; 14 is the rotary valve, and 3 is the rotary valve seat; 18 is the equalizing discharge valve, which controls the trainpipe exhaust *m* and *n*. The action of the discharge valve has already been explained under "service application position."

As cavity D above the discharge valve is very small, it is necessary to have a greater volume of air to control it than the cavity alone will contain, and this greater volume is supplied by a little drum, or equalizing reservoir, which holds about 500 cubic inches of air, and is located, usually, under the footboard of the cab. It is connected to the brake valve at T (Fig. 1), and from T to cavity D there is a connecting passage, as shown by *s* in Figs. 2 and 3, and as the little drum is always charged equally with cavity D, whenever the pressure in cavity D is reduced it is also reduced in the little drum. This greater volume is needed above the discharge valve to compensate for the volume in the trainpipe.

When the handle of the brake valve is placed in service position the rotary shuts off the main reservoir and also cavity D from the trainpipe, and allows the air to escape from cavity D by way of port *e*, groove *p* and preliminary exhaust port *h* to the atmosphere through the main exhaust *k*, and when the handle is moved to lap it closes the preliminary exhaust, and thus holds the little drum pressure at whatever it was reduced to, as shown by the black hand of the gauge, and when the trainpipe has exhausted until it becomes less than the pressure in cavity D the discharge valve is forced to its seat by the pressure in the little drum, and stops any further flow of air from the trainpipe.

Nos. 34 to 46 in Fig. 3 of plate 19 all refer to the old style feed valve attachment as used on the F-6 brake valve. The essential parts are the supply valve (34), valve spring (35), diaphgram piston (37), regulating spring (39), regulating nut (41).

When the rotary is in running position the operation of the feed valve is as follows: the regulating spring being set at seventy pounds tension, it forces the piston up against the stem of the supply valve and raises it off its seat, causing the main reservoir pressure to flow from the top of the rotary down through port *j* in the rotary (Fig. 4 plate 20), and through port *f* in the rotary seat (Fig. 3, plate 19), through a passage (*f*), and under the supply valve to the top of the diaphragm piston, then through a port (shown by

ITS USE AND ABUSE

dotted lines, and marked i, Fig. 2, plate 20), which leads off the top of the piston into the trainpipe by way of the main supply port, as shown by dotted lines in Fig. 2. As the rotary is now in position so that the large cavity (c), as shown in Fig. 4, plate 20, connects the main supply port with the equalizing port g (which passes through the rotary seat into cavity D), the air that is passing from the top of the rotary through the feed valve into the trainpipe, is also filling cavity D, and the little drum, by way of ports g and s, as shown in Fig. 3, plate 19 (while plate 19 shows full release position, still ports s and g are plainly shown, and if the handle was moved to running position the port through the rotary that registers with port e in Fig. 3, would be in register with port f; port g is indicated by dotted lines).

In running position, when the trainpipe and little drum are charged up to seventy pounds there is also seventy pounds on top of the diaphram piston, and as the regulating spring is set at a fraction less than seventy, the air pressure forces it down and allows the supply valve to seat and shut off the main reservoir from the trainpipe. But as soon as the pressure in the trainpipe falls below seventy, the piston is again

MODERN AIR-BRAKE PRACTICE

forced up by the regulating spring and keeps the supply valve open until the pressure is again restored in the trainpipe.

The feed valve attachment is in operation *only* when the handle of the brake valve is in running position.

The course of the air through the brake valve in full release position is as follows: the return pipe from the main reservoir is connected to the brake valve at X, and passes directly to the top of the rotary through passage A, then through port a in the rotary into cavity b in the rotary seat and under a bridge in the rotary (which now stands midway over cavity b), and on over the seat of the rotary, through large cavity c, direct into the main supply port (1) to the trainpipe. In passing over the rotary seat the air also passes down through the equalizing port g, into cavity D, and from cavity D through port s into the little drum; and as the feed valve is cut out when the handle is in full release, both the little drum and trainpipe pressure would charge up to main reservoir pressure if the rotary was left in full release. In full release position, port j in the rotary registers with port e in the seat, so that cavity D charges faster in full release than in running position.

ITS USE AND ABUSE

Always remember that the little drum is simply an enlargement of cavity D, and the same pressure is in both.

The Warning Port, through which the air is heard escaping as long as the handle remains in full release, is a small port through the rotary about the size of a pin, which allows the main reservoir air to whistle through it to warn the engineer that he is liable to overcharge his trainpipe. It should always be kept clean.

The black hand of the gauge is piped to the little drum at W (Fig. 1, plate 19), as stud 17 is tapped into pipe 15 which connects the little drum with cavity D by way of port s.

The red hand of the gauge and also the pump governor are piped to the main reservoir pressure at R.

To make an emergency application the handle must be moved to the extreme right, when the large cavity (c) in the rotary will connect the main supply port (l) of the trainpipe with the main exhaust port (k), and allow the air in the trainpipe to exhaust directly into the atmosphere.

PLATES 21 AND 22—THE G-6 BRAKE VALVE AND NEW SLIDE VALVE FEED VALVE

The G-6 brake valve is identical with the F-6, with the exception of the feed valve. In the

MODERN AIR-BRAKE PRACTICE

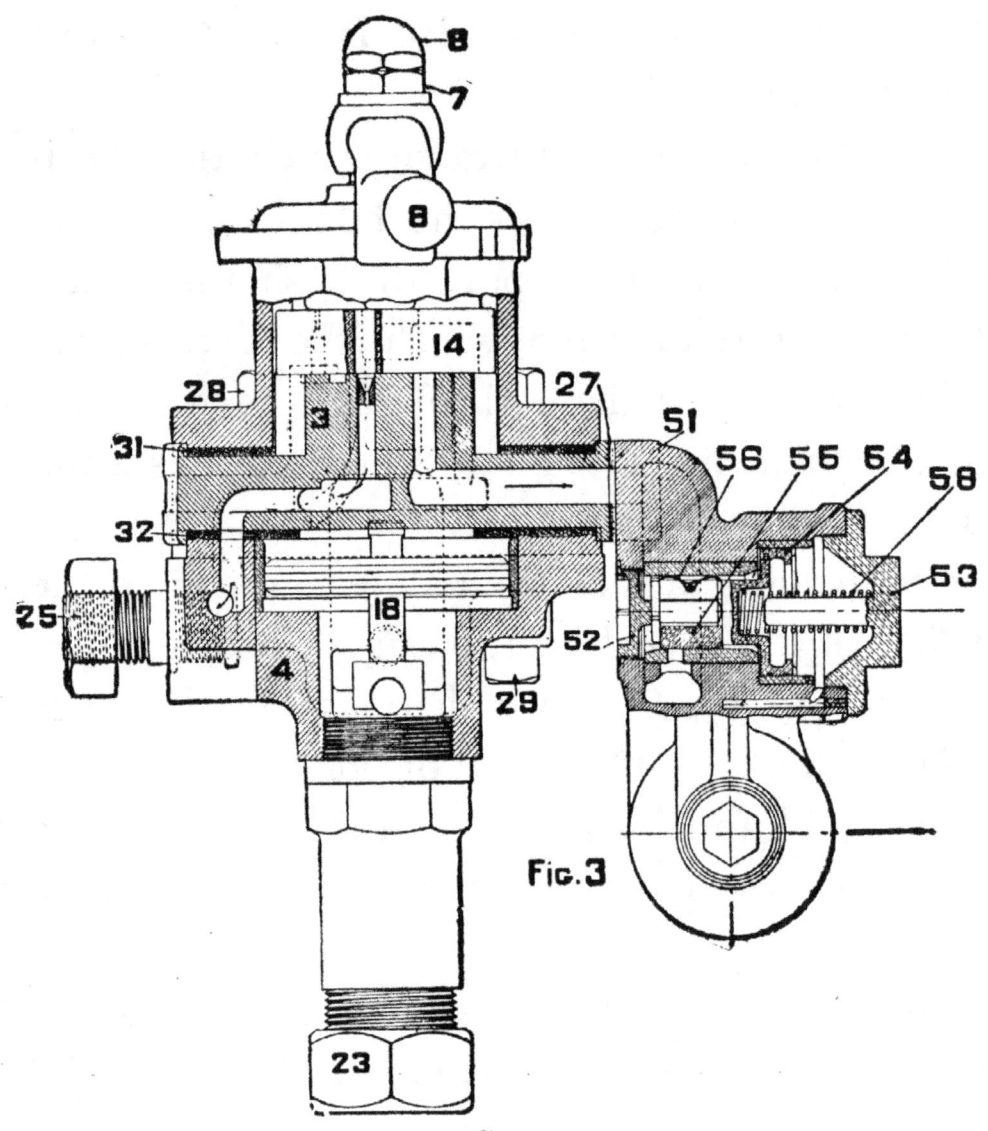

PLATE NO. 21—G-6 BRAKE VALVE.

DESCRIPTION OF PLATE 21—G-6 BRAKE VALVE.

Fig. 3 shows how the new slide valve feed valve is attached.

ITS USE AND ABUSE

new slide valve feed valve the only material change is that a slide valve controls the flow of air from the main reservoir into the trainpipe, which allows the pressure to be raised much quicker than it can be with the old style feed valve.

The working parts of the new slide valve feed valve are as follows: all of the essential parts of the old style feed valve are retained, as shown by plate 22, with slight modification, for 64 is the diaphragm piston, which instead of having a rubber diaphragm has two sheet-brass diaphragms (57) on the piston head, supported by a ring (63); 67 is the regulating spring; 65 the regulating nut; 59 a small valve corresponding exactly with supply valve 34 in the old style feed valve, and 60 is the spring which controls valve 59.

By plate 21, Fig. 3, you will see that there is a slide valve (55) attached to a piston (54), and this piston is forced forward by a spring (58).

The action of the new slide valve feed valve is as follows: when the handle of the rotary is in running position, main reservoir pressure drives the slide valve and piston back, which uncovers a port in the slide valve seat that connects with feed port *i*, and as the slide valve does not move until the trainpipe is fully charged, it causes the

MODERN AIR-BRAKE PRACTICE

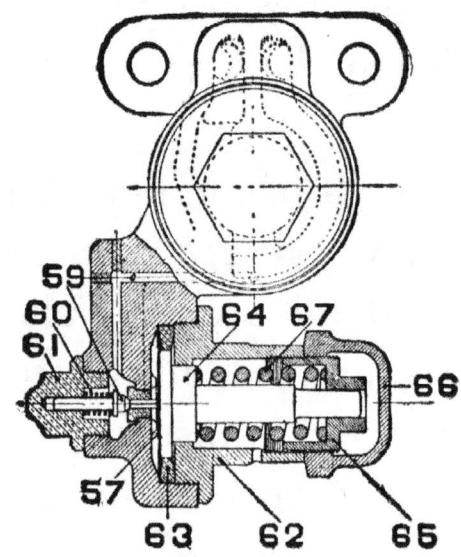

PLATE NO. 22—SLIDE VALVE FEED VALVE.

DESCRIPTION OF PLATE 22 — SLIDE VALVE FEED VALVE

57. Diaphragm piston.
59. Cut-off valve.
67. Regulating spring.
65. Regulating nut.

The slide valve is shown in plate **21**.

ITS USE AND ABUSE

pressure to be restored very quickly after it has been reduced from any cause.

The reason the slide valve does not move until the pressure is restored is because the piston has no packing rings, and the air is allowed to circulate by it through a small passage that leads to the supply valve chamber, from which it passes under the cut-off valve across the diaphragm into feed port i, and when there is a pressure of seventy pounds on the diaphragm it moves away from the supply valve and allows it to seat, when the circulation by the piston is stopped, causing the pressure to equalize on both sides of the slide valve piston, when spring 58 moves the slide valve and closes communication between the main reservoir and the trainpipe. Whenever trainpipe pressure falls below seventy the diaphragm forces valve 59 off its seat and the same action is repeated as before.

As the new Westinghouse Automatic Brake Valve which is used in connection with their new Distributing Valve is not as yet in general use throughout the country, a full description of it will be sent by applying to the Railway Publications Society, Chicago, Ill.

MODERN AIR-BRAKE PRACTICE

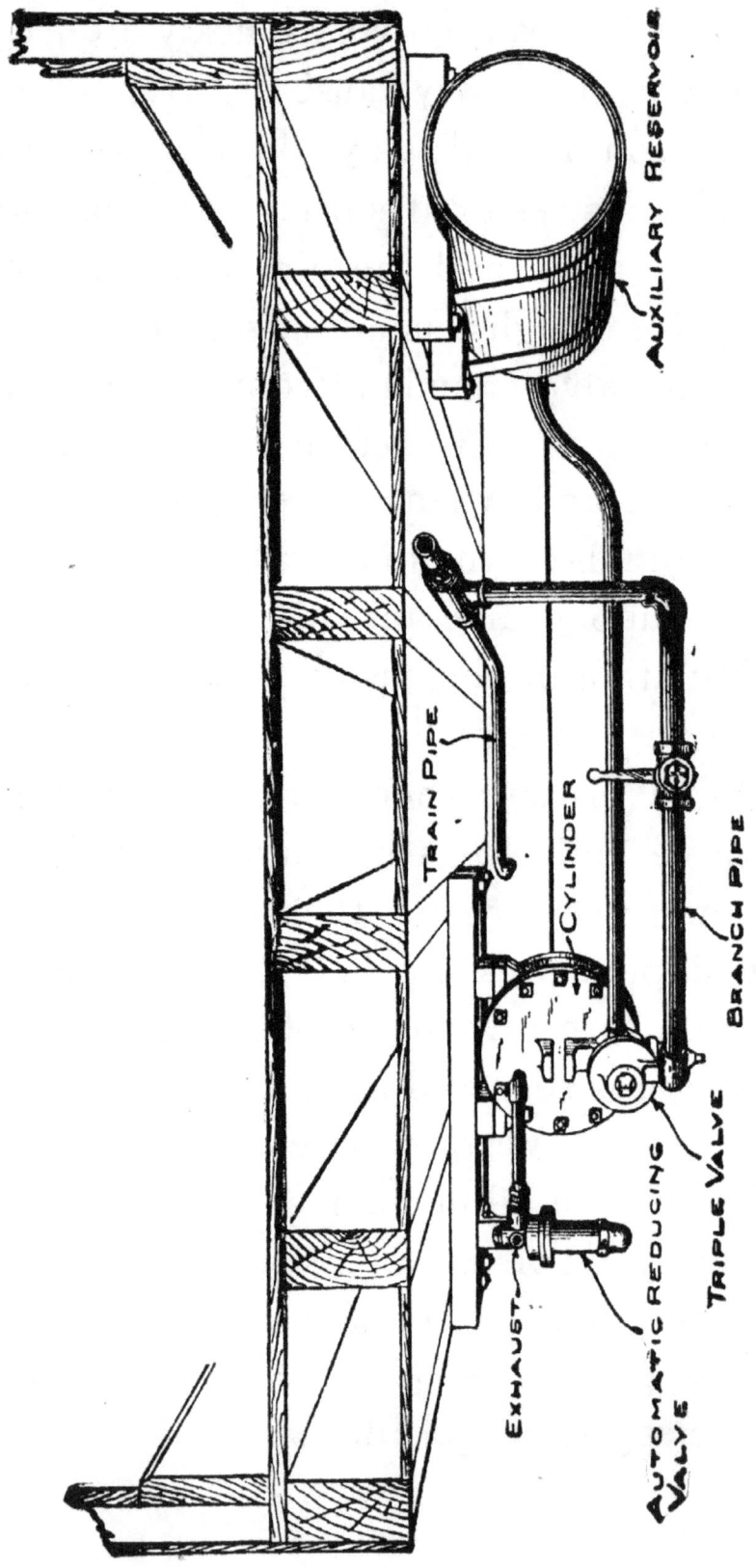

PLATE NO. 23.—HIGH-SPEED BRAKE.

DESCRIPTION OF PLATE 23—HIGH-SPEED BRAKE AS ATTACHED TO CAR

This illustration shows how the reducing valve is attached to a car and piped to the pressure head of brake cylinder.

THE HIGH-SPEED BRAKE

Briefly stated, the high speed brake is an apparatus which enables the engineer to apply a very high pressure to the brake cylinders while running at a high speed, which automatically reduces as the train slows down.

When a train is equipped with the high-speed

brake a pressure of 110 pounds is carried in the trainpipe and auxiliaries and 120 in the main reservoir.

The equipment for the high-speed brake is the same as the ordinary quick-action brake, except that there is a duplex pump governor, an additional slide valve feed valve, a quick action instead of a plain triple on the tender, a specially designed plain triple for the driver and truck brakes, and an automatic reducing valve attached to the cylinder 8 under the locomotive and each car, as shown in plate 23.

As the high pressures are only to be used on trains which run at a very high speed, there are cut-out cocks on the pump governor and feed valves so that the regular seventy and ninety pounds can be carried when required.

When it is desired to change the locomotive equipment from the quick-action to the high-speed brake it is only necessary to turn two handles, that of the reversing cock of the feed valve and that of the quarter-inch cut-out cock on the pipe leading to the governor. These handles must be turned at right angles to the position occupied when the quick-action brake is being used.

The duplex pump governor consists merely of

ITS USE AND ABUSE

two diaphragm portions of the ordinary pump governor (only one of which is in use at a time) connected with one steam valve portion.

The principle of the high-speed brake is as follows: As the friction between the shoe and the wheel is lessened as the rapidity of rotation of the wheel increases, and as the adhesion between the wheel and rail remains practically the same regardless of speed, a greater cylinder pressure can be used while the train is moving at a high speed without danger of sliding wheels, but as the train slows down the cylinder pressure must be correspondingly reduced. This is done by what is called the automatic reducing valve.

PLATE 24—THE AUTOMATIC REDUCING VALVE FOR THE HIGH-SPEED BRAKE

Attached to the brake cylinder on each car there is an automatic reducing valve. Fig. 2 shows how the air passes in at Z, through a strainer (17), and, if the pressure is above sixty pounds, it overcomes the tension of regulating spring 11, and piston 4 is forced down, which carries the slide valve (8) with it, so that port b in the valve registers with port a in the seat, and allows the surplus pressure to escape to the atmosphere until the cylinder pressure is down

MODERN AIR-BRAKE PRACTICE

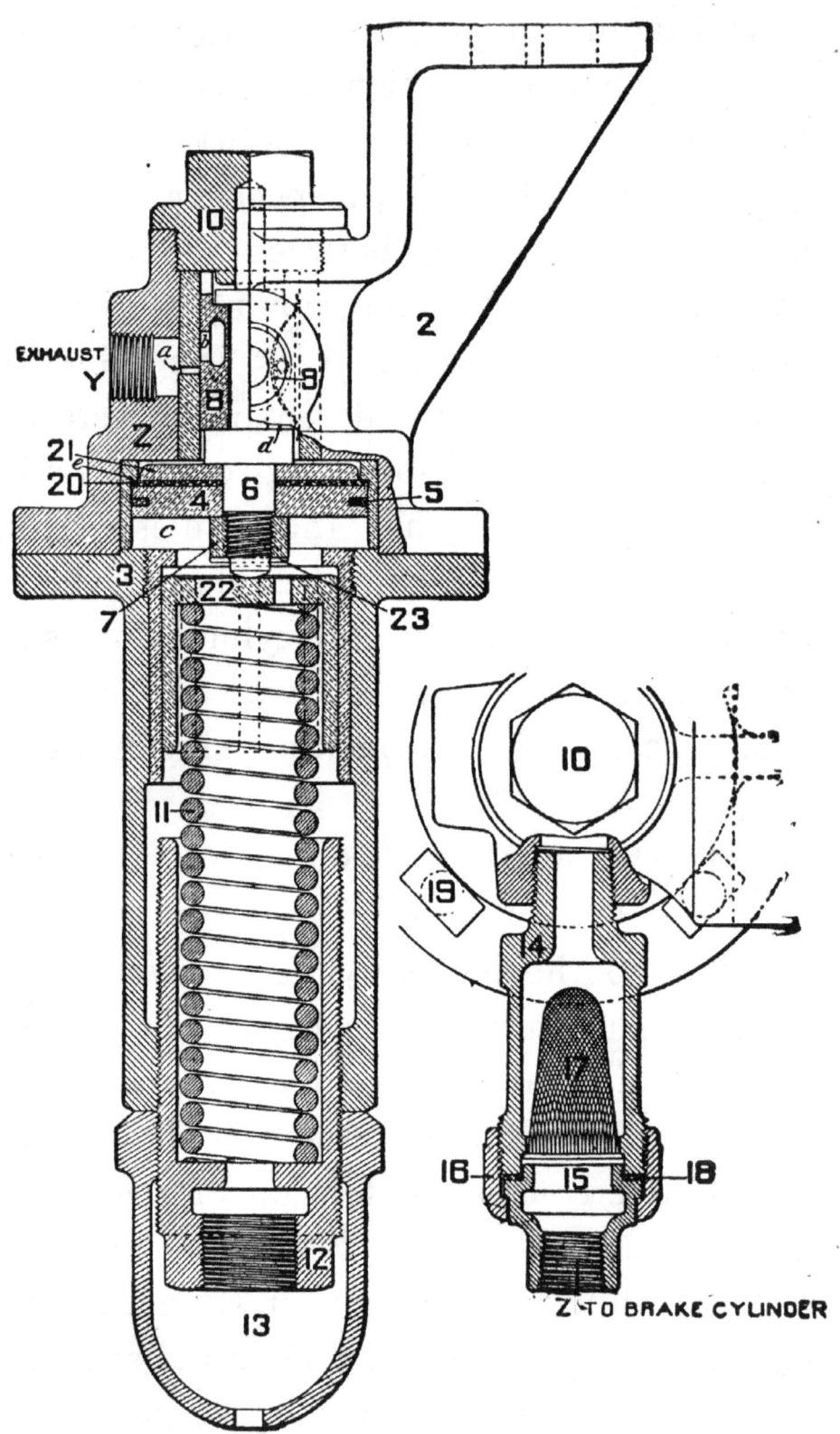

PLATE NO. 24—THE AUTOMATIC REDUCING VALVE.

DESCRIPTION OF PLATE 24—AUTOMATIC REDUCING VALVE

10. Cap nut.
9. U spring of slide valve.
8. Slide valve.
6. Slide-valve piston.
11. Regulating spring.
12. Regulating nut.

MODERN AIR-BRAKE PRACTICE

to sixty pounds, when the regulating spring forces the slide valve up and thereby closes the exhaust port *a*, and holds the sixty pounds in the cylinder until the engineer releases the brake in the usual way.

Plates 25, 26 and 27 illustrate the positions of the ports in the valve seat and slide valve of the reducing valve when making a service stop, an emergency stop, or when there is sixty pounds or less in the cylinder.

The opening *d* in the side of the slide valve always admits cylinder pressure to port *b*, and, as port *b* is triangular in form, when a service stop is made the largest end of port *b* is in register with port *a*, to allow the air to reduce as rapidly as possible from the cylinder, but when an emergency application is made the slide valve is forced down so that the small end of port *b* is in register with port *a*, and as the surplus cylinder pressure is gradually exhausted the regulating spring gradually raises the slide valve until, when there is a fraction less than sixty pounds left in the cylinder, port *b* is beyond port *a*, and the exhaust is closed.

The air remaining in the cylinder is released in the usual manner, by way of the triple exhaust.

The reducing valve should be examined occa-

ITS USE AND ABUSE

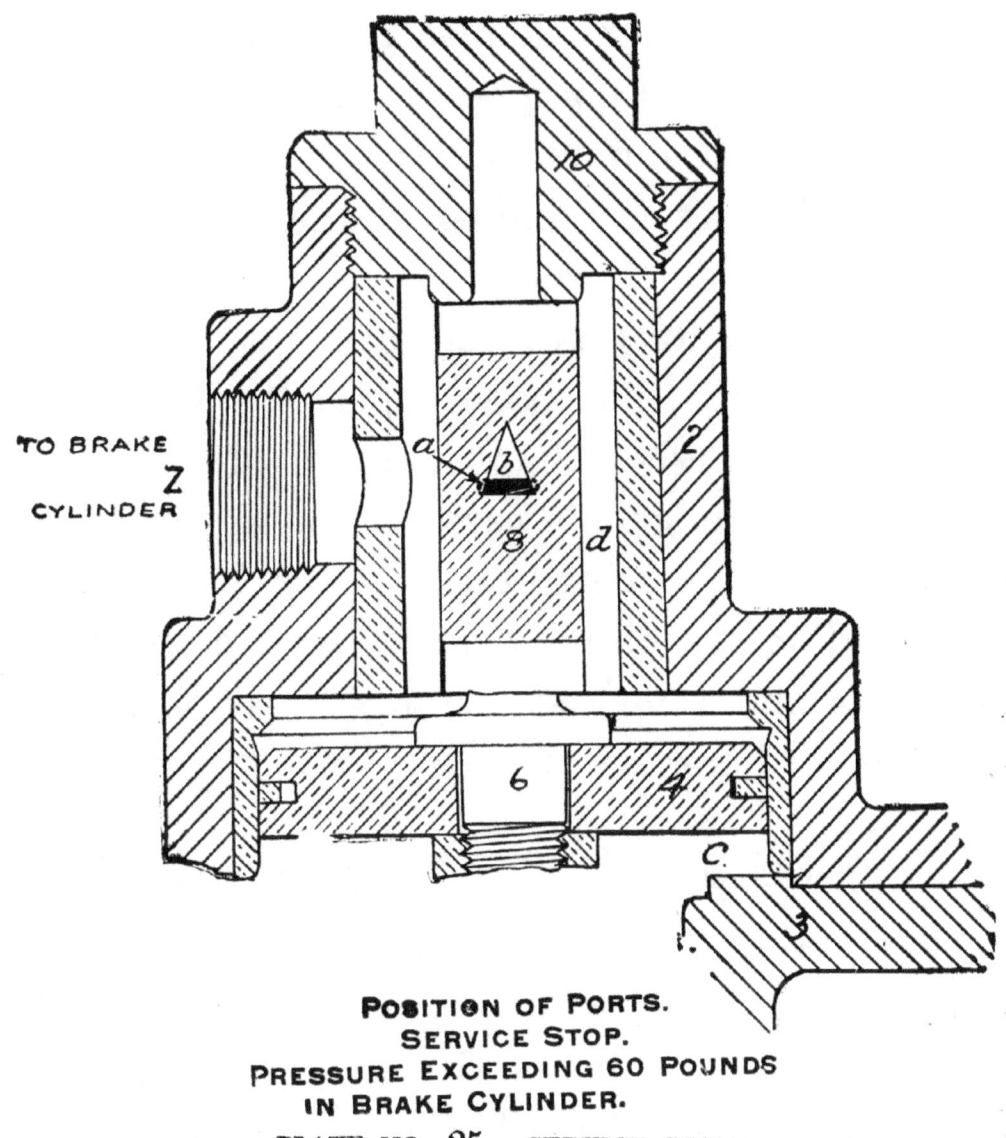

POSITION OF PORTS.
SERVICE STOP.
PRESSURE EXCEEDING 60 POUNDS
IN BRAKE CYLINDER.
PLATE NO. 25.—SERVICE STOP.

DESCRIPTION OF PLATE 25—SERVICE STOP

8. Face of slide valve, showing large end of port *b* to be in register with exhaust port *a*.

MODERN AIR-BRAKE PRACTICE

sionally in order to detect and overcome any possible leak through the discharge port.

Cars that are not equipped with the automatic reducing valve should never be attached to trains employing the high-speed brake, unless the brake cylinders are equipped with the safety valve provided for temporary use in such cases. The safety valve has been especially designed to prevent a higher than standard pressure in the brake cylinders of cars not equipped with the automatic reducing valve. It may be quickly screwed into the oiling hole of the brake-cylinder head and removed when the car is again placed in ordinary service.

HIGH-PRESSURE CONTROL OR SCHEDULE U

The purpose of the high-pressure control equipment is to enable enginemen to safely handle freight trains which are hauled out empty and brought back loaded.

For example, all freight-brake rigging is supposed to be adjusted so that the brake power exerted will be equal to only seventy per cent of the light weight of the car with a seventy-pound auxiliary pressure, and when you load a car, you of course change its weight; consequently if the brake power on an empty car

ITS USE AND ABUSE

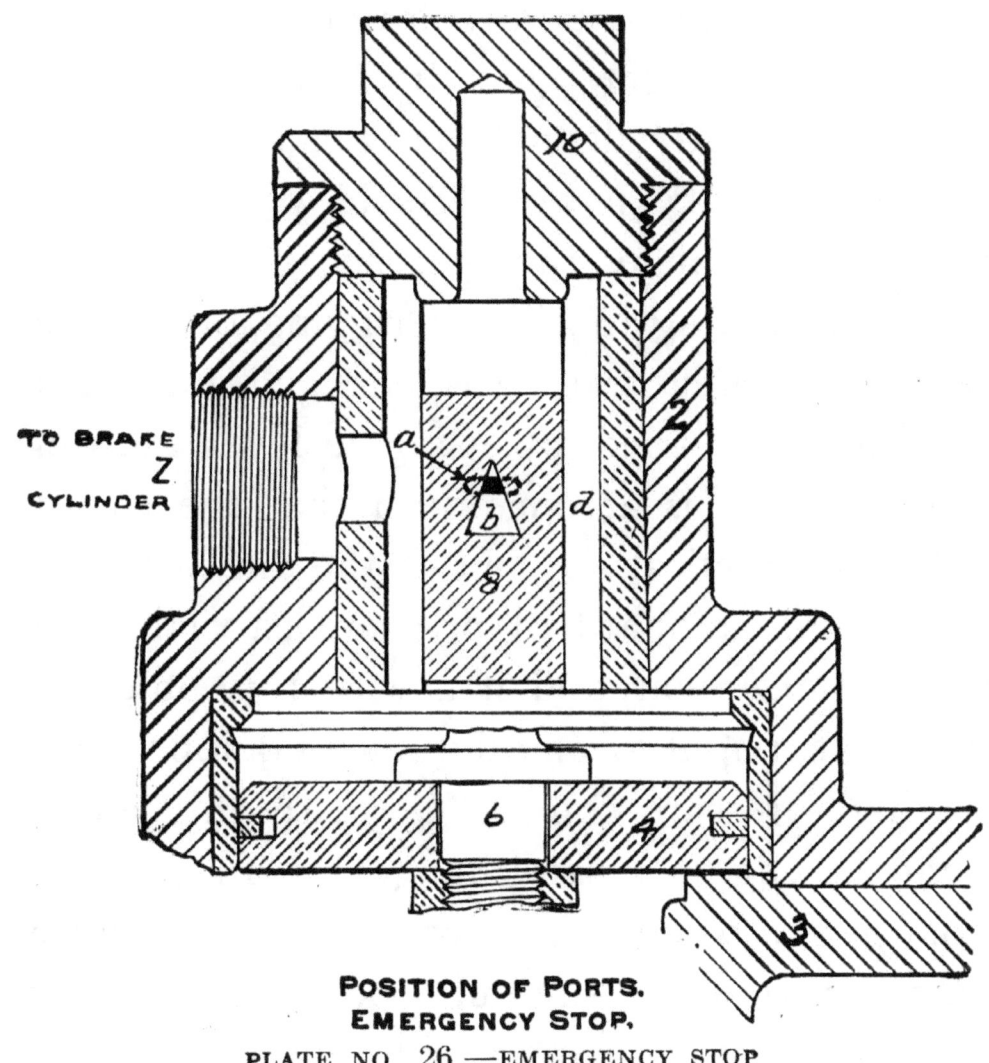

POSITION OF PORTS.
EMERGENCY STOP.

PLATE NO 26.—EMERGENCY STOP

DESCRIPTION OF PLATE 26—EMERGENCY STOP

8. Face of slide valve, showing small end of port *b* to be in register with exhaust port *a*.

should be only seventy per cent, that percentage would be very materially lowered when you increase the weight by loading the car. Even a very light load will materially change the percentage of brake power. As it would be very difficult to change the percentage of brake power by altering the brake rigging every time the weight of a train was changed (although this has been tried by using a lever shifting attachment) it is at once seen that the easiest and most practical way out of the difficulty is to change the standard of pressure carried in the auxiliary reservoir, and it is with this object in view that freight locomotives are equipped with the high-pressure control, for with this equipment an engineer can change his air pressure from 70 and 90 pounds to 90 and 110 pounds by simply turning a cut-out cock, and thereby increasing the percentage of brake power on his train.

To make this plain to you I will explain by saying that if the brake-piston travel on a car is eight inches, and you make a service application of thirty pounds from a seventy-pound auxiliary and trainpipe pressure you would simply get fifty pounds in your brake cylinder, and would be wasting ten pounds of your trainpipe pres-

ITS USE AND ABUSE

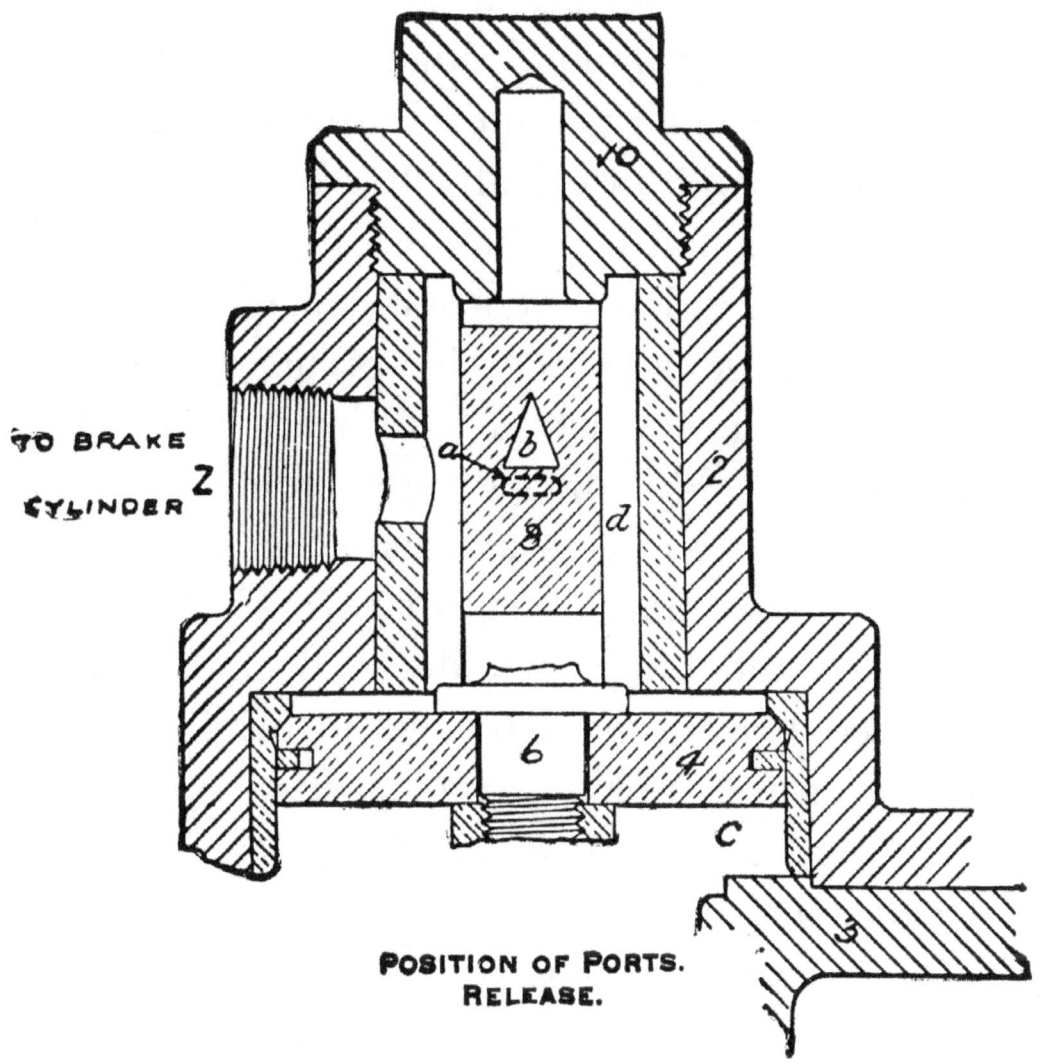

PLATE NO. 27.—RELEASE POSITION.

DESCRIPTION OF PLATE 27—RELEASE POSITION

8. Face of slide valve, showing port *b* closed to exhaust port *a*.

sure, because the auxiliary and brake cylinder would have equalized at fifty pounds with a twenty pound trainpipe reduction. Now, if you should have ninety pounds in the auxiliary you would have to draw off about twenty-six pounds in order to equalize the auxiliary and brake cylinder pressures, but they would equalize at about sixty-seven pounds, thereby giving you much more brake power with a full service application than you would ordinarily get from a seventy-pound trainpipe pressure with an emergency application. The reason for this is because the auxiliary and brake cylinder will equalize at a point ($\frac{2}{7}$) two-sevenths below the original auxiliary pressure. For example, a 20-pound reduction from a 70-pound auxiliary pressure will equalize at 50, and 20 is two-sevenths of 70.

By this arrangement an engineer can greatly increase the brake power on his train so that he has it under better control in descending grades, and with little or no chance of sliding wheels, for the reason that the increased load not only makes the increased cylinder pressure safe, but absolutely essential.

As a precaution against sliding wheels on the engine and tender, there is attached to them safety valves which automatically let out all

ITS USE AND ABUSE

but fifty pounds of the brake cylinder pressure when an application is made. An additional safeguard against sliding or heating of engine tires is the Dukesmith Driver Brake Control Valve, furnished by the Dukesmith Air Brake Company, Pittsburg, Pa. With this new valve an engineer can release any part or all of the brake pressure on the locomotive without interfering with the train brakes. (See Driver Brake Control.)

The difference between the high pressure control and the high-speed brake is as follows: the cars require no additional parts when using the high-pressure control; safety valves are used on the engine and tender instead of automatic reducing valves, and plain triple valves are used on both the engine and tender brakes, whereas a quick-action triple is used on the tender with the high-speed brake. The duplex pump governor is piped to both the main reservoir and slide valve feed valve with the high pressure control, whereas with the high-speed brake the governor is piped direct to the main reservoir.

Owing to the fact that the ninety pound pump governor is piped to the feed valve and because the feed valve is automatically cut out by the action of the rotary whenever the handle of the brake-valve is in any other position but running or release, it will be seen that when the handle of the brake valve is in any other position the 110-pound governor controls the pump, thereby causing it to quickly pump up the excess pressure.

With the high speed brake the governor is piped direct to the main reservoir, the same as with the quick-action equipment, consequently the cutting in or out of the ninety-pound governor by the quarter-inch cut-out cock on the governor pipe will give you the low or high pressure as desired. The reason for having but one cut-out cock for the two governors with the high-speed brake is because if you cut in the ninety-pound governor the steam valve will be closed at ninety pounds, and if you cut out the ninety-pound governor it will require 120 pounds to unseat the diaphragm valve in order to let the air shut off the steam valve. The tension of the steam valve spring is, of course, always the same, no matter which governor is in use, but the tension of the diaphragm spring (41) is regulated by nut 40, so that one diaphragm valve will be lifted by ninety and the other by 120 pounds, or, if you are using the high pressure control, at ninety and 110 pounds.

COMBINED STRAIGHT AIR AND AUTOMATIC ENGINE BRAKE

A very good addition, indeed to the air-brake

ITS USE AND ABUSE

system has recently been made by what is known as the Combined Straight Air and Automatic Engine Brake. Besides the regular apparatus used with the automatic brake, the Westinghouse and New York Straight Air equipment consists of the following parts: a double check valve for the purpose of automatically shifting the connection from the cylinder to either the triple valve or the straight air-brake valve, as the case may require; a straight air-brake valve, having three positions, release, lap and application; a slide valve feed valve, set at forty-five pounds, and attached to the straight air-brake valve, to reduce the main reservoir pressure when using straight air. The double check valve is used on both the engine and tender brakes. The Dukesmith Combined Automatic and Straight Air Brake Valve does not require the use of double check valves, and has an additional advantage in that the engineer can release the locomotive brakes with this valve even though the trainpipe hose should be bursted. For full description of this valve see Driver Brake Control.

PLATE 28—THE DOUBLE CHECK VALVE

The double check valve consists of a casing (2-3) with two end and two side openings, and has inside a loose, spool-shaped piece with a leather seat

MODERN AIR-BRAKE PRACTICE

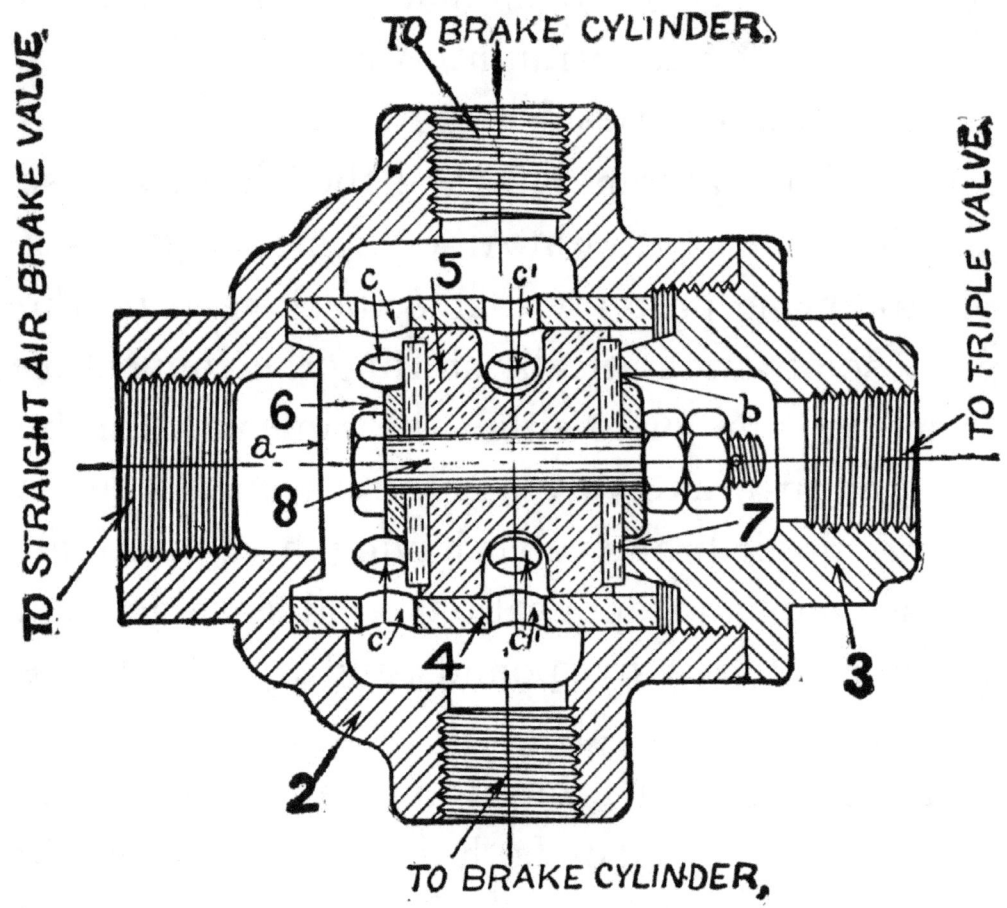

PLATE NO. 28.—DOUBLE CHECK VALVE.

DESCRIPTION OF PLATE 28—DOUBLE CHECK VALVE

4. Bushing.
5. Check valve.
a and b. Valve seat.
c, c. Ports for "straight air."
c_1, c_1. Ports for "automatic."
7. Leather gasket.

ITS USE AND ABUSE

on each end (6) for the purpose of making a joint with the valve seat (*a-b*) at either of the end openings, against which it is driven by the air pressure entering at the other.

The pipe leading from the straight air-brake valve is connected to one end opening of the double check valve, and the pipe from the triple is connected to the other end opening, and the connection with the brake cylinder is made by a pipe leading from either of the side openings, and to the other side opening is attached a safety valve set at about fifty pounds.

Plate 28 shows the double check valve when straight air is being used, for as the air from the brake valve strikes the check valve it is forced against seat *b*, which shuts off the triple and opens port *c*, which allows the air to rush into the cylinder.

To release the brake the engineer simply places the handle of the brake valve on release position and the air in the cylinder returns through the same ports in the check valve and escapes to the atmosphere by way of the release port in the brake valve.

To apply the brakes with the automatic, the old style straight air brake valve *must* be in release position, and when using the straight air the

MODERN AIR-BRAKE PRACTICE

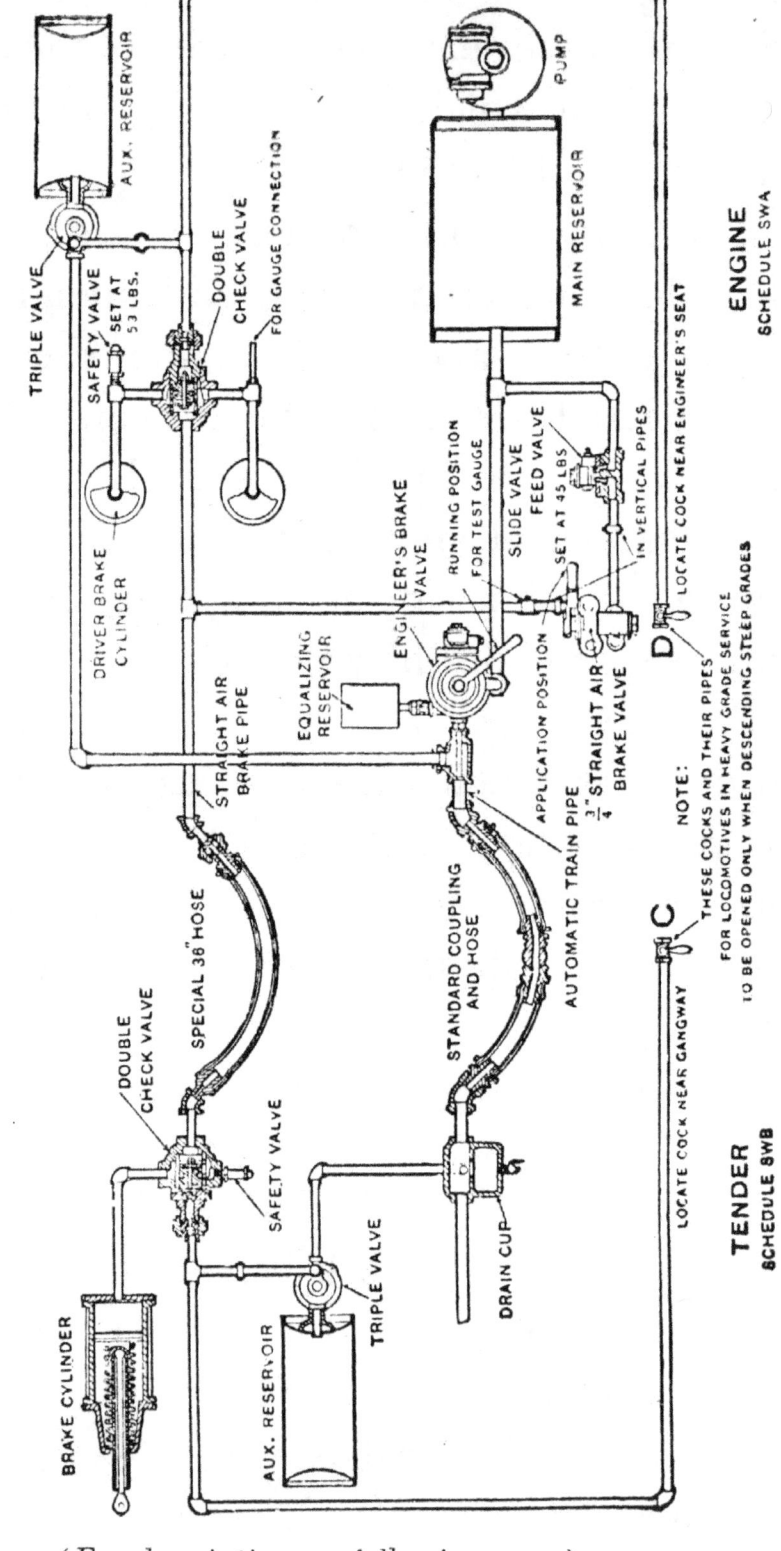

PLATE 28-A. METHOD OF PIPING THE WESTINGHOUSE COMBINED AUTOMATIC AND STRAIGHT AIR BRAKE.

(For description see following page.)

ITS USE AND ABUSE.

DESCRIPTION OF PLATE 28A.

[Plate 28A is a diagrammatic illustration showing the method of piping the Westinghouse Combined Automatic Straight Air Valve. The main features to be remembered are that the hose between the engine and tender, marked special 36 inch hose, should be one continuous piece in order to avoid possible leakage; that with this arrangement two double check valves are needed and two exhaust valves marked D and C, are used for the purpose of enabling the engineer to reduce the pressure on the locomotive when descending heavy grades, or when the wheels are sliding; as shown in the chart, there must also be a safety valve attached to the brake cylinders on both the engine and tender, and there should also be a pressure gauge for indicating what the brake cylinder pressure is at all times, and this is very important, for the reason that should the reducing valve become defective it is liable to allow a much too heavy pressure to get into the cylinder. The exhaust valves C and D, as shown in the diagram, are located on the tender and engine respectively, and should be within easy reach of the engineer, for when they are needed they are needed in a hurry.]

MODERN AIR-BRAKE PRACTICE

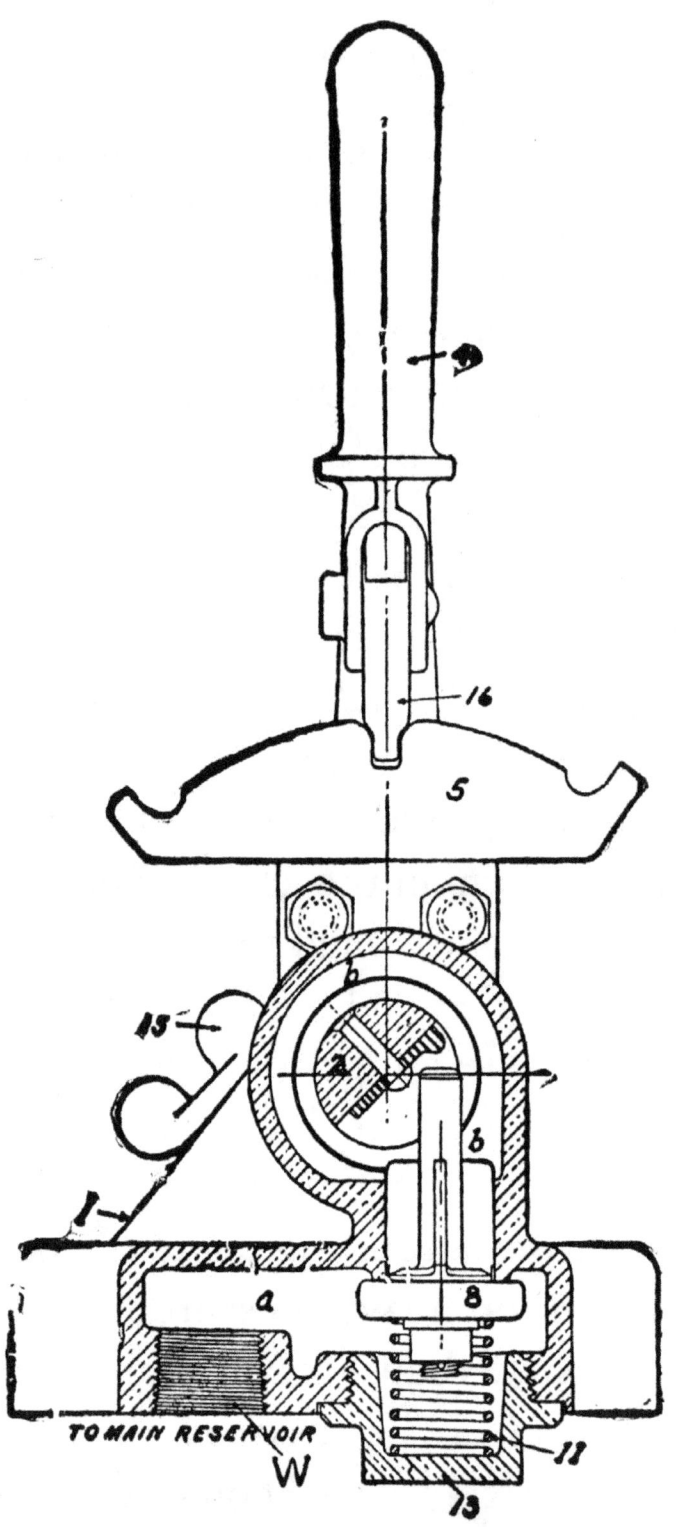

PLATE NO. 29A.—BRAKE VALVE AND COMBINED STRAIGHT AIR AND AUTOMATIC ENGINE BRAKE.

(*For description see following page.*)

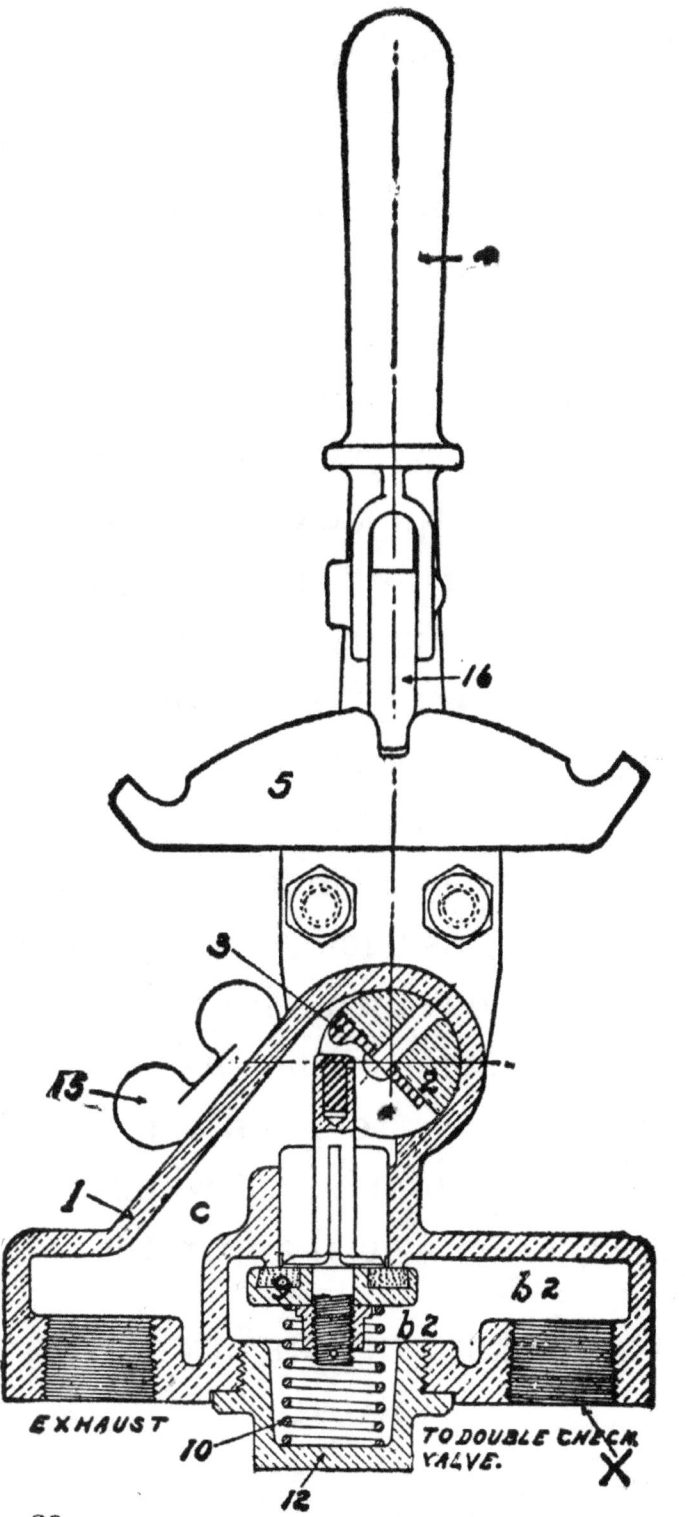

PLATE NO. 29B.—BRAKE VALVE AND COMBINED STRAIGHT AIR AND AUTOMATIC ENGINE BRAKE.

Description of Plates 29A *and* 29B.—2. Shaft attached to handle (4) for operating valve 8 and release valve 9. The handle is on lap position.

MODERN AIR-BRAKE PRACTICE

automatic brake valve must be left in running position. The engine brakes cannot be released with the Westinghouse or New York Straight Air Brake Valve if the triple is in application position.

When a reduction is made on the trainpipe pressure in the usual way, with the automatic brake valve, the air from the auxiliary forces the check valve against seat a, and thereby opens ports c_1, which allows the auxiliary air to rush into the cylinder. The brake is released in the usual way, for when the automatic brake valve is placed in full release position the triple piston reverses the slide valve, and the exhaust being thus opened the air in the cylinder flows back through ports c_1 in the check valve and out through the triple exhaust.

When an engineer wishes to do so he can keep his train brakes released and still have his engine and tender brakes set, when his engine is equipped with this special apparatus.

PLATES 29A, 29B AND 30—THE STRAIGHT AIR-BRAKE VALVE

The Westinghouse straight air-brake valve has three positions: release, lap and application. Plates 29 and 30 show it on lap. It is very simple, as the essential parts are the handle (4); the shaft (2), to which the handle is fastened, which oper-

ITS USE AND ABUSE

ates two check valves (8 and 9). Check valve 8 contains the supply of air from the main reservoir to the brake cylinder, and valve 9 controls the exhaust from the cylinder.

Look at plate 29A and imagine that you have moved the handle to the right, which would cause the shaft to force valve 8 down and allow main reservoir pressure, which is always in chamber a, to flow under the valve into passage b and through b_1 (plate 30), b_2 and X (plate 29B) to the double check valve and and on into the cylinder, as explained under plate 28.

To release the brake, the handle is moved back to the extreme left, which causes the shaft to allow valve 8 to reseat, and forces valve 9 down, when the air from the cylinder passes back through X, b_2, under valve 9, through passage c to the exhaust.

The slide valve feed valve is attached to the pipe leading to the double check valve, and when the handle is thrown to application position the flow of air from the main reservoir to the cylinder is shut off automatically at forty-five pounds. Should the feed valve leak, or be set too high, the safety valve will allow the surplus pressure to escape, and should the safety valve not seat properly it would allow the cylinder pressure to leak off when

MODERN AIR-BRAKE PRACTICE

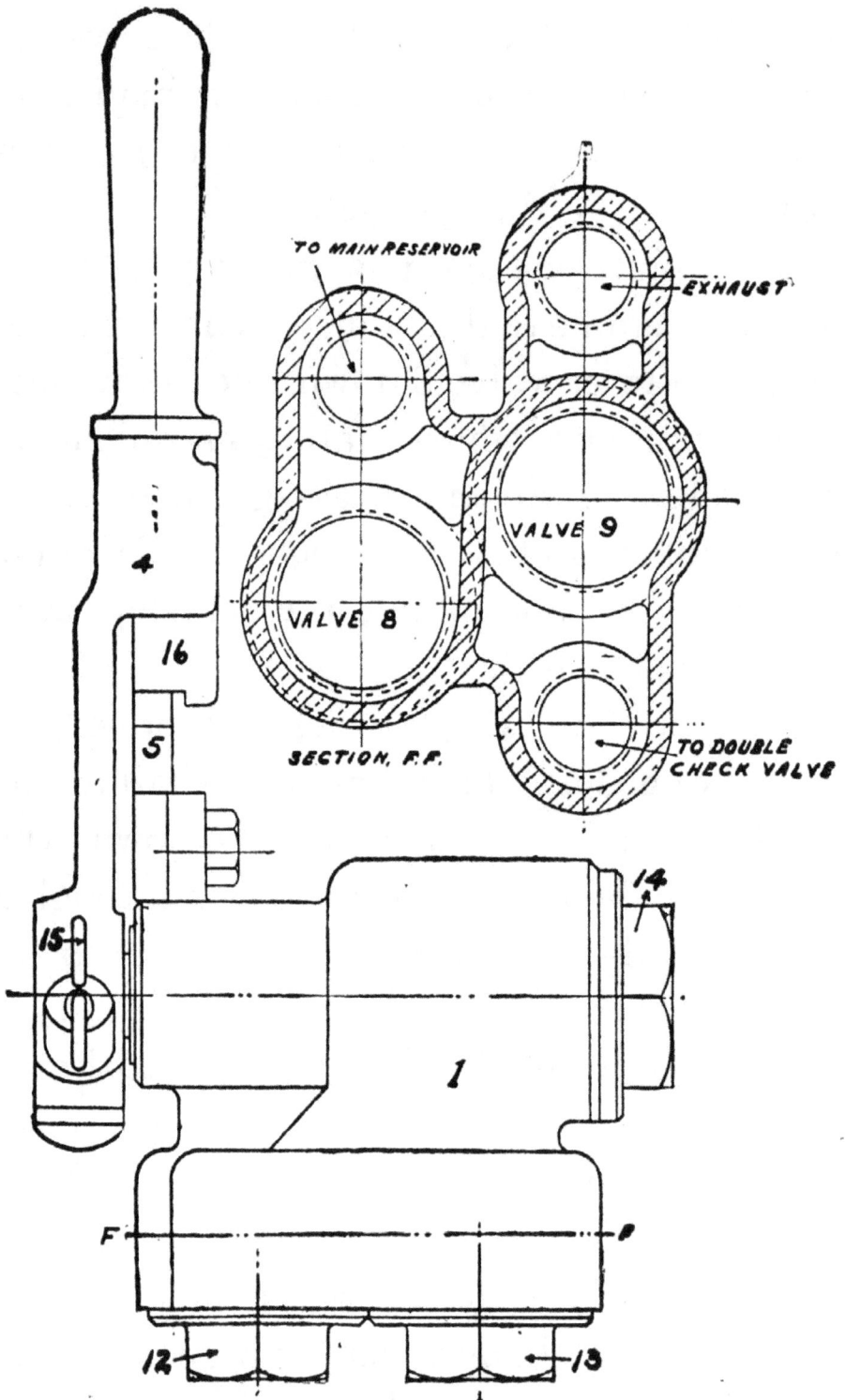

PLATE NO. 30A.—BRAKE VALVE FOR COMBINED STRAIGHT AIR AND AUTOMATIC ENGINE BRAKE.

(For description see following page.)

ITS USE AND ABUSE

DESCRIPTION OF PLATE 30

Section F.F. shows how the air passes from the main reservoir by way of valve 8 to the double check valve, and how in returning from the double check valve it passes under valve 9 to the exhaust.

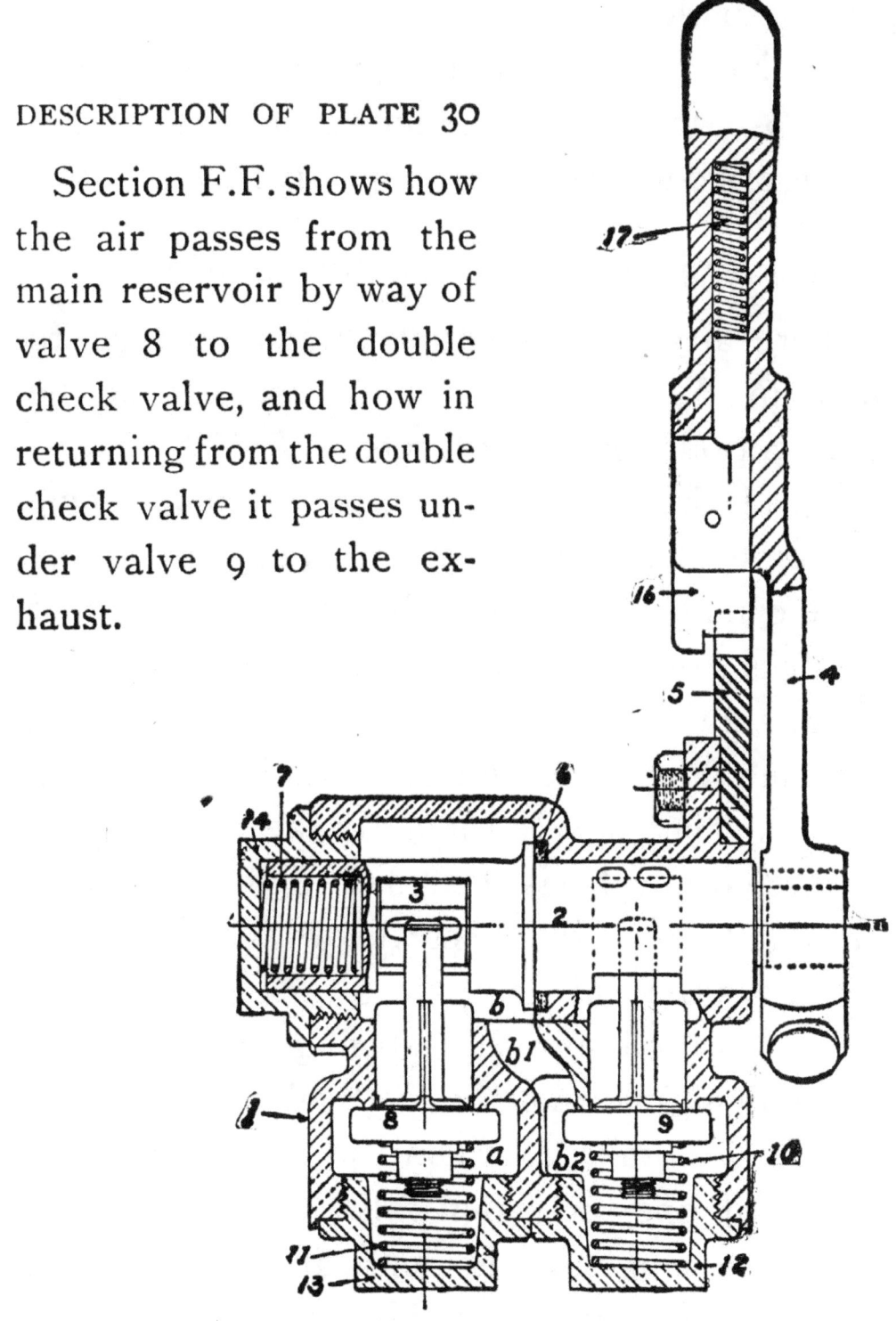

PLATE NO. 30.—BRAKE VALVE FOR COMBINED STRAIGHT AIR AND AUTOMATIC ENGINE BRAKE.

MODERN AIR-BRAKE PRACTICE

either a partial straight air or automatic application was made.

The New York Straight Air Brake Valve has four positions and the Dukesmith has five, and are described under their respective headings.

PLATE 31—THE WHISTLE SIGNAL SYSTEM

There are four essential things that go to make up the air-signal system, aside from the pipes, cut-out cocks, cords, etc.

Fig. 1 is the signal valve, and stands in the same relation to the whistle as the auxiliary does to the brake cylinder, for it is in the signal valve that the air is stored for use in blowing the whistle.

Fig. 2 is the car discharge valve, and stands in the same relation to the air signal as the conductor's valve does to the air brake, for when the car discharge valve is opened the air escapes from the signal pipe and causes the whistle to blow.

Fig. 3 is the whistle.

Fig. 4 is the improved reducing valve, which is to the air-signal what the feed-valve attachment is to the air brake, as it controls the pressure in the signal pipe and signal valve.

The reducing valve is identical in its operation with the old style feed-valve attachment, and when you understand one you know the other,

ITS USE AND ABUSE

for as the regulating spring 13 is set at forty pounds, the diaphragm piston (10) will keep the supply valve (4) off its seat until the main reservoir pressure (which flows in at A) has filled the signal pipe (B) to a fraction over forty pounds, when the piston is forced down and allows the supply valve to shut off the main reservoir pressure until the signal-pipe pressure is again reduced, when the piston will again raise and unseat the supply valve to allow the main reservoir to quickly restore the pressure in the signal pipe, when the valve will again seat by the piston being forced away from it.

The signal valve is attached to the main signal pipe by a short branch pipe at Y, and whatever pressure is in the pipe the same is in chambers A and B, for as air passes through port d into chamber A, it also passes down passage C and raises the diaphragm stem (10) so that the small groove cut around the stem at f is above bushing 9, and as the side of the stem is flat as far up as the groove, when the stem is raised the air is free to enter chamber B, and when it equalizes with A the stem drops to its seat (7) by its own weight and closes port e. The stem is attached to a rubber diaphragm (12), and as the whistle is piped to the signal

MODERN AIR-BRAKE PRACTICE

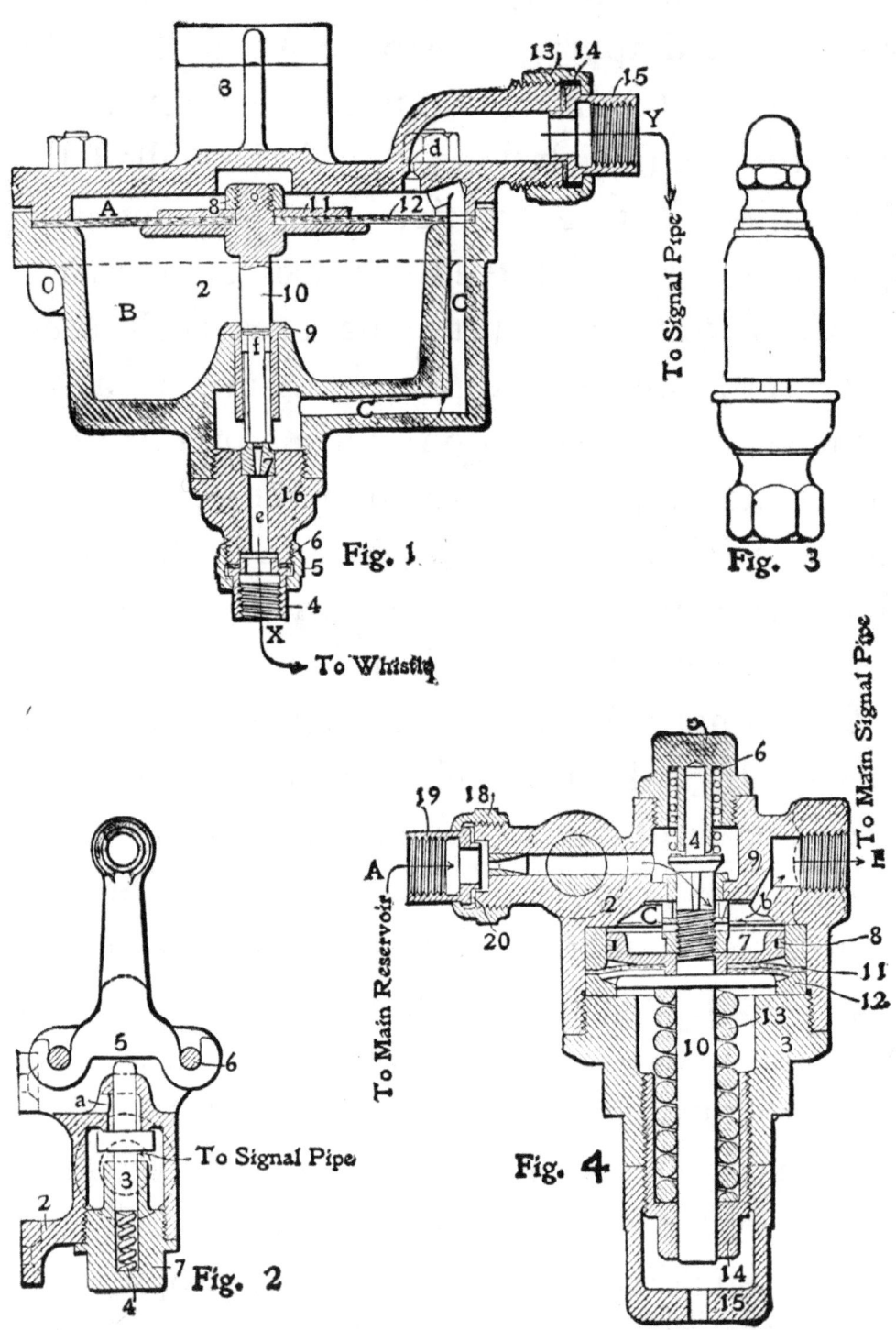

PLATE NO. 31.—WHISTLE SIGNAL SYSTEM

ITS USE AND ABUSE

DESCRIPTION OF PLATE 31—WHISTLE SIGNAL SYSTEM

Fig. 1. Whistle signal valve.

Fig. 2. Car discharge valve.

Fig. 3. The whistle.

Fig. 4. Reducing valve.

valve at X, whenever the lever (5, Fig. 2) of the car discharge valve is moved either to the right or left the small valve (3) is forced off its seat to allow the air to escape from the signal pipe, and when the pressure is thus reduced the air in chamber A is also reduced, and as the volume of B is so much greater than A the rubber diaphragm is forced up, which unseats the stem and allows the air in B and some of the signal-pipe air to rush out through the bell-shaped whistle and cause it to blow.

In order to insure the whistle giving the proper blast it is necessary to make a sudden reduction, and as it is the air in the signal valve that blows the whistle, at least two seconds must be allowed between each pull of the cord to let chamber B fully recharge, and on a long train four seconds is better.

Plates 32 and 33 are diagrammatic illustrations showing (32) the Quick-Action Automatic Brake, and (33) the High-Speed Brake Equipment. These plates are remarkably complete in detail, and the reader will at once see the exact relation each part is to the other.

QUESTIONS AND ANSWERS TO SECTION 2

THE PARTS OF THE WESTINGHOUSE AUTOMATIC, HIGH SPEED AND STRAIGHT AIR-BRAKE EQUIPMENT AND THEIR DUTIES

1. What is meant by an automatic brake?

Ans.—A brake that is self-acting.

2. When an engine is equipped with the Westinghouse automatic quick-action air brake, what are the essential parts, and what are their duties?

Ans.—A steam-driven pump to compress the air; a main reservoir in which the air is stored ready for use; an automatic air-controlled governor for stopping and starting the pump according to the amount of compressed air required in the brake apparatus; a duplex gauge for registering the pressure in the main reservoir and trainpipe; an engineer's brake valve for controlling the flow of air from the main reservoir into the trainpipe and from the trainpipe to the atmosphere; an equalizing reservoir attached to the brake valve for controlling the equalizing discharge valve within the brake valve; pipe connections between the pump and the main reservoir, between the main reservoir and the brake valve, between the main reservoir and the governor, between the main reservoir and the red hand of the gauge, between the equalizing reservoir and the black hand of the gauge, representing trainpipe pressure; and to the brake valve is attached the trainpipe, in which is located a cut-out cock just below the brake valve for the purpose of closing communications between the brake valve and trainpipe as occasion demands, as in double heading; branch pipes leading from the trainpipe to the triple valve, from the triple valve to the auxiliary reservoir, from the triple to the brake cylinders, as there are two brake cylinders on the engine the pipe leading from the triple to the cylinders is teed so that one branch leads to the right and the other to the left-hand cylinder; there are cut-out cocks on the branch pipe leading from the trainpipe to the triple and from the triple to the brake cylinder, and from the triple to the auxiliary; there is an auxiliary reservoir for supplying air to the brake cylinder and a plain triple valve for charging, setting, and releasing the brake. When an engine is said to be fully equipped there is also a truck brake equipment consisting of an additional auxiliary reservoir of

ITS USE AND ABUSE

smaller capacity, a truck brake cylinder, an automatic slack adjuster, and a Driver Brake Control Valve, with Automatic Release Signal, in which case there would be no triple on the tender.

3. What additional apparatus is required on a passenger engine from that of a freight engine?

Ans.—The whistle signal equipment, consisting of a reducing valve set at forty pounds, a whistle signal valve, and an air whistle, together with a signal pipe and suitable connections between the main reservoir and reducing valve, and from the reducing valve and signal valve, and from the signal valve to the air whistle.

4. What additional apparatus is needed to change the quick-action equipment into a high-speed equipment?

Ans.—A duplex pump governor, an extra slide valve feed valve and bracket, and an automatic reducing valve.

5. What are the parts required on a tender in ordinary freight or passenger service, with a standard Westinghouse equipment?

Ans.—A trainpipe, brake cylinder, auxiliary reservoir and a plain triple valve with branch pipes, cut-out cocks, an angle-cock and hose. In passenger service there is, in addition, the signal pipe with its angle-cock and hose. But when an engine is equipped with the Dukesmith Driver Brake Control Valve no triple valve or auxiliary reservoir is required on the tender.

6. When equipped for high-speed brake, what additions are needed on a tender?

Ans.—An automatic reducing valve set at sixty pounds, and the quick-action triple is substituted for the plain triple. To be fully equipped the tender in any kind of service should also have an automatic slack adjuster and a release signal.

7. What apparatus is required on a freight car?

Ans.—A trainpipe with angle-cocks and hose at both ends; a quick-action triple valve; a brake cylinder; a branch pipe leading from the trainpipe to the triple in which is a cut-out cock; a release rod leading from the release valve on the auxiliary to either side of the car; a pressure-retaining valve clamped to the end of the car near the top along side the staff of the hand brake; a pipe connecting the retaining valve with the triple exhaust so that when it is desired to allow the engineer to recharge the auxiliary reservoir on descending grades the handle of the retainer can be turned up and thereby retain a pressure of fifteen or fifty pounds in the brake cylinder while recharging, (at the foot of the grade, or sooner if desired, the handle must be turned down again in order to

ITS USE AND ABUSE

permit the engineer to release the brakes); if the car control valve is used instead of the ordinary retaining valve there should also be a release signal, which is clamped to the end of the car just below the top and piped direct to the brake cylinder, the purpose of which is to signal the train crew every time the brake sets, releases, leaks off, has too much piston-travel or sticks. It also enables the trainmen to detect a "kicker," or brake that flies into emergency with a service application. When a brake sticks so that it cannot be released from the engine the brakeman can release it from the top of the car by simply moving the handle of the car control valve which is on the end of the car until the signal drops into its pocket. On a dark night when the brakes are felt to be dragging, the brakeman will not have to drop off and watch the brakes as the train passes in order to find the defective brake, but can, when the release-signal is used, run back over the top of the train, and by the light of his lantern, see the release-signal as it appears above the top of the car, as it is a foot square, and having thus quickly and surely found it, has only to open the valve and let the brake off without having to take any personal risk, as he does when dropping off the train, and without causing a dangerous delay to the train, as is frequently the case when brakes get to dragging and have to be bled off by the auxiliary release valve. Whenever the air is out of the brake cylinder the release-signal will automatically drop into its pocket below the top of the car. To find a "kicker" in a train, have the engineer make a five-pound reduction, and on all cars on which the triple valves are properly working the signal will show itself just a little way above the top of the car. In case the "kicker" should be caused by a weak graduating spring in one of the first seven cars, it would throw the whole train in emergency on the first light reduction, but on the car which has the "kicker," if it is not caused by a weak graduating spring in the first seven cars, the signal will not move with the first five-pound reduction, so that when the next five-pound reduction is made, the signal which did not move at all on the first reduction will jump up, showing that the defective triple is on that particular car. Having thus found the "kicker," cut that brake out, card the car and report as usual. As uneven piston-travel is one of the worst evils railroads have to contend with, all air brake cars should also be equipped with the automatic slack-adjuster.

8. On a passenger coach, in ordinary service,

ITS USE AND ABUSE

what additional apparatus is required from that of a freight car?

Ans.—Either a car control valve or the ordinary conductor's valve attached to the end of a pipe which leads from the trainpipe to within the body of the coach, (usually in the toilet room), by means of which the conductor can stop the train, if desired, by letting the air out of the trainpipe either gradually or suddenly, according to circumstances. For an ordinary stop it should be opened gradually, but for an emergency it should be pulled wide open quick and held open until the train comes to a full stop, when it should be again closed. On a passenger car there is also a whistle signal pipe from which there is a branch pipe leading to the car discharge valve, and as there is a cord attached to the discharge valve, a sudden jerk of the cord will open the valve and let out signal-pipe pressure, thereby causing the whistle to blow on the engine; a lapse of at least two seconds should be allowed between pulls in order to insure the correct signal, and on long trains four seconds is better. All passenger coaches should have the automatic slack-adjuster and the release-signal.

9. To equip a passenger car for the high-speed brake, what extra apparatus is needed?

Ans.—Simply the automatic reducing valve.

Where a car is temporarily used in a train equipped with the high-speed brake, a safety valve must be screwed into the oil hole of the brake cylinder, and when it is returned to ordinary service the safety valve should be removed and the plug replaced in the oil hole of the cylinder. Ordinary passenger coaches do not usually have retaining valves on them, but all Pullman and most private cars do, and are placed on the end of the car in the vestibule.

10. Now, to return to the engine equipment. How many main pistons are there in the pump?

Ans.—Two; the main steam piston and the main air piston, the former in the top and the latter in the bottom section of the pump.

11. Are these pistons connected together?

Ans.—Yes. One piston rod operates both.

12. What is the principal difference in the construction of the eight-inch pump from that of the $9\frac{1}{2}$-inch pump?

Ans.—The eight-inch pump has its steam valves on the side and top, while the $9\frac{1}{2}$-inch pump has its steam valves all at the top, and where in the eight-inch pump the flow of steam is controlled by pistons with packing rings, in the $9\frac{1}{2}$-inch pump the flow of steam is controlled by a common D slide valve actuated by two pistons of unequal

ITS USE AND ABUSE

diameter. Both pumps contain a reversing valve with reversing valve rod which operates within the hollowed-out main piston rod.

13. What difference is there in the two pumps in regard to the air valves?

Ans.—The eight-inch pump has its discharge and receiving valves all on one side, whereas in the 9½-inch pump there is a receiving and discharge valve on either side of the pump. The eight-inch pump has two air inlets; the 9½-inch pump has only one.

14. Is there any difference in the lift of the air valves in the two pumps?

Ans.—Yes; in the eight-inch pump the receiving valves have a lift of $\frac{1}{8}$ of an inch, and the discharge valves have $\frac{3}{32}$, whereas in the 9½-inch pump all air valves have a lift of $\frac{3}{32}$.

15. Can you explain the operation of the steam end of the eight-inch pump?

Ans.—When the steam enters at the side of the pump it flows into a chamber in which is contained two pistons of unequal diameter, which, in combination, is known as the main steam valve of the pump; leading from near the top of this chamber there is a steam passage which conducts the steam to the top of the reversing valve and from thence through a small passage into another chamber in which is contained the reversing piston (see plate 14), the steam having thus passed from the main valve chamber to the reversing valve chamber and into the chamber of the reversing piston, and as the reversing piston and the bottom piston of the main valve combine to make a much larger area than the top piston of the main valve, it naturally forces the main valve down so that the steam from the main valve chamber can pass through the bottom ports in the main valve chamber to the steam cylinder of the pump, and thereby force the main steam piston up; as the main steam piston rises, the reversing plate strikes the shoulder of the reversing valve rod and thereby changes the position of the reversing valve, so that the steam in the chamber over the reversing piston can pass through the second passage in the head through the cavity in the reversing valve, through the lower passage in the head, to the exhaust passage, which begins at the bottom of the reversing piston chamber and ends at the main exhaust. The pressure having now been removed from the top of the reversing piston, the large piston in the main valve chamber is forced up, causing the small bottom valve to close the bottom supply ports to the steam cylinder and at the same time open the bottom exhaust ports of the steam cylinder, thereby allowing the steam to exhaust from the under side of the main piston. While the main valve is in this position, the exhaust port, from the top side of the piston, is closed, and the supply port from the main valve chamber to the top of the steam piston is open, so that the steam can pass from the main valve chamber to the top of the main steam piston and thereby force it down; in doing so the reversing plate engages the button on the end of the reversing valve rod, which again changes the position of the reversing valve, thereby allowing the same action to take place as in the beginning.

16 How do the air valves in the eight-inch pump operate?

Ans.—On the up-stroke of the main air piston a partial vacuum is formed in the air cylinder, and as the atmospheric pressure is then greater on the outside of the pump, it enters the air inlet and forces the receiving valve off its seat until the air cylinder is filled with atmospheric pressure. As the reversing valve causes the main pistons to reverse just before reaching the top of the cylinders, the compression of the air, which begins immediately that the piston starts down, causes the receiving valve to be firmly closed, and as the compression in the air cylinder is increased over the pressure in the main reservoir, it causes the discharge valve to be lifted and allow the air from the pump to be forced into the main reservoir. As the piston starts up again the main reservoir air holds the discharge valve to its seat in the same manner that the air cylinder pressure held the receiving valve to its seat on the down-stroke. In making the up-stroke, the upper receiving and discharge valves operate in the same manner as did the lower valves.

17. Can you explain the operation of the steam end of the 9½-inch pump?

Ans.—The steam entering the pump at the main steam connection is conveyed to a chamber in the top head of the pump in which is contained two pistons of unequal diameter, on the piston rod of which is a common D slide valve moving over a seat having three ports. One of these ports leads to the under side of the main steam piston, one leads to the top of the main steam piston, and one to the main exhaust, and as the cavity in the slide valve can only connect two of these ports at any one time, it naturally follows that when the steam enters between the two pistons of unequal diameter that the slide valve is moved towards the end of the chamber containing the large piston. In doing so it

ITS USE AND ABUSE

uncovers the port leading to the under side of the main steam piston, which causes the piston to move up, which in doing so operates the reversing valve in the same manner as previously explained in the eight-inch pump. As the piston nears the top it changes the position of the reversing valve in order to allow steam to pass from between the two unequal pistons, through a port in the bushing, to the outer side of the large piston. By this action the pressure on both sides of the large piston is equalized, and as there is no pressure on the outer side of the small piston, the expansion of the steam forces the slide valve to the end of the chamber in which is contained the smaller piston. This action causes the slide valve to connect the port leading from the bottom side of the main steam piston with the port leading to the main exhaust, and while the steam is exhausting from the under side of the piston, live steam is being admitted to the top of the main steam piston through the port in the slide valve seat which is now uncovered, thereby forcing the main steam piston down, which in doing so causes the reversing valve to be reversed the same as in the eight-inch pump, which action exhausts the steam from the outer side of the large piston so

MODERN AIR-BRAKE PRACTICE

that it is again forced in the same direction as described in the beginning of the stroke. The action of the air valves is the same as in the eight-inch pump except that the lift of the valves in the 9½-inch pump is $\frac{3}{32}$ of an inch all around, whereas in the eight-inch pump the receiving valves have $\frac{1}{32}$ of an inch greater lift than the discharge valves.

18. What is the name of the pipe leading from the pump to the main reservoir?

Ans.—The main reservoir discharge pipe.

19. What is the name of the pipe leading from the main reservoir to the engineer's brake valve?

Ans.—Main reservoir return pipe.

20. What prevents the main reservoir pressure from flowing back into the pump?

Ans.—There are two valves, known as the discharge valves, which are held to their seat by the main reservoir pressure, but when the pump compresses air to a higher pressure than that contained in the main reservoir, the discharge valves are lifted from their seat until the pressures equalize, when the valves drop to their seat by their own weight.

21. If the main reservoir pressure begins at the pump, where does it end?

ITS USE AND ABUSE

Ans.—If the engineer's brake valve is on lap, it ends on top of the rotary valve and at the pump governor and at the red hand of the gauge.

22. If the brake valve is on lap, why does not the main reservoir pressure end on top of the brake valve?

Ans.—Because there is a branch pipe leading from the main reservoir pipe just before it reaches the brake valve, which carries the air to the red hand of the gauge and to the pump governor.

23. Are there any other attachments which might consume main reservoir pressure?

Ans.—Yes; the bell ringer and air sander, and if it is a passenger engine, the whistle signal pipe. Should a leak occur in any of these connections, it would be a main reservoir leak.

24. Where should the main reservoir on the engine be located?

Ans.—While circumstances regulate the location of the main reservoir, it should, however, be always placed in such a position that it will be lower than the pump, so that all oils and condensations may settle in it, and should be piped so that the discharge pipe, or the one which connects with the pump, is separated as far as possible from the return pipe leading to the engineer's brake valve.

25. What is the engineer's brake valve for?

Ans.—For the purpose of enabling the engineer to properly charge, set and release brakes and control the flow of main reservoir and trainpipe pressure.

26. What are the essential parts of the engineer's brake valve?

Ans.—The rotary valve and the handle which controls it, the equalizing discharge valve, the feed valve attachment, or trainpipe governor, and the equalizing reservoir.

27. What is the purpose of the rotary valve?

Ans.—To open and close the ports in the brake valve.

28. What is the handle of the brake valve for?

Ans.—To control the movement of the rotary valve.

29. What is the equalizing discharge valve for?

Ans.—To open and close the trainpipe exhaust port according to the pressure above or below it.

30. What is the equalizing reservoir intended for?

Ans.—To maintain a large volume of air on the upper side of the equalizing discharge valve, in order to compensate for the volume of air in

ITS USE AND ABUSE

the trainpipe, which is on the under side of the equalizing discharge valve.

31. What is the feed-valve attachment, or trainpipe governor for?

Ans.—As its name implies, it is for the purpose of controlling the pressure in the trainpipe.

32. In what way does it control the pressure in the trainpipe?

Ans.—As there is a regulating spring (see illustration) which is set at seventy pounds, it requires an air pressure of a little over seventy pounds to compress the spring and allow the feed valve to close and shut off the flow of air from the main reservoir to the trainpipe.

33. How many kinds of feed valves are there in use?

Ans.—Two. The old style feed valve is merely a poppet valve, while the new slide valve feed valve contains, in addition to the poppet valve feature, a slide valve, which is actuated by both air and spring pressure.

34. Which feed valve is preferable and why?

Ans.—The new slide valve feed valve; for the reason that with it the trainpipe pressure can be quickly raised and more evenly maintained, whereas with the old style feed valve the flow of air into the trainpipe is materially retarded after the pressure has reached fifty pounds, on account of the gradual closing of the poppet feed valve, while with the new slide valve feed valve the trainpipe port virtually remains wide open until the full seventy-pounds pressure is in the trainpipe. While the slide valve controls the port leading to the trainpipe, it in turn is controlled by the small poppet valve, for the reason that when the trainpipe pressure of seventy pounds forces the diaphragm of the regulating spring away from the poppet valve, the latter is allowed to seat, which prevents the circulation of the air through the feed-valve attachment, and the pressure thus becomes equalized on both sides of the slide-valve piston. The spring behind the piston forces it forward and causes the slide valve to close the trainpipe port. It is because the air is intended to circulate freely on both sides of the slide-valve piston that there are no packing rings on this piston.

35. In what position must the brake handle of the brake valve be in order to have the feed-valve attachment in operation?

Ans.—In running position, as that is the only position of the brake-valve handle in which you can get air from the main reservoir to the trainpipe through the feed valve.

ITS USE AND ABUSE

36. Can you trace the course of the air from the pump to the brake cylinder?

Ans.—In order to get the air from the main reservoir to the brake cylinder, if the handle of the engineer's brake valve is on lap at the beginning, it is necessary to make at least two movements of the handle of the brake valve, which I will explain in a moment. The pump having compressed the air, it is forced through the discharge valves and through the discharge pipe into the main reservoir, from thence it passes through the return pipe to the top of the rotary valve in the engineer's brake valve. When the handle of the brake valve is thrown to the left, or full release position, the main reservoir pressure can then pass through the largest ports in the brake valve direct into the trainpipe. From the trainpipe it passes through the branch, or cross-over pipe, to the trainpipe side of the triple piston, and in forcing the triple piston forward there is opened a feed groove in the casing of the triple piston cylinder, which allows the trainpipe pressure to flow over the piston and over the top of the slide valve into the auxiliary reservoir. The air has now been carried from the pump through the main reservoir, through the engineer's brake valve, through the trainpipe cross-over pipe, through the triple valve and into the auxiliary reservoir. When a sufficient pressure has been stored in the auxiliary reservoir and it is desired to set the brakes, the engineer must move the handle of the brake valve to at least service application position, which action causes the preliminary exhaust port in the brake valve to open and allow the pressure from the top of the equalizing discharge valve to escape to the atmosphere, which causes the trainpipe pressure, which is on the under side of the equalizing discharge valve, to force the valve up and open the trainpipe exhaust port. With the trainpipe exhaust port open the air rushes out from the trainpipe, and as the triple piston stands between the trainpipe and auxiliary pressure, it naturally follows that when the trainpipe pressure has been made lower than the auxiliary pressure the triple piston is forced towards the weaker pressure by the auxiliary pressure, and as it carries with it the slide valve, the ports in the slide valve and in the slide-valve seat are thereby opened, which allows the auxiliary pressure to flow into the brake cylinder against the brake piston, which is connected with the brake levers, which forces the shoes up against the wheels and the brake is then set.

ITS USE AND ABUSE

37. Can you explain how the brakes are released?

Ans.—By the excess pressure in the main reservoir. When it is desired to release the brakes the handle of the brake valve is placed in full release position, in order that the great volume of air contained in the main reservoir may pass quickly through the trainpipe and strike the triple piston a hammer blow, in order to overcome the pressure in the auxiliary reservoir, thereby causing the slide valve to be moved, so that the exhaust port of the triple valve will be opened and permit the brake cylinder pressure to pass out into the atmosphere, and the pressure having thus left the brake cylinder the return spring in the brake cylinder forces the brake piston back, thereby moving the brake levers to their original position, which allows the brake shoes to drop away from the wheels.

38. Can the brakes be released with the handle of the brake valve in any other position than that of full release?

Ans.—Yes. They can be released sometimes in running position, but it is a very dangerous practice for an engineer to do so, for the reason that when it is necessary to release the brakes in a train they should be all released at the same moment, if possible, because if some brakes release and others do not, it is very liable to pull the train in two.

39. How many positions are there on the brake valve?

Ans.—Five. Full release, running, lap, service and emergency.

40. What are these positions intended for?

Ans.—Full release is for charging and releasing the brakes. Running position is to enable the engineer to maintain an even pressure in the trainpipe and auxiliary reservoir while running along, and keep up the excess pressure, because as the air in the trainpipe escapes through leaks of any kind, the feed-valve attachment automatically opens to allow main reservoir pressure to flow into the trainpipe, but automatically closes when the pressure has been restored. Lap position, which is the third on the brake valve, is for the purpose of closing all ports, so that no air can flow into or out of the trainpipe. Service application position is for the purpose of making a gradual application of brakes, and emergency application position is for the purpose of allowing the trainpipe pressure to rush out as quickly as possible, in order that all brakes in the train may be

ITS USE AND ABUSE

set instantly, or nearly so. Emergency position should never be used except in case of actual or probable danger, and should never be used when an engine is on the turntable.

41. What main reservoir and trainpipe pressures should be carried with the quick-action brake equipment?

Ans.—Ninety pounds in the main reservoir, which is shown by the red hand, and seventy pounds in the trainpipe, which is shown by the black hand. In 1904 the Air Brake Association in National Convention at Buffalo recommended ninety pounds as the standard trainpipe pressure, but suggested that its use be gradually adopted, as circumstances would permit.

42. When the high-speed brake equipment is used what pressure should be carried?

Ans.—One hundred and twenty pounds in the main reservoir and 110 pounds in the trainpipe.

43. When the high pressure control is used, what pressure should be used on the engine?

Ans.—With a light train ninety and seventy pounds, but with a loaded train a hundred and ten, and ninety pounds.

44. What is meant by excess pressure, and what is it used for?

Ans.—Excess pressure is the amount of air

carried in the main reservoir over and above what is carried in the trainpipe. If the trainpipe governor is set at seventy pounds and the main reservoir or pump governor at ninety pounds, there would be an excess pressure of twenty pounds in the main reservoir. The object in carrying this extra or excess pressure is to enable the engineer to quickly recharge the trainpipe after making a reduction, in order to strike the triple pistons a hammer blow to drive them to release position.

SECTION 3

CHAPTER III

WESTINGHOUSE AIR-BRAKE DEFECTS—HOW TO TEST FOR AND REMEDY THEM

While a great many defects are constantly found in the air-brake equipment, it must be borne in mind that they arise more from abuse and neglect than from wear and tear.

When it is taken into account that the equipment is handled by such a great variety of men, and is required to perform its function under such varying conditions, it is really amazing that it will remain in service as long as it does without having to be renewed. But, as good as it is, it can easily get out of order, and the growing demand for greater safety in the running and handling of trains requires that the equipment be kept in as nearly perfect condition as possible, to reduce to a minimum the recurrence of the terrible wrecks and accidents that will continue to happen as long as railroads exist.

There are many once happy homes now shrouded in black despair as the result of some air brake defect that was either neglected or overlooked until it was too late. For instance, an engineer on a certain road in Pennsylvania was pulling a heavy freight train over a mountain division, and having neglected to keep his air pump in proper condition, could not pump sufficient air to overcome the trainpipe leaks and still maintain the proper pressure in the auxiliaries, and as a consequence the braking power of the train gradually fell down and down until, upon reaching a very heavy grade, the train got the start of the brakes, when, like a crazed monster it rushed down the mountainside until, like a flash, it left the rails and piled up a mass of wreckage beneath which lay the crushed remains of the engineer and fireman—and as a result two once happy wives were thus made widows, not because "the brakes failed to work," but because the engineer failed to maintain his brake equipment in the condition it should have been.

I shall take up the various defects in the same order in which the several parts of the equipment have been described.

The Triple Valve. The duties of the triple valve being to charge, set and release the brake, if it fails to do any one of these things it is because there is a defect somewhere, and if the trainmen expect the equipment to be kept in

ITS USE AND ABUSE

working order, they must be able to make an intelligent report to the car repairers.

Failure to charge the auxiliary may be on account of any of the following reasons: The strainer being clogged, feed groove clogged, bad leak under the slide valve, bolts loose on triple, bad gaskets, or if the release valve on the auxiliary does not seat properly. A very bad leak by the emergency valve will cause the brake to set while the auxiliary is being charged, and the air will be heard blowing out of the retainer.

Failure to set the brake may result from not having sufficient pressure in the auxiliary; triple piston packing ring worn so that auxiliary pressure reduces as fast as trainpipe pressure is reduced; very dirty strainer preventing reduction to be made quick enough to close the feed groove in triple; leaky cylinder; and sometimes the supply port in the triple valve seat becomes clogged up, preventing the auxiliary pressure from getting into the cylinder. It has happened that the supply pipe in the auxiliary of freight equipments has become clogged so that no air can get into the brake cylinder, but this is very rare.

Failure to release may be caused by not raising the trainpipe pressure above the auxiliary pressure quick enough, as the pressure will equalize and fail to move the slide valve if the triple piston packing ring is badly worn or gummed up, or if the strainer is clogged and retards the flow of air, either of which will cause the brake to remain set. This frequently happens after an emergency application, for as the auxiliary pressure is then very high, it is necessary that the trainpipe pressure should be raised *suddenly* against the plain side of the triple piston, otherwise a leak by the packing ring would allow the auxiliary to charge without releasing the brake, and as a consequence the wheels would be slid, or bursted, or a drawhead pulled out. If the retainer is turned up, or any dirt clogs in the retainer pipe, or should port h in the triple valve seat become clogged, the brake cannot be released from the engine, and must be bled off by letting the air escape through the auxiliary release valve, unless the car is equipped with a release signal, when it can be used to let the brake off. But when a car is not so equipped, and should the auxiliary release valve become clogged before the triple moves to release position, take out the drain plug in the auxiliary. Should the hand brake be set, on a freight car, the push rod would not follow the piston back

ITS USE AND ABUSE

when the air was released, nor if the brake rigging was caught on a bolt head, or anything. A great amount of oil on the slide valve seat will prevent a brake being bled off on a detached car, as the oil forms a suction so that the cylinder pressure can't lift the slide valve to let the air out.

Blow at the triple exhaust, or at the retainer, is caused by a leak from either the auxiliary or trainpipe side of the triple piston, and may be that the slide valve is off its seat, or the gasket between the triple and auxiliary may be leaking, either of which would be a leak from the auxiliary side of the piston; sometimes, on freight cars only, the blow may be caused by a leak in the supply pipe (*b*) between the triple and brake cylinder, but this is rare; a blow from the trainpipe side on the triple piston would be caused by a leak under the emergency valve, or there may be a leak by check gasket 14. To tell where the blow is coming from, cut the brake out and if it sets itself the leak is from the trainpipe side of the triple piston; if the brake don't set when you cut it out, the trouble is an auxiliary leak, and to tell if it is the triple gasket or the slide valve, cut the brake in and make a reduction on the trainpipe, and if the blow stops while the brake is set but starts again when the brake is released, it is the gasket, but if the blow continues while the brake is set or released the slide valve is causing the trouble.

Quick action, or going into emergency when only a service application was made, is caused by either a sticky triple, weak or broken graduating spring, or broken graduating valve pin. The latter trouble and a sticky triple both act alike, for on the first light reduction, if it is a sticky triple, the slide valve fails to move and open the port into the brake cylinder, and a broken pin would prevent the graduating valve from unseating, and in either case when the second reduction was made the graduating spring could not prevent the triple from going to emergency position. This action would be the same no matter in what part of the train the defective triple was located, but if the emergency was caused by a weak or broken graduating spring it would have to be within seven cars from the engine, and would show itself on the first light reduction. To find which car it is, cut out a part of the train and have the engineer make a very light reduction, and if you find a brake not set, watch it while the second reduction is being made and you will see it fly on, when of course

ITS USE AND ABUSE

you will cut it out. If you cut a part of the train out and the brakes fail to go in the emergency, you know that the trouble is not in that part of the train, when you will cut in more cars and try again, until you find the bad triple. When a train is equipped with the release signal it is not necessary to cut out a part of the train to find the "kicker," as a five-pound reduction will cause all the good brakes to show a portion of their signal, and the bad brake will show no signal until the second light reduction, when it will fly into the emergency. This can be seen from the ground or on the top of a freight train.

Eight-inch Pump. Where the same defect is possible in the 9½ as in the eight-inch pump, it will be explained under the 9½-inch pump.

If the stem of the reversing piston gets broken the pump will sometimes fail to reverse, until the piston is jarred down again by lightly tapping over the cap nut.

If the stop pin becomes broken the main valve will drop down and allow the packing ring of the small piston to catch and prevent the pump from reversing.

Should the packing rings on either of the main valve pistons, or the reversing piston, become so badly worn as to allow free passage of steam, the pump would not work, and if sufficient oil does not reach the reversing piston the pump is liable to stop.

Nine and One-half Inch Pump. Constant attention and careful management is required to keep an air pump in proper working order. It is always best to work a pump to its proper capacity, but it ruins it to overwork it. Your pump is your best friend, so take care of it.

Pump Running Hot. This may be caused from leaky packing rings on the main air piston; the piston-rod packing being too tight; not enough lift to the discharge valves; choked air passages or choked discharge pipe; leaky discharge valves; too small a main reservoir with long train, and fast running. To prevent choking up air passages or discharge pipe, or causing valves to leak don't use too much oil in the air cylinder, and never allow it to be sucked through the air inlet. To prevent overheating never run the pump faster than sixty full strokes a minute, and never run it with leaky valves or packing rings.

To test for leaky discharge valves, pump up to ninety pounds in the main reservoir and shut off the pump, then open the oil cup and hold your finger over it, and if top discharge valve is leak-

ITS USE AND ABUSE

ing the air will blow out continuously. For testing bottom discharge valve remove bottom plug. The plug should be removed before making oil-cup test, as a leaky bottom valve and leaking packing rings would let air blow by the piston.

To test for leaky packing rings, on the air piston, open the oil cup, run the pump about forty-five strokes a minute, and if they leak you will feel a gush of air through the oil cup as the piston makes the down-stroke.

Jiggling, or dancing, of the main piston is usually caused by too much oil getting under the seat of the reversing valve. To take the valve out shut off the pump until you get the cap nut off, then give it enough steam to raise the piston, when you can catch the valve. In putting it back be sure you get the groove over the guide pin. A bent reversing valve stem will sometimes catch on the reversing plate, or the latter, having a burr on it, will cause the pump to jiggle.

Pounding may be caused by any of the following defects: Too much lift in the air valves; pump loose from frame or frame loose from boiler; a worn shoulder on reversing valve not allowing piston to reverse quick enough; bottom end of piston rod worn too far into piston head, which allows piston to strike before reversing; nuts loose on main piston. If the pump is started up fast it will pound if the condensation is not drained from the steam end, or if there is no air cushion for the piston to strike against.

The Eleven-inch Pump, being made after the same pattern as the nine-inch pump, the same rules will apply to both pumps.

Pump Governor. If the governor don't shut off at all and the diaphragm valve port is not closed, it is because the pressure has equalized on both sides of the air valve, and the vent port and waste pipe need to be opened. If the governor shuts off at a low pressure, it is either because the regulating spring is too loose, or the diaphragm valve dirty or battered so it won't seat, or the valve has been filed off too short so it can't seat.

The Gauge. If, with the brake valve in full release, the red hand shows less than the black hand, it is because the gauge pipes have been crossed or the hands have become twisted on the pinion.

The gauge should be tested once a month by attaching a test gauge to the trainpipe hose on the tender, and placing the brake valve in **full**

ITS USE AND ABUSE

release for the red hand and in running position for the black hand.

Engineer's Brake Valve. The only thing that can get wrong with a brake valve is a leak somewhere which will let the pressures run together or escape to the atmosphere.

Pressures Equalize in Running Position. If the handle is in running position and both hands show the same pressure, the trouble is one of three things: either the body gasket, rotary or feed valve is leaking. To tell which it is, place the handle in service position, and if the body gasket is cracked so that main reservoir pressure flows into cavity D as fast as it passes out of the preliminary exhaust, there will be no discharge from the trainpipe exhaust, and the brakes will not set. If the brakes do apply in service position, but release when the handle is brought to lap, the trouble is a leaky rotary. But if the brakes apply in service and remain set on lap, and when the handle is again brought to running position the pressures again equalize at main reservoir pressure, the feed valve needs attention. If it is a D-8 brake valve, any one of these leaks will allow the governor to shut off the pump at seventy pounds, as the governor is controlled by trainpipe pressure. In making these tests the angle-cock behind the tender must be closed, as they cannot be satisfactorily made if the train brakes are cut in.

To test if the supply valve of the feed valve is leaky, draw off all trainpipe pressure, lap the valve and remove the diaphragm piston, place the handle on running position and if the supply valve leaks you will feel the air blowing out by holding your finger under the valve. If no blow is felt, the trouble is either in the gasket between the feed valve and brake valve, or else the rubber diaphragm buckles on account of the spring box being screwed up too tight. If it is a new slide valve feed valve, the trouble may be caused by spring 58 being gone, or the small valve 59 having a bad seat or too short.

Never oil any part of the old style feed valve, and use only high grade machine oil for the rotary and the slide valve of the new style feed-valve.

Blow at Trainpipe Exhaust. This is caused either by a leak from cavity D or its connection with the little drum or black hand of the gauge, or dirt under the seat of the equalizing discharge valve. If it is dirt causing the blow, it can generally be knocked out by closing the cut-out cock under the brake valve and making a

ITS USE AND ABUSE

reduction of about fifteen pounds and then throwing the handle to full release, which will cause the short trainpipe temporarily to have a higher pressure than cavity D, and of course the discharge valve is forced up and the air rushes out the trainpipe exhaust and blows the dirt out. After trying this and the blow continues, then look over the pipe connections.

If the pipe to the little drum gets broken, plug it up and also the trainpipe exhaust and use the emergency position in making an application; but be very careful to place the handle gradually on emergency, and just as carefully bring it back to lap, to prevent releasing the brakes by the surging of the air.

Failure to open the trainpipe exhaust when the handle is placed in service position, is on account of the discharge valve not raising. This may be due either to a broken body gasket, letting main reservoir pressure into cavity D, or the packing ring around the discharge valve may be letting the trainpipe pressure up on top of it. To tell which is causing the trouble, lap the brake valve, and if it is the body gasket, the trainpipe and main reservoir pressure will equalize.

Whistle Signal. A leak in the signal pipe will

cause the whistle to blow. If the supply valve of the reducing valve leaks and allows main reservoir pressure to equalize with the signal pipe, whenever a trainpipe reduction is made and the brakes released, the whistle will blow, because the main reservoir air going into the trainpipe allows the signal pipe air to flow back into the main reservoir, which thus makes a reduction on the signal pipe and blows the whistle.

If the whistle fails to respond it is more than likely on account of the rubber diaphragm in the signal valve being baggy, or the whistle needs adjusting, or it is cut out at the reducing valve, or an angle-cock is turned.

If more than one blast is heard when but one pull was made, it may be that the diaphragm stem needs filing off to allow it to drop further down, or there may be dirt holding it up.

Should the signal pipe fail to charge up it is either cut out at the reducing valve or there is some dirt lodged in the small opening which admits the main reservoir pressure into the reducing valve.

When releasing brakes on a passenger train, if the whistle blows it is because the reducing valve is letting the signal pipe air back into the

ITS USE AND ABUSE

The engine air gauge can be used in setting the reducing valve by drawing the main reservoir pressure down to forty pounds, and slacking off the regulating spring until the whistle fails to blow when the main reservoir pressure is reduced below forty pounds.

QUESTIONS AND ANSWERS TO SECTION 3

THE CAUSE OF WESTINGHOUSE AIR-BRAKE DEFECTS AND HOW TO DETECT AND REMEDY THEM

The following questions will start with the engine equipment and be carried right through the train.

45. What effect is produced by leaky packing rings in the air end of the pump?

Ans.—It prevents the pump from producing the proper amount of air within the required time and causes it to run hot, for the reason that if the packing rings are leaking, on the downstroke of the pump the air which is being compressed in the lower end of the cylinder would be forced to the upper end and prevent the receiving valve from letting in the required amount of fresh air, thereby lowering the efficiency of the pump. The same action will cause the pump to run hot for the reason that on a warm summer's day the air in a pump working against a ninety-pound pressure in the main reservoir is raised to a temperature of 550

ITS USE AND ABUSE

degrees, and naturally if the free air is saturated with a portion of the compressed air which is already made hot by compression, it follows that a second compression of it greatly increases the temperature, thereby causing the pump to run hot.

46. How should you test for leaky packing rings in the pump?

Ans.—First ascertain if the discharge valves are leaking, which is done by shutting off the steam to the pump and opening the oil cup and removing the bottom plug, and holding your finger slightly above the oil cup to see if any air is blowing out. If the air blows out of the oil cup the top discharge valve is leaking, and if it blows out of the bottom plug hole the lower discharge valve is leaking. If no blow is felt either at the oil hole or plug hole, then replace the plug, leave the oil cup open and start up the pump at about forty strokes a minute, and if the packing rings are leaking you will feel a gush of air through the oil hole as the pump makes the down-stroke.

47. What will cause a pump to jiggle or dance?

Ans.—Too much oil getting under the seat of the reversing valve, or if the reversing valve

MODERN AIR-BRAKE PRACTICE

stem catches on the reversing plate, or if the reversing plate has a burr on it, it has a chance to jiggle.

48. What will cause a pump to pound?

Ans.—Too much lift in the air valves; pump being loose from the frame, or frame loose from the boiler; a worn shoulder on the reversing valve which would prevent the piston from reversing quick enough; the bottom end of the piston-rod worn too far into the piston head will allow the piston to strike before reversing; loose nuts on the main piston; or if the pump is started to running fast before the condensation has been properly drained, or if there is no air cushion for the piston to strike against.

49. When should the air end of the pump be oiled?

Ans.—Every time the engine is started on a trip and oftener if required, but great care must be taken not to get too much oil in the air end as it will cause the valves to gum up and make the pump run hot. Automatic oil cups should be used.

50. What kind of oil should be used in the pump?

Ans.—Cylinder oil· engine oil should never be used in the pump.

51. How often should a pump be cleaned?

ITS USE AND ABUSE

Ans.—At least every six months by running a solution of potash through it, and in doing so the connections between the main reservoir and the tender should be broken so that no potash can work into the brake equipment.

52. If the trainpipe and main reservoir pressure equilizes while the handle of the brake valve is in running position, what might be causing the leak?

Ans.—Any one of three things; either the rotary valve, the body gasket, or feed-valve attachment. To tell which it is, place the handle in service position and if the body gasket is cracked so that main reservoir pressure flows into cavity D as fast as it passes out of the preliminary exhaust, there will be no discharge from the trainpipe exhaust and the brakes will not set. If the brakes apply in service position, but release when the handle is brought to lap, the trouble is a leaky rotary. But if the brakes apply on service and remain set on lap, and when the handle is again brought to running position the pressure again equalizes, it is the feed valve that needs attention. To tell if it is the supply valve of the feed valve, draw off all trainpipe pressure, lap the valve, and remove the diaphragm piston, place the handle in running position, and if the supply valve leaks you will feel the air blowing out by holding your finger under the valve. If no blow is felt the trouble is either in the gasket between the feed valve and the brake valve, or else the diaphragm buckles on account of the spring box being screwed up too tight. If it is a new slide valve feed valve, the trouble may be caused by spring 58 being gone, or the small supply valve 59 having a bad seat, or else too short. If you are testing a D-8 brake valve, any one of these leaks will allow the governor to shut off the pump at seventy pounds for the reason that the governor is controlled by trainpipe pressure with this kind of a valve.

53. Is there anything else that would prevent the trainpipe exhaust from opening in service position besides a cracked body gasket?

Ans.—Yes. If the packing ring around the equalizing valve leaks badly it will allow the trainpipe pressure to get on top of it as fast as the preliminary exhaust port lets the air out of cavity D.

54. Should you lose your equalizing reservoir, or damage it so that it leaked, how would you handle your train?

Ans.—Plug the trainpipe exhaust and also the

ITS USE AND ABUSE

pipe leading to the equalizing reservoir and use the emergency position for applying brakes, but be very careful to go slowly to the emergency position and also slow in bringing the handle to lap.

55. If the pump governor doesn't shut off at all, what is the trouble?

Ans.—It is because the pressure has equalized on both sides of the air valve, and the vent port and waste pipe need to be opened. If it shuts off at a low pressure it is caused by the regulating spring being too loose or the diaphragm valve is dirty or battered so that it wont seat, or else the valve has been filed off so that it is too short to seat.

56. What would cause the black hand of the gauge to show more pressure than the red hand?

Ans.—Either the pipes have been crossed or the hands have become twisted on the pinion, or stuck.

57. What will cause the whistle to blow when the brake valve handle is thrown to full release?

Ans.—A leak in the supply valve in the reducing valve, which allows the signal whistle pressure to flow back into the main reservoir.

58. What will prevent the whistle from responding when the whistle cord is pulled?

Ans.—The rubber diaphragm in the signal valve being baggy, or the whistle not being properly adjusted. Of course, if it is cut out at the reducing valve, or a cut-out cock is turned it will not whistle.

59. What will cause a blow at the triple exhaust?

Ans.—A leak from either the auxiliary or trainpipe side of the triple piston.

60. How many places are there at which such a leak might occur?

Ans.—Four. Under the slide valve, or the gasket between the triple and auxiliary, under the emergency valve, or by check gasket 14. To tell where the blow is coming from, cut the brake out and if it sets itself the leak is from the trainpipe side of the piston; if the brake don't set when you cut it out the trouble is an auxiliary leak, and to tell if it is the triple gasket or the slide valve, cut the brake in and make a trainpipe reduction, and if the blow stops while the brake is set but starts again when the brake is released it is the gasket; but if the blow continues while the brake is either set or released, it is the slide valve that is causing the trouble.

61. What causes a brake to fly into the emergency when a service application is made?

ITS USE AND ABUSE

Ans.—It is either because of a sticky triple, weak or broken graduating spring, or broken graduating valve pin. The latter trouble and a sticky triple both act alike, for on the first light reduction if it is a sticky triple, the slide valve fails to move, and consequently no air gets into the brake cylinder, and a broken pin would prevent the graduating valve from unseating, so that in either case, when the second reduction was made the graduating spring could not prevent the full travel of the triple piston, and the brake would, of course, go into the emergency. The action would be the same no matter in what part of the train the defective triple was located, but a weak or broken graduating spring would cause an emergency application only when the defect is within seven cars from the engine, in which case the brakes would fly into emergency on the first light reduction.

SECTION 4

CHAPTER IV

NEW YORK AIR-BRAKE EQUIPMENT—THE PARTS AND THEIR DUTIES

As the reader has already thoroughly informed himself on the construction, operation and handling of the Westinghouse air-brake equipment, it will not be necessary to repeat here many of the things which have already been gone over.

The question of leverage applies the same with one air-brake system as with another. The general instructions regarding train handling covers the New York Air Brake the same as it does the Westinghouse, with a few minor changes which will be explained in their proper order.

The operation and care of the steam and air cylinders, air valves, etc., is the same with both systems. The specific differences are mentioned in their proper places.

With this explanation you will at once see that aside from the information contained in the previous pages of this book it will only be necessary for me to explain the construction, operation and handling of the several parts of the New York Air-Brake Equipment in order for the

ITS USE AND ABUSE

reader to be as fully informed on the New York air-brake system as he is on the Westinghouse.

The order in which we will describe the New York Air-Brake Equipment is as follows: The Air Pump, Pump Governor, Engineer's Brake Valve, Triple Valve, Straight Air-Brake Valve, and Whistle Signal Apparatus.

After describing the several parts of the equipment I shall then treat of the defects and diseases and their remedies, the same as I have with the Westinghouse equipment.

THE NEW YORK DUPLEX AIR PUMP

The New York Duplex Air Pump is now being made in four sizes, No. 1, No. 2, No. 6, and No. 5. In general principle they are all the same. This pump has four cylinders, two of which are for steam and two for air. The steam cylinders of the No. 1 pump are five inches in diameter; the high pressure air cylinder is five inches in diameter; and the low pressure air cylinder is seven inches in diameter.

The steam cylinders of the No. 2 pump are seven inches in diameter, the high pressure air cylinder is seven inches and the low pressure air cylinder is ten inches in diameter; the steam cylinders of the No. 6 are seven inches, the high

MODERN AIR-BRAKE PRACTICE

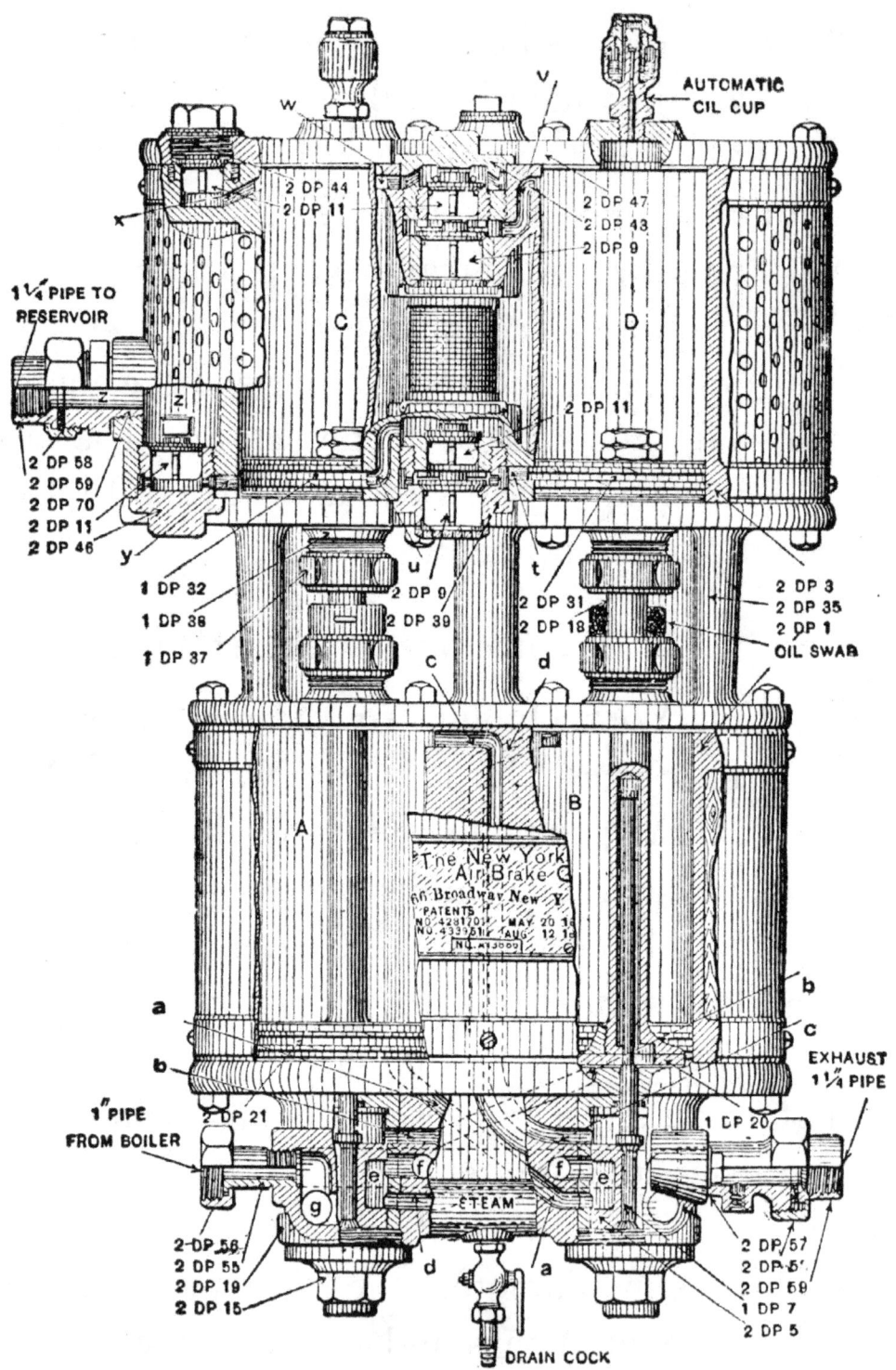

CHART 35—FIG. 1. NEW YORK DUPLEX PUMP, PISTONS AT REST.

pressure cylinder is seven inches, and the low pressure cylinder eleven inches; the steam cylinders of the No. 5 pump are eight inches in diameter, the high-pressure air cylinder is eight inches and the low-pressure air cylinder is twelve inches in diameter.

The stroke of the Nos. 1 and 2 pump is nine inches, while the stroke of the No. 6 is ten inches, and the No. 5 is twelve inches. The No. 6 is meant to take the place of the No. 2, as it has better proportioned cylinders and a longer stroke.

The air cylinders of the New York pumps are located above the steam cylinders, whereas with the Westinghouse pumps the air cylinders are located on the bottom end. The low pressure cylinder of the New York pump has a volume capacity of about double that of the high pressure cylinder. The diameter of the steam cylinders is always the same as the diameter of the high pressure air cylinder.

A feature you should thoroughly impress yourself with in regard to the action of the New York pump is that there is only one steam piston in motion at any one time, for whenever a piston has made its stroke it waits for the other piston to make a stroke before it moves again. This is

MODERN AIR-BRAKE PRACTICE

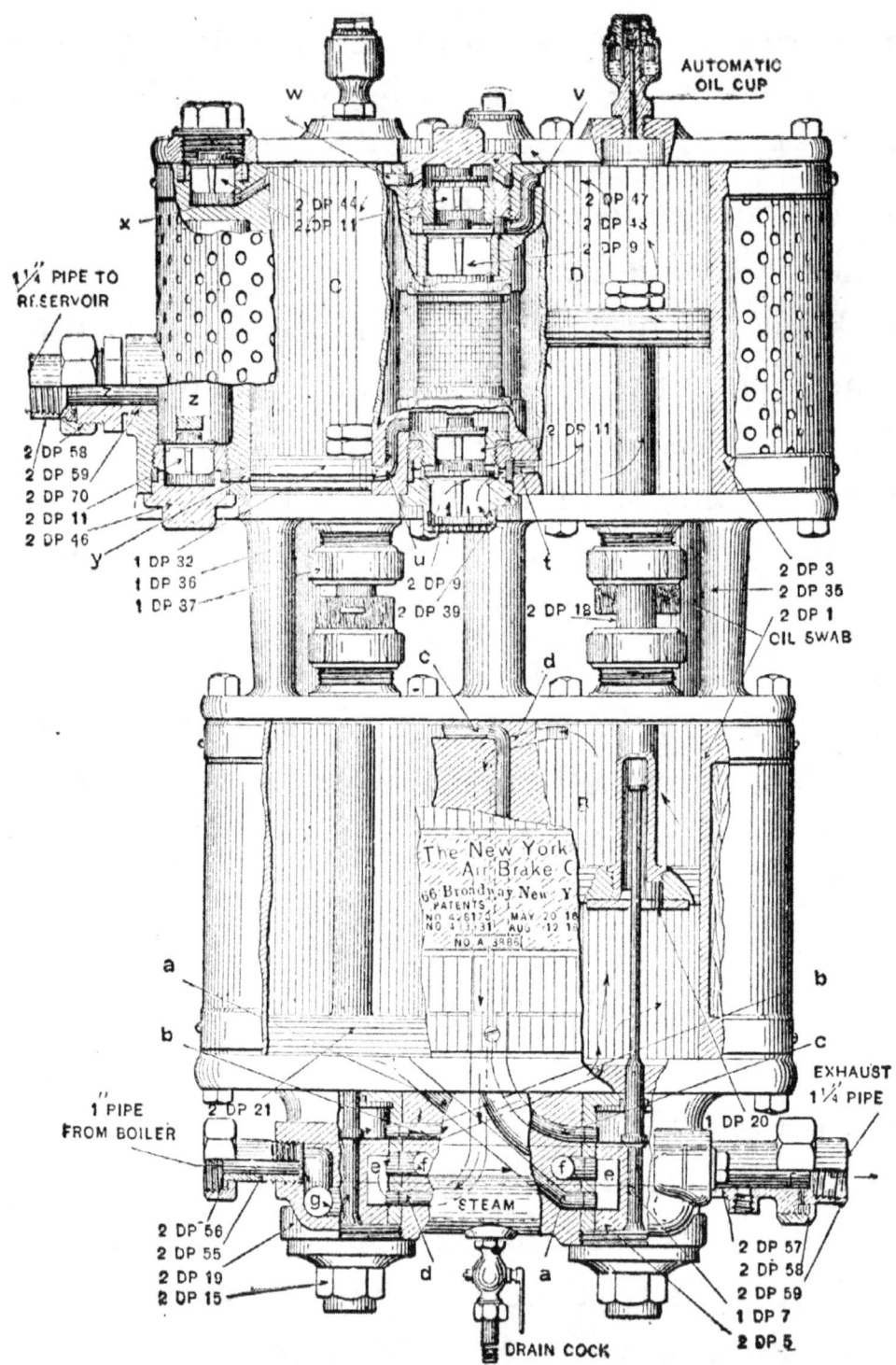

CHART 35—FIG. 2. NEW YORK DUPLEX AIR PUMP, LOW PRESSURE PISTON ON THE UP STROKE.

ITS USE AND ABUSE

brought about by reason of the fact that the reversing-valve on one piston controls the action of the opposite piston.

As you are already familiar with the reversing-valve used in the Westinghouse pump, this knowledge will enable you to understand the action of the steam end of the New York pump, for the reason that the valve-gear of the steam end of the New York pump is very similar to an ordinary reversing-valve, in fact the general appearance is the same.

Like the steam piston in the Westinghouse pump the piston-rod is hollow, and on the reversing-valve side of the steam piston there is a plate bolted which performs the same function in the New York pump as the reversing-valve rod plate does in the Westinghouse pump, that is, it moves the reversing-valve up or down.

By referring to Fig. 1, on Chart 35, this will be made very clear to you if you will but look at the engraving up-side-down, for in this position you will see that the reversing-valve rod extends down into the hollow steam piston-rod. You will also notice that this reversing-valve rod has a shoulder on one end and a button on the other end for the purpose of controlling the movement of the small D slide valve which is connected to

MODERN AIR-BRAKE PRACTICE

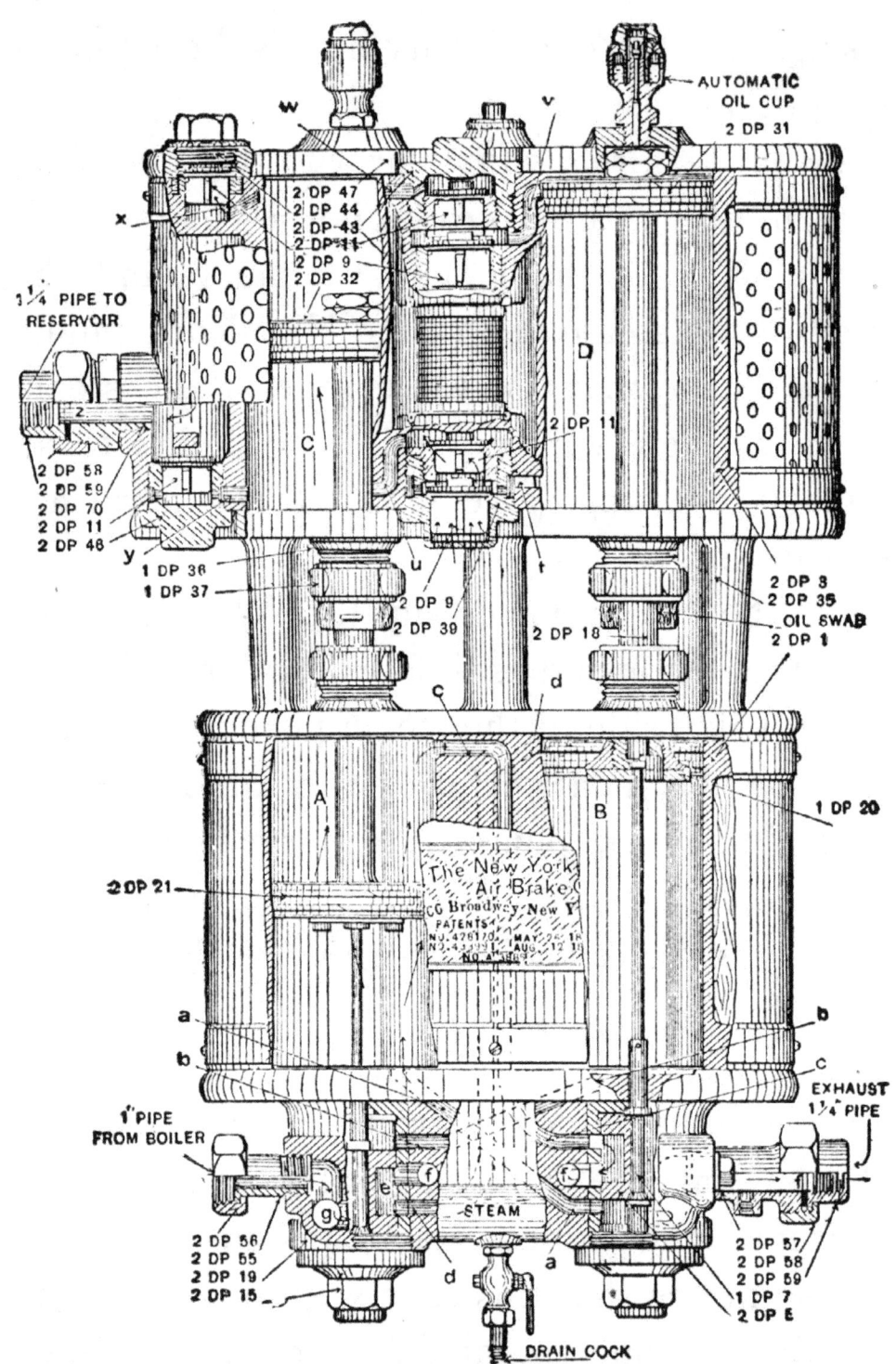

CHART 35 - FIG 3   NEW YORK DUPLEX PUMP, UP-STROKE, HIGH PRESSURE PISTON.

ITS USE AND ABUSE

the reversing rod, the same as the reversing-valve is connected to its rod in the Westinghouse pump.

Fig. 1 shows both steam pistons at rest. Now if you will turn the engraving right-side-up again you will notice that the boiler connection is made at the left-hand side. Now remember that the slide valve which controls the right steam cylinder is located under the *left* steam cylinder, and the valve that controls the left steam cylinder is located under the *right* steam cylinder. This will be made plain to you when you notice that the ports are crossed, as shown by dotted lines.

As the pump is shown with the pistons at rest, you will notice that both slide-valves are in their bottom positions, so that as the steam enters the pump from the boiler it fills the slide-valve chamber under the left-hand piston and at the same time passes through port G to the slide-valve chamber under the right-hand piston, and live steam also passes through the port marked b to the under side of the right-hand piston and at the same time passes up through the port c to the top side of the left-hand piston; this causes the right-hand, or low pressure piston, to be forced up, as shown in Fig. 2, of Chart 35. Just

MODERN AIR-BRAKE PRACTICE

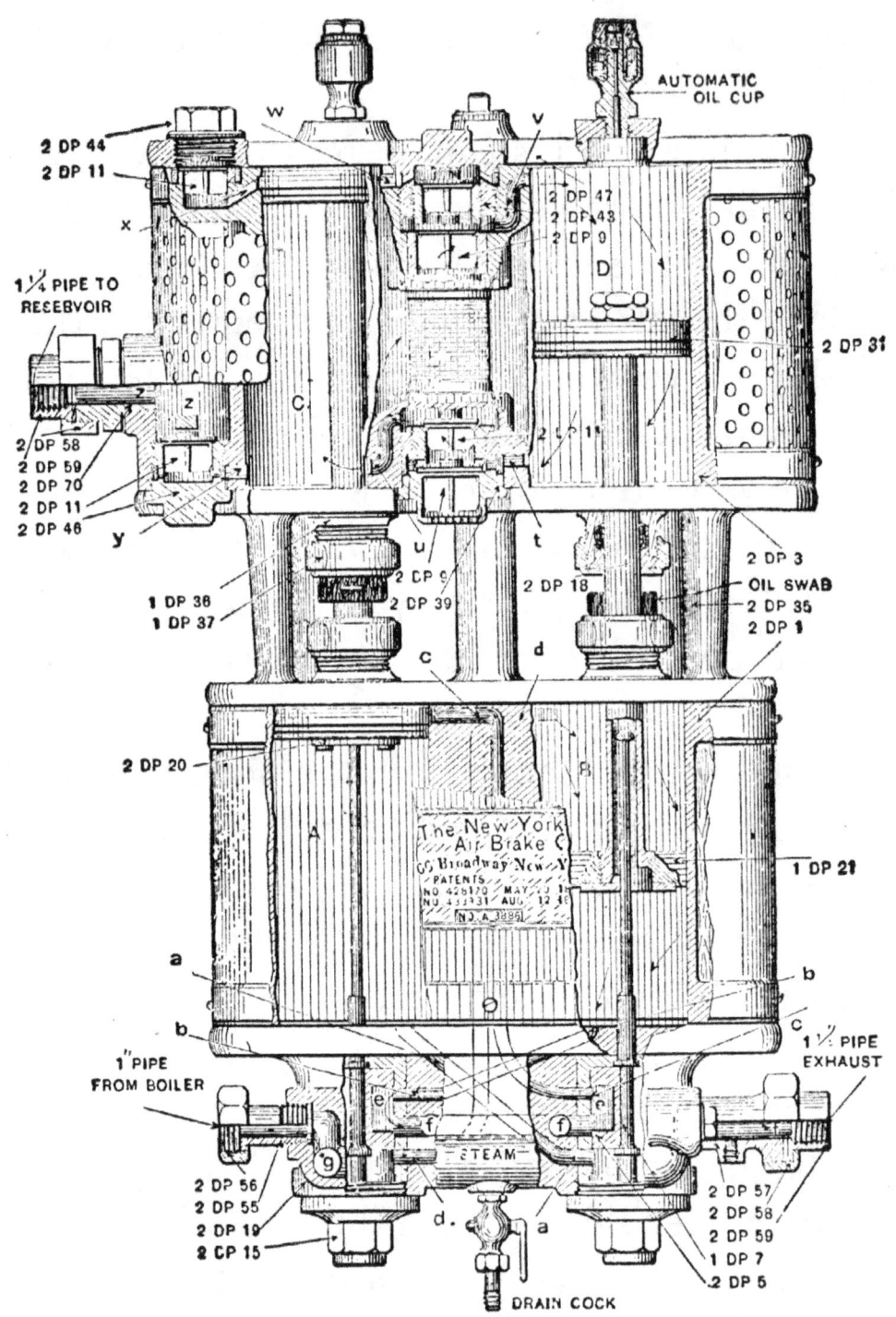

CHART 35—FIG. 4. NEW YORK DUPLEX PUMP, DOWN-STROKE, LOW PRESSURE PISTON.

ITS USE AND ABUSE

as the piston reaches the end of its top stroke the reversing-rod plate engages the button on the end of the reversing-rod and pulls it up. This action connects port c with the exhaust cavity F and at the same time live steam from the reversing-valve chamber under the right-hand piston passes through port a to the under side of the left-hand piston, so that while the steam is exhausting from the top side of this piston live steam on the under side is pushing it up, as shown in Fig. 3, of Chart 35. As port b leads from the under side of the low pressure piston that port is still closed to the exhaust cavity by the left-hand slide-valve, so the right-hand piston is held up by the steam which is confined under it, but as soon as the left-hand piston reaches the end of its up stroke the reversing-rod is pulled up by the reversing-rod plate, thereby connecting port b with port F, by way of cavity E in the slide-valve, so that the steam can exhaust from the under side of the right-hand piston, and at the same time live steam from the left-hand slide-valve chamber passes through port d to the top of the right-hand piston, which forces it down, and when it reaches the end of its down stroke it reverses the slide-valve, thereby exhausting the steam from the under side of the

MODERN AIR-BRAKE PRACTICE

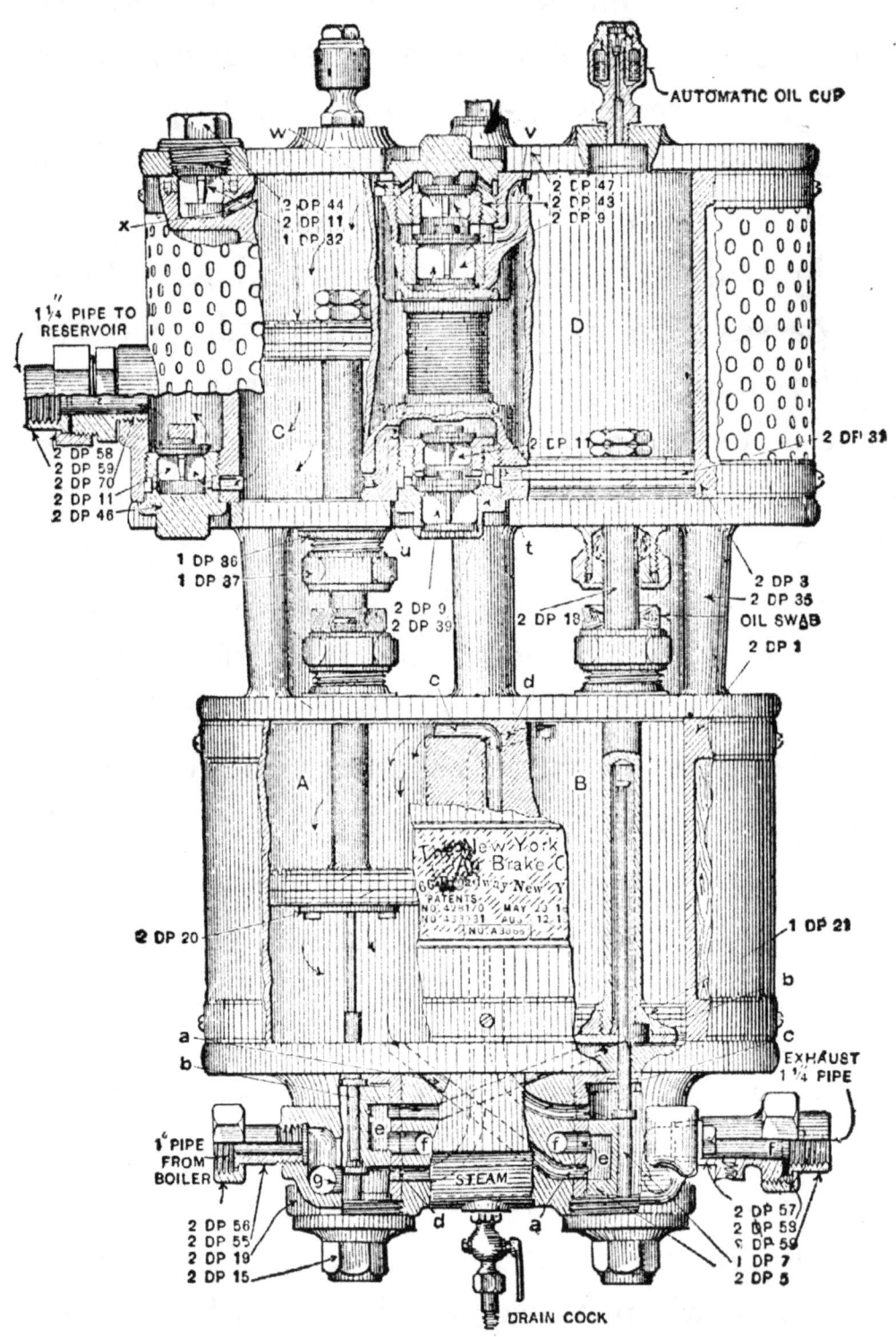

CHART 35—FIG. 5. NEW YORK DUPLEX PUMP, DOWN-STROKE, HIGH PRESSURE PISTON.

ITS USE AND ABUSE

left piston, and at the same time lets live steam get on the top side of that piston which forces it down, so that both pistons have now made a full stroke up and down.

This valve arrangement is certainly very simple and very effective. There are no packing rings to contend with in the reversing-valve arrangement, and if too much oil is not allowed to get into the slide-valve chambers there will be very little trouble result from this mechanism. If too much oil is allowed to get into the slide-valve chamber it will naturally cause the valve to be forced off its seat, and thereby disarrange the port connections.

The drain cock should always be left open when the pump is not working, and should always be opened before starting the pump in order to let the condensation pass away.

The air end of the New York pumps, except the No. 6 and No. 5, contains six air valves; two of them are ordinary receiving valves; two are intermediate valves and two are ordinary discharge valves. On the No. 1 and No. 2 pump the receiving valves and intermediate valves are located between the two cylinders, as shown in Chart 35 and the two discharge valves are located on the left side of the pump, above and below

MODERN AIR-BRAKE PRACTICE

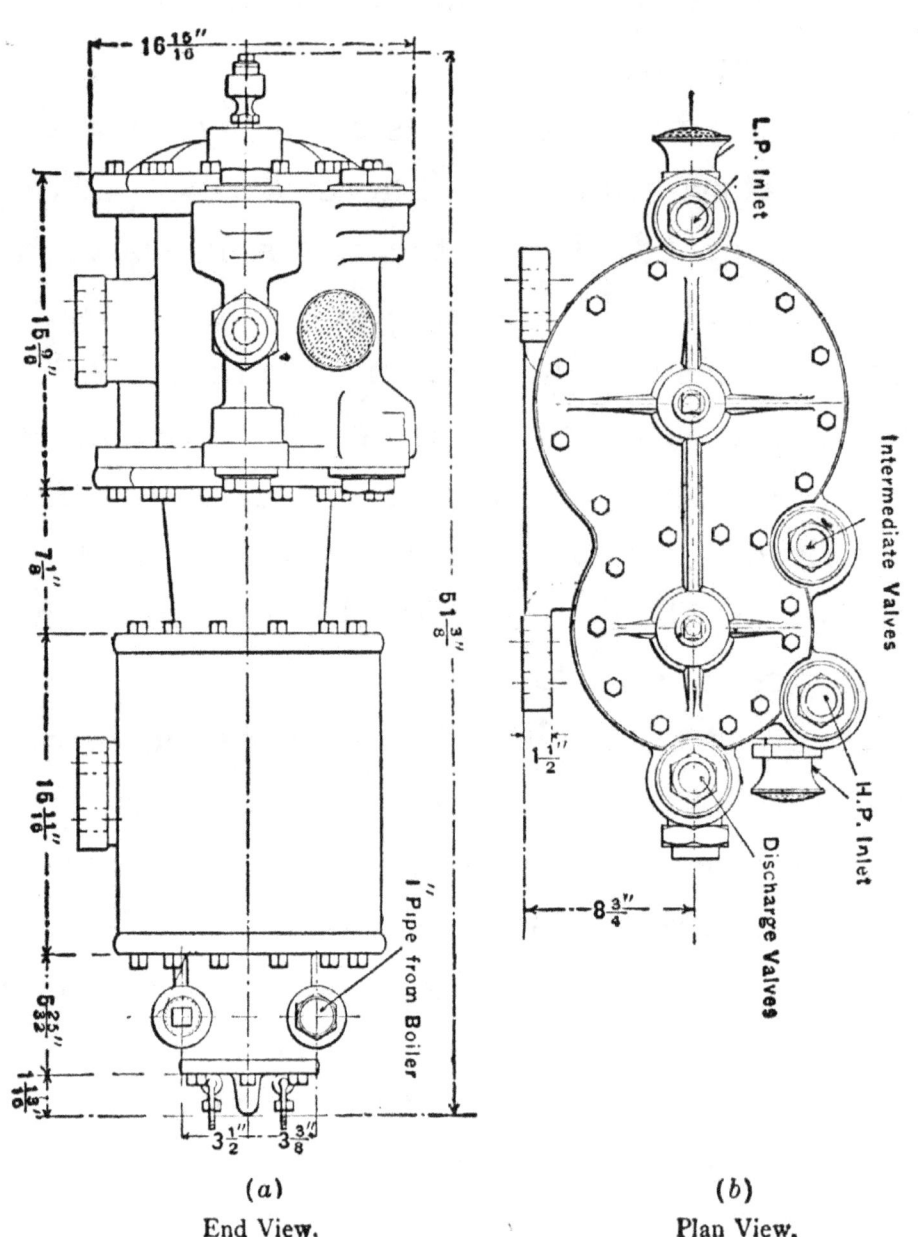

(a) End View. (b) Plan View.

CHART 35—FIG. 6. NEW YORK NO. 5 DUPLEX AIR PUMP

ITS USE AND ABUSE

the pipe leading to the main reservoir, but the air valves on the No. 6 and No. 5 are arranged a little different, as will be explained further on, and are eight in number, instead of six as in the other pumps. (See plate 35, Fig. 6.)

The reason the intermediate valves are so called is because of the fact that the low pressure cylinder discharges its air into the high pressure cylinder through these valves, and they are therefore intermediate between the low and high pressure air cylinders.

The action of the air valves in the No. 1 and No. 2 pump is as follows: As the low pressure piston starts on its upward stroke a partial vacuum is created on the under side of it and atmospheric pressure forces the bottom receiving valve up (this valve is marked 9) and allows the low pressure cylinder to receive a charge of atmospheric pressure, and as this piston does not move again until the piston in the high pressure cylinder makes its stroke, you will at once see that when the piston of the high pressure cylinder moves up, and creates a partial vacuum, the bottom receiving valve and also the bottom intermediate valve (marked No. 11) are both forced off their seats by atmospheric pressure rushing into the high pressure cylinder, and

MODERN AIR-BRAKE PRACTICE

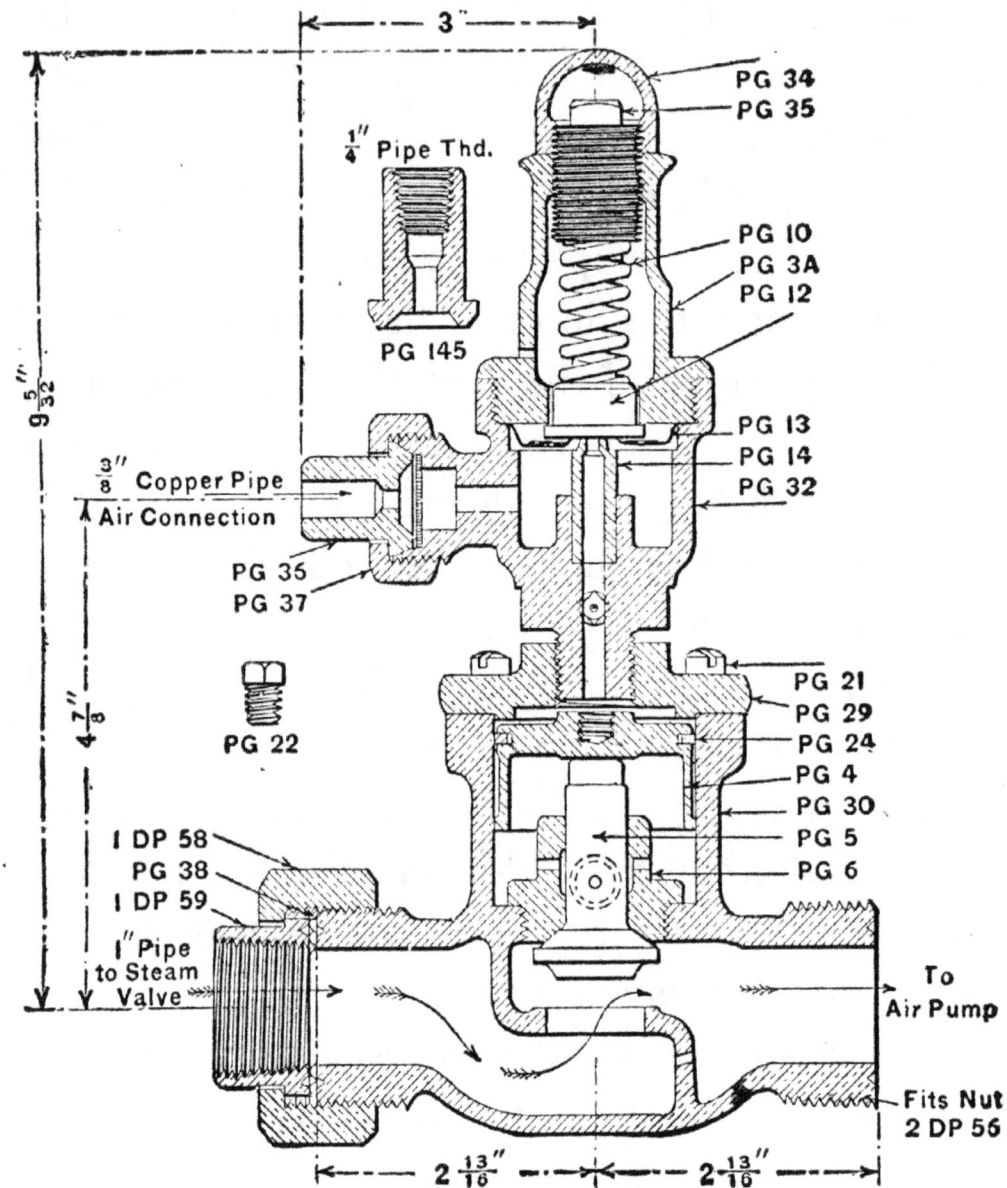

CHART 36—FIG 1. STYLE C, NEW YORK PUMP GOVERNOR, STEAM VALVE OPEN.

when the atmosphere has equalized in both cylinders then both valves drop to their seats. As the low pressure piston starts down, as shown in Fig. 4 on Chart 35, the lower intermediate valve (11) is forced from its seat so that the compressed

ITS USE AND ABUSE

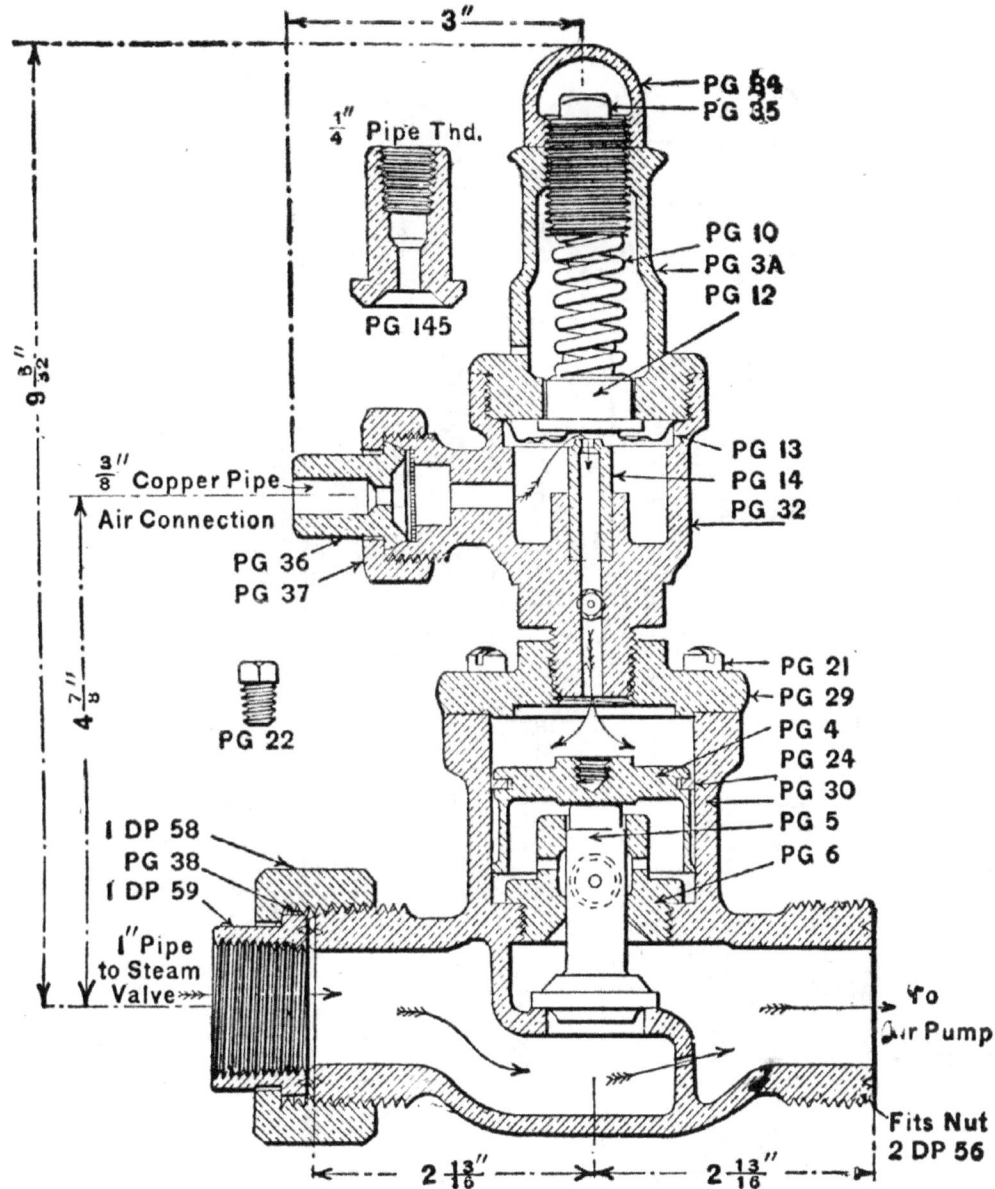

CHART 36—FIG. 2. STYLE C, NEW YORK PUMP GOVERNOR, STEAM VALVE CLOSED.

air in cylinder D passes into cylinder C, which was previously charged with atmospheric pressure, so that the cylinder C now contains three measures of air, for the reason that cylinder D is twice the size of the high pressure cylinder, C.

MODERN AIR-BRAKE PRACTICE

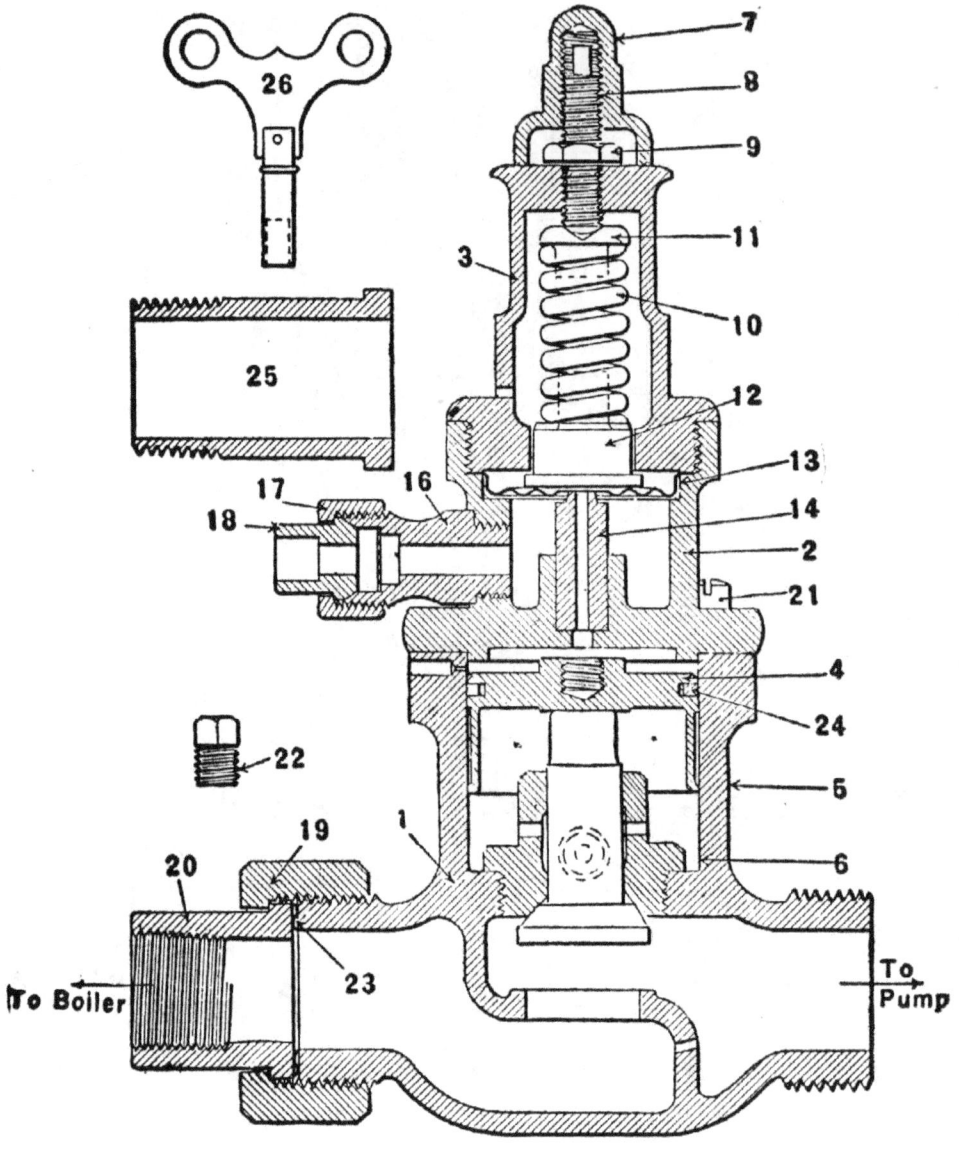

CHART 36—FIG. 3. STYLE A, NEW YORK PUMP GOVERNOR.

While the low pressure piston is moving down the top receiving-valve (9) is forced from its seat by atmospheric pressure rushing into cylinder D.

When the high pressure piston starts down, a partial vacuum is created in the high pressure

ITS USE AND ABUSE

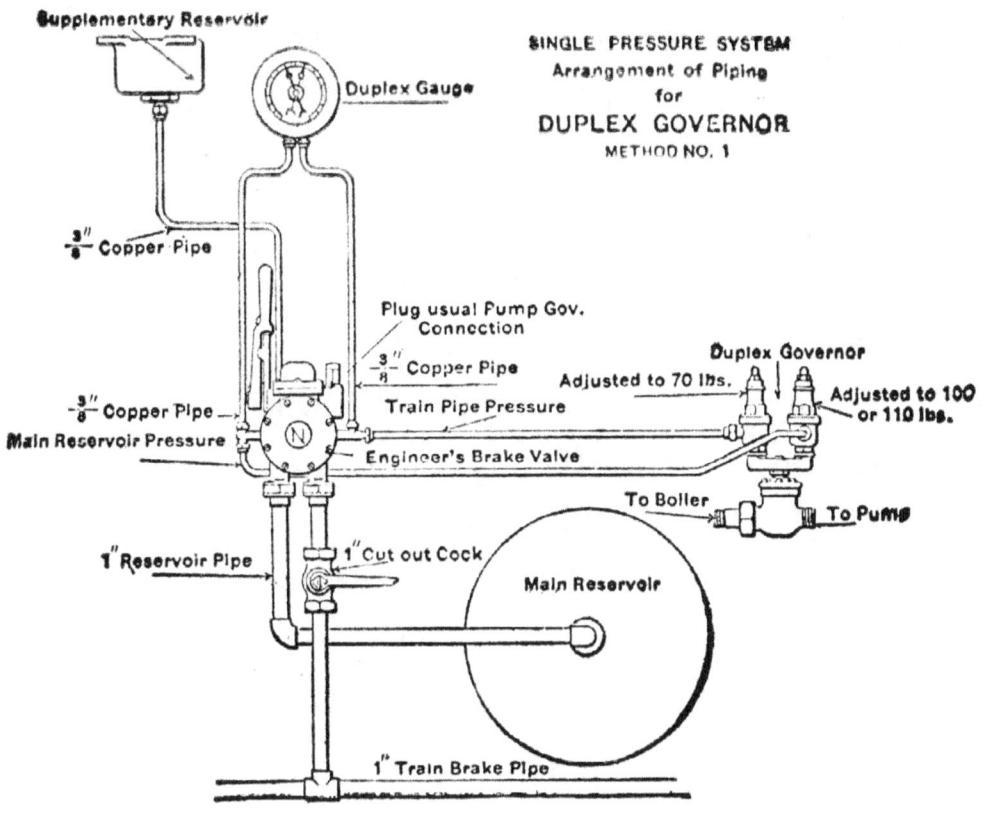

CHART 36—FIG. 4.   NEW YORK AIR BRAKE.

cylinder and atmospheric pressure forces the top receiving-valve and top intermediate valve from their seats, and when the pressure in both cylinders has equalized these valves drop to their seats the same as did the bottom valves. Now as the low pressure piston starts up, the air on the top side of it is compressed, which causes the top receiving-valve to be held to its seat while the top intermediate valve is forced from its seat and the air from cylinder D is forced into cylinder C, as shown in Fig. 2, Chart 35.

From the above description it is seen that the high pressure cylinder at every stroke of its piston discharges into the main reservoir, through either the top or bottom discharge valve, three full measures of air, one received from the atmosphere direct and a double-sized one from the low pressure cylinder, so that this pump is not only duplex but it is also a compound pump.

Referring to the No. 5 duplex pump, Fig. 6, Chart 35, you will notice that there is an air inlet shown on both the right and left-hand side, the same as the No. 6 pump, and these are shown in Fig. 6. One air inlet is for the low pressure cylinder and one for the high pressure cylinder. The intermediate valves in the No. 5 and No. 6 pumps are located the same as in the No. 1 and No. 2, but the No. 5 and No. 6 pumps have a separate set of receiving valves as shown in Fig. 6. Fig. 6 plan view b, is the only way in which both air inlets can be seen together.

The general instructions regarding oiling, speed, drainage, etc., apply to the New York pumps the same as they do to the Westinghouse pumps.

The illustration of the New York pump shown on Chart 35 shows the automatic oil cup. This is a very simple device, as the method by which the oil is passed into the air cylinder is very much

ITS USE AND ABUSE

after the same manner in which air is passed into the pump cylinder.

The Westinghouse Automatic oil cup consists of a brass body, in the main chamber of which the oil is contained, and extending through this chamber there is a regulating valve, the end of which is pointed so that if it is desired to increase or decrease the flow of oil the pin-valve can be moved up or down by means of a regulating nut, and kept in position by a small lock nut. In the body of the valve below the pin-valve there is a ball valve, and the operation of this automatic oil cup is as follows: As there is a small port in the cap nut, atmospheric pressure is always admitted to the top of the oil in the main oil chamber, consequently when the pump piston is moved down the partial vacuum in the pump cylinder causes the ball valve to be lifted off its seat so that the oil, which has previously passed from the main oil chamber around the point of the regulating valve into the passage controlled by the ball valve, is drawn into the pump in the form of a fine spray. When the piston makes its up stroke the compressed air holds the ball valve to its seat, thereby preventing the oil from being blown out of the oil cup chamber. This refers to the No. 1 oil cup.

The No. 2 Westinghouse oil cup consists merely of a brass body having a chamber in which the oil is contained and, instead of having a ball valve and a regulating valve there is a small check valve to which is attached a needle-rod of very small diameter, and which extends up through a small opening in the bottom of the oil chamber. On the under side of the check valve there is a spring, so that the operation of the No. 2 automatic oil cup is as follows: As the pump piston makes a down stroke the partial vacuum in the pump cylinder causes the check valve to be unseated, thereby allowing the oil to be drawn from the oil cup into the air cylinder of the pump. The up stroke of the piston causes the check valve to be held to its seat, thereby preventing the oil from being blown out of the cup.

In the body of both of these oil cups there are suitable heating chambers for the purpose of allowing the warm compressed air to surround the oil chamber, and thereby keep the oil in a liquid state.

The New York automatic oil cup is made in two styles, A and B. Style A consists of a brass body in which there is an oil chamber, and in the center of the body there is a regulating valve which can be moved up or down for the purpose

ITS USE AND ABUSE

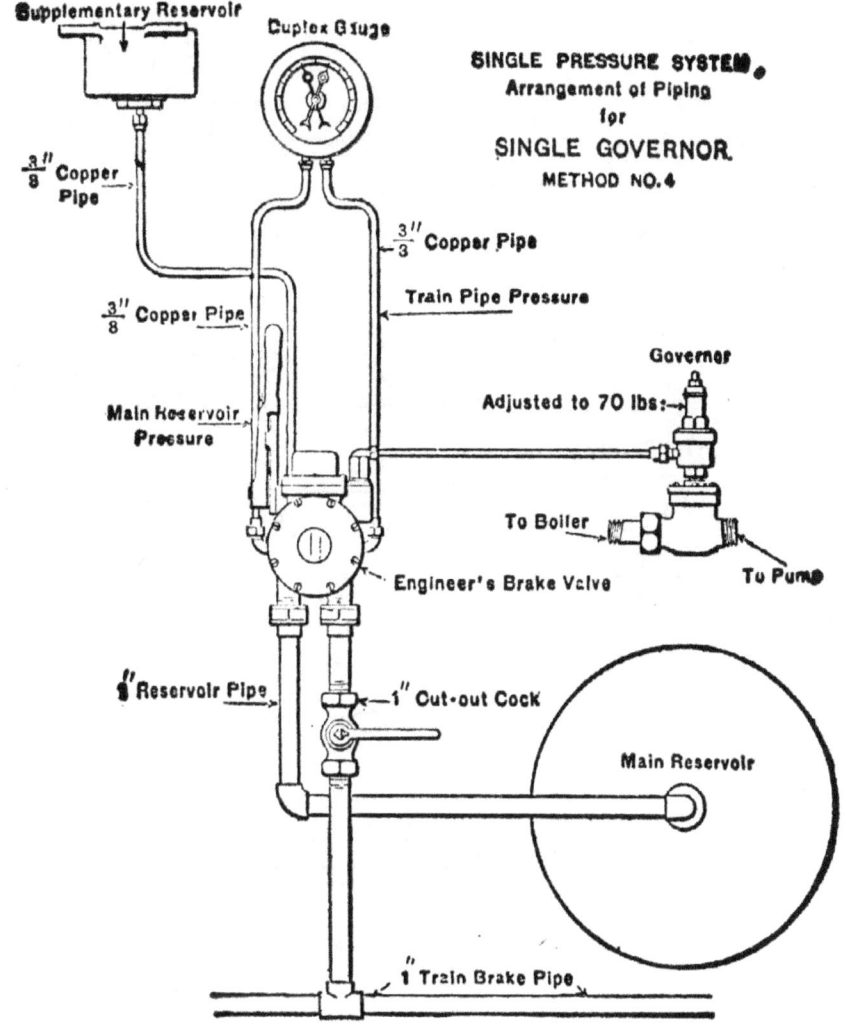

CHART 36—FIG. 5.   NEW YORK AIR BRAKE.

of increasing or decreasing the amount of oil to be fed to the pump. The operation of this cup is as follows: When the piston in the air cylinder moves up, compressed air is forced through the oil to the top of it, so that when the pump piston makes a down stroke the partial vacuum in the cylinder combines with the compressed air

MODERN AIR-BRAKE PRACTICE

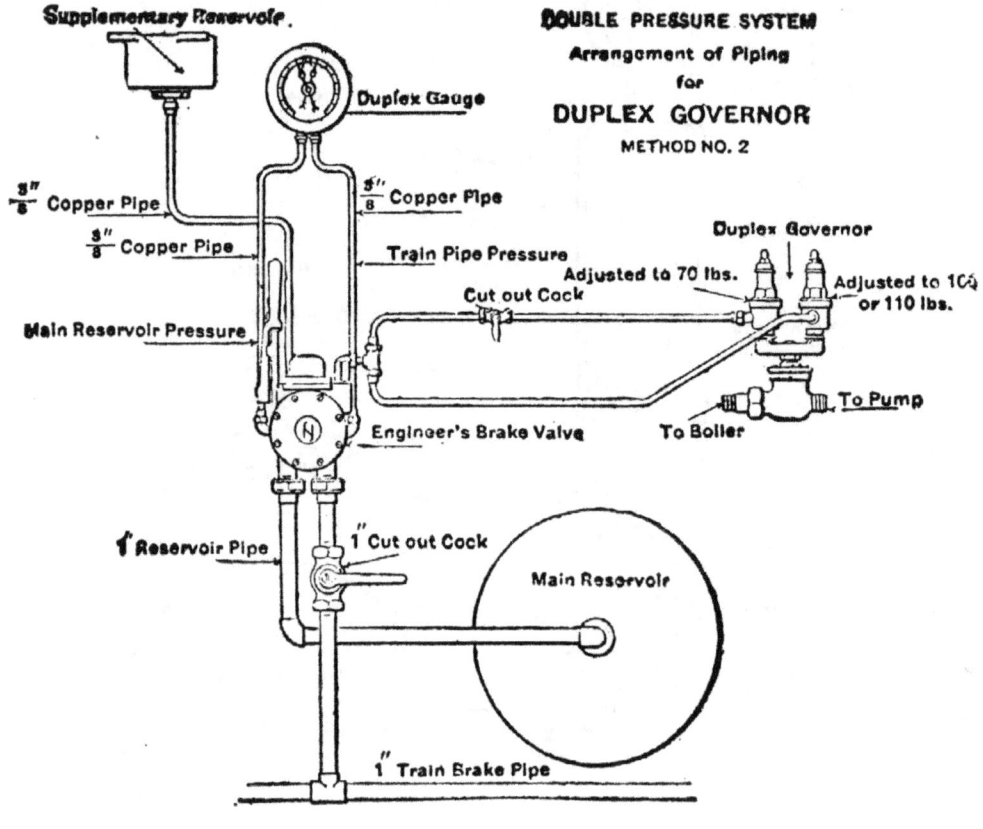

CHART 36—FIG. 6. NEW YORK AIR BRAKE.

on top of the oil and causes a small portion of oil to be drawn into the air cylinder of the pump and sprayed around the walls.

The New York style B automatic oil cup has no adjustable feed, but has a very small port through the body of the oil cup which permits a small amount of oil to be drawn into the air cylinder every time the piston makes a down stroke.

ITS USE AND ABUSE

NEW YORK PUMP GOVERNORS

The principle of the New York pump governor is the same as the Westinghouse, and when you understand one you can operate the other. There is a slight difference in the construction, but not enough to make it necessary to re-describe the entire governor here. It has a diaphragm, regulating spring and a regulating nut, just the same as the Westinghouse, but instead of having a diaphragm pin-valve like the Westinghouse, the diaphragm-valve in the New York governor

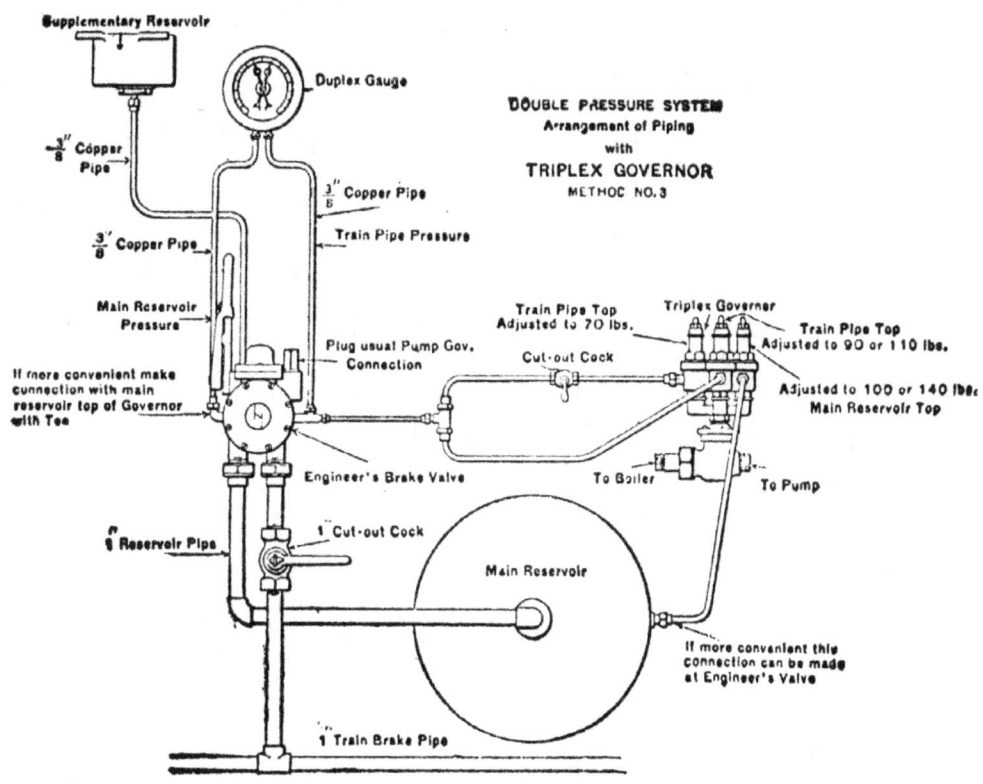

CHART 36—FIG. 7.   NEW YORK AIR BRAKE.

closes the port leading from the diaphragm chamber to the steam piston. Another difference is that there is no spring under the steam piston as in the Westinghouse, so that the steam valve and the steam-valve piston is forced up by steam pressure alone, whereas with the Westinghouse both steam and spring are used to force the steam-valve piston up. There is an outlet port from the steam piston chamber to allow any back pressure to escape to the atmosphere, which is the same as in the Westinghouse. This vent port prevents the pressure from accumulating under the steam piston, for if not allowed to escape the air pressure on top would not be able to force the steam valve to its seat and shut off the pump, whereas, when steam is operating against the steam-valve alone it only requires about one-third as much air pressure on the large area of the top of the piston to overcome the steam pressure and force the steam valve down.

Fig. 1, in Chart 36, shows style C of the New York Pump Governor with the steam-valve open, and Fig. 2 shows this same style C governor with the steam-valve closed.

The old style A New York governor requires a key with which to set the regulating spring, and is shown in Fig. 3 of Chart 36.

ITS USE AND ABUSE

The duplex pump governor is simply a governor with one steam portion but having two air portions, as shown in Fig. 4, Chart 36. A duplex governor consists of one steam-valve body, steam-valve, steam-valve piston, and a Siamese fitting to which is attached two pressure taps, or diaphragm-valve portions.

The method of piping the governor differs according to whether or not a single or double pressure system is to be controlled.

The method of piping used in the single pressure system, that is, the ordinary automatic air-brake, is shown in Fig. 4, Chart 36. In this diagram you will notice that the trainpipe pressure is connected to only one side of the governor, and is adjusted to seventy pounds, while the other side is connected to the main reservoir, and may be adjusted at ninety or a hundred pounds, or whatever number of pounds is considered standard on the particular road operating the device. You will notice there are no cut-out cocks on the pipe leading to the Duplex governor, for the reason that with this method of piping the movement of the brake valve handle determines just which side of the governor is operative. This will be explained more fully when we describe the engineer's brake valve.

MODERN AIR-BRAKE PRACTICE

Fig. 5 shows the method of piping a single governor for a single pressure system. With the New York Brake Valve the single governor would be connected to the trainpipe, whereas with the Westinghouse system the single pump governor is always connected to the main reservoir.

Fig. 6 in Chart 36 shows a cut-out cock in the pipe leading from the low pressure side of the governor to a T connection with the trainpipe, so that by cutting out the low pressure governor the trainpipe pressure will be raised to one hundred or a hundred and ten pounds, as the case may be, according to what the high pressure governor is set at. But it must be remembered that with this method of piping there is no separate governor control for the main reservoir pressure. (See New York Brake Valve.)

Fig. 7, Chart 36, shows the triplex governor by which the main reservoir pressure is controlled in addition to giving two other degrees of control to the trainpipe pressure as may be desired. With the triplex pump governor the main reservoir pressure can be adjusted at any point that may be considered safe. If it were the high pressure control system or the high speed brake it would be proper to set the main reservoir

ITS USE AND ABUSE

governor at one hundred and ten or one hundred and twenty pounds, and the middle pressure-top would be set at ninety or one hundred and ten pounds, and the remaining top at seventy pounds.

With the New York Brake Valve on lap, service or emergency position there is no connection between the brake valve and the trainpipe pump governor, so that if the brake valve had only one governor, as shown in Fig 5, and the handle of the brake-valve should be left on lap, service, or emergency position, there would be nothing to stop the pump, and consequently it would continue to operate as long as the steam in the boiler could move it, but when the Duplex governor is used and one governor pipe is connected to the trainpipe and one to the main reservoir, as shown in Fig. 4, then if the brake valve handle should be on lap, service or emergency position the pump would have to shut off when the main reservoir pressure reached the point at which the main reservoir governor was set.

Pump and Pump Governor Defects and Remedies will be treated in their proper order in a chapter devoted to that subject.

THE NEW YORK ENGINEER'S BRAKE VALVE

The student of air-brakes who possesses a previous knowledge of the Westinghouse Engineer's

MODERN AIR-BRAKE PRACTICE

Brake Valve, must remember when studying the New York Engineer's Brake Valve that when a service application is made with the Westinghouse valve the first escape of air is from off the top of the equalizing discharge valve, whereas with the New York valve the first escape of

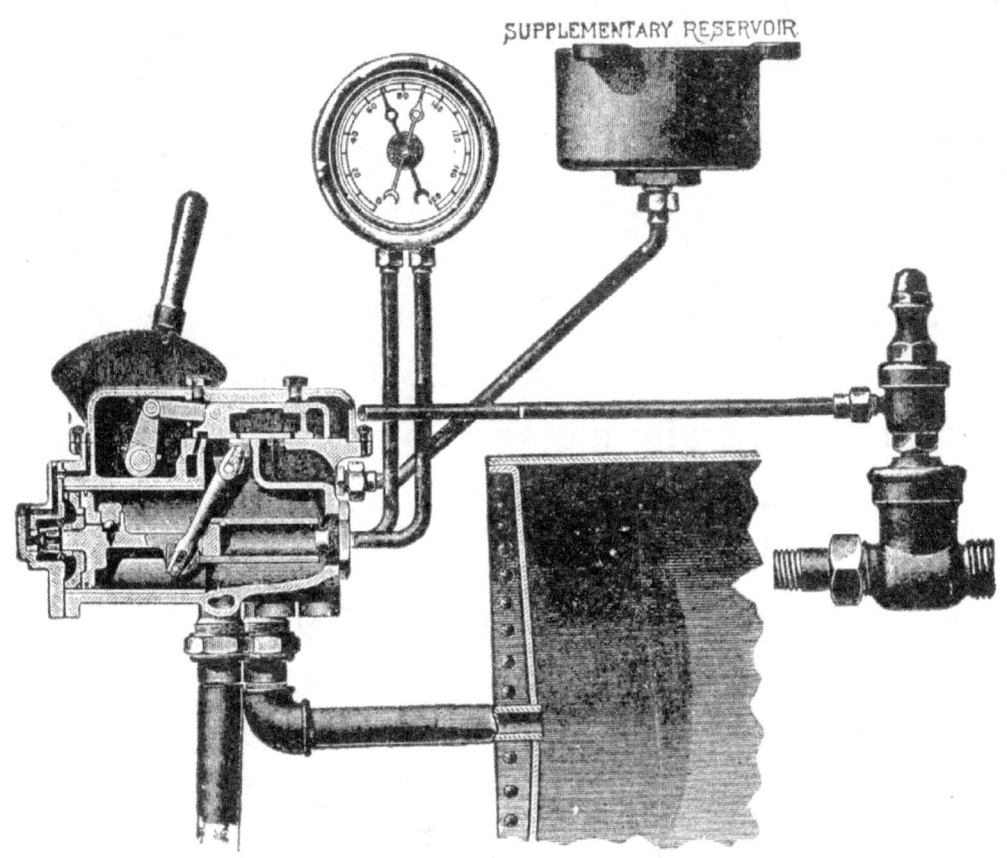

General Arrangement of Brake Valve, Supplementary Reservoir, Etc.

CHART 37—FIG. 1. NEW YORK AIR BRAKE.

ITS USE AND ABUSE

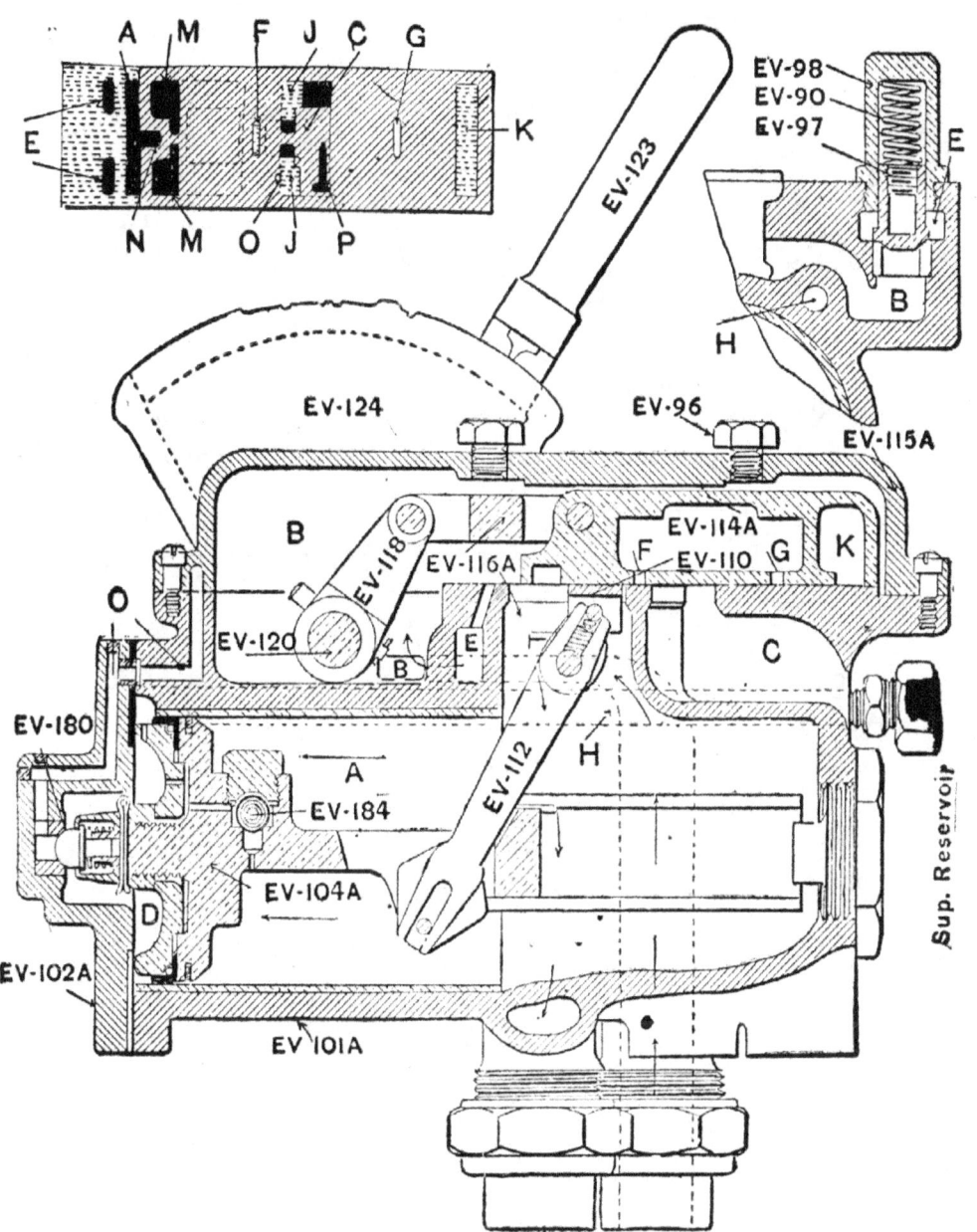

CHART 37—FIG. 2. FULL RELEASE POSITION OF BRAKE VALVE.

air in making a service application is direct from the trainpipe, but that the port opening from the trainpipe to the atmosphere is much smaller in making a service application than it is when making an emergency application.

MODERN AIR-BRAKE PRACTICE

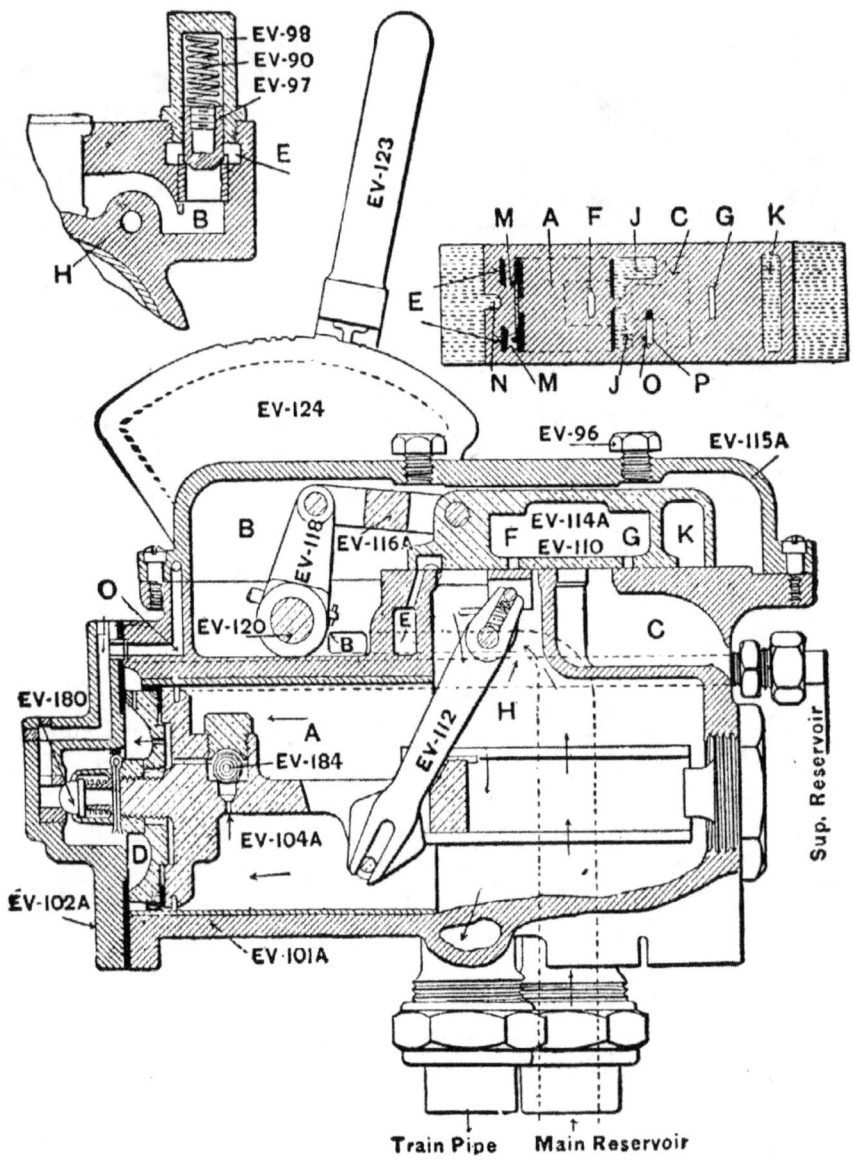

CHART 37—FIG. 3.   RUNNING POSITION NEW YORK BRAKE VALVE.

Another important feature that the student should impress himself with at the beginning is that the service position on the New York Brake Valve is divided up into five notches, and that if he is handling a train of four cars or less with the New York Brake Valve he should always begin a service application by placing the handle in the

ITS USE AND ABUSE

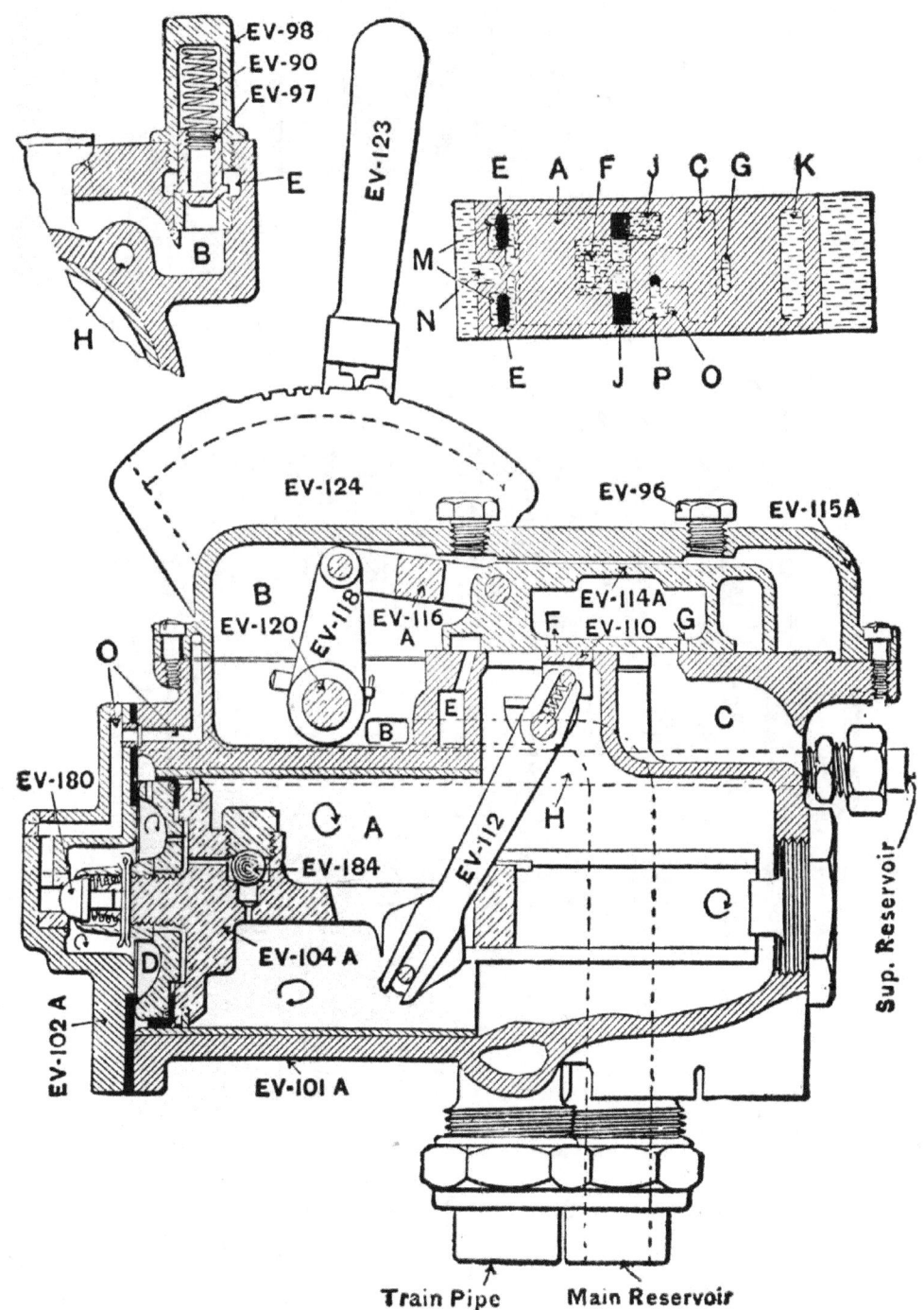

CHART 37—FIG. 4. LAP POSITION, NEW YORK BRAKE VALVE.

first notch, because of the fact that the service port gradually becomes wider as the handle is moved over the quadrant, and with a short train

MODERN AIR-BRAKE PRACTICE

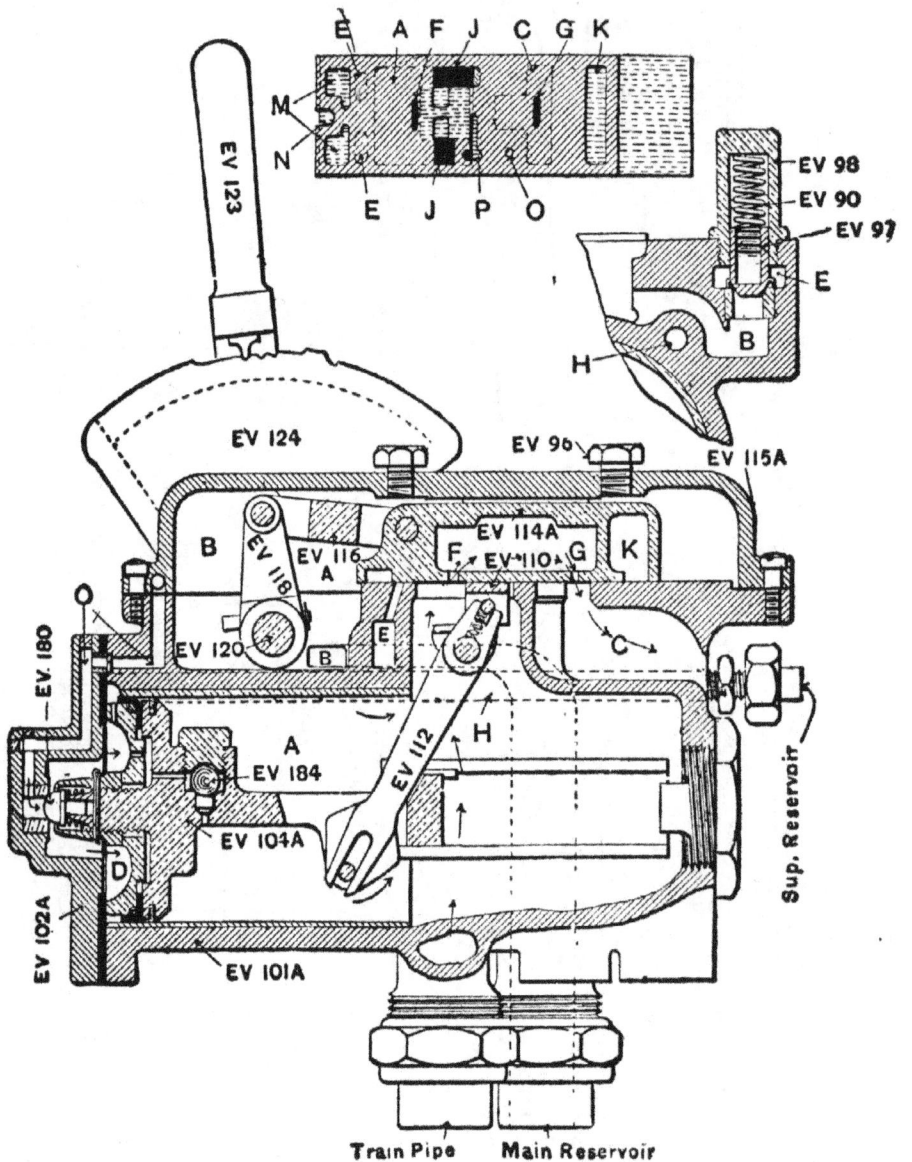

CHART 37—FIG. 5.   SERVICE POSITION, NEW YORK BRAKE VALVE.

of four cars or less the trainpipe volume is so small that. if the handle were moved past the first notch it is very likely to produce an emergency application, as it allows the trainpipe pressure to be reduced too suddenly.

With this explanation, let us now begin a careful study of the New York Engineer's Automatic

ITS USE AND ABUSE

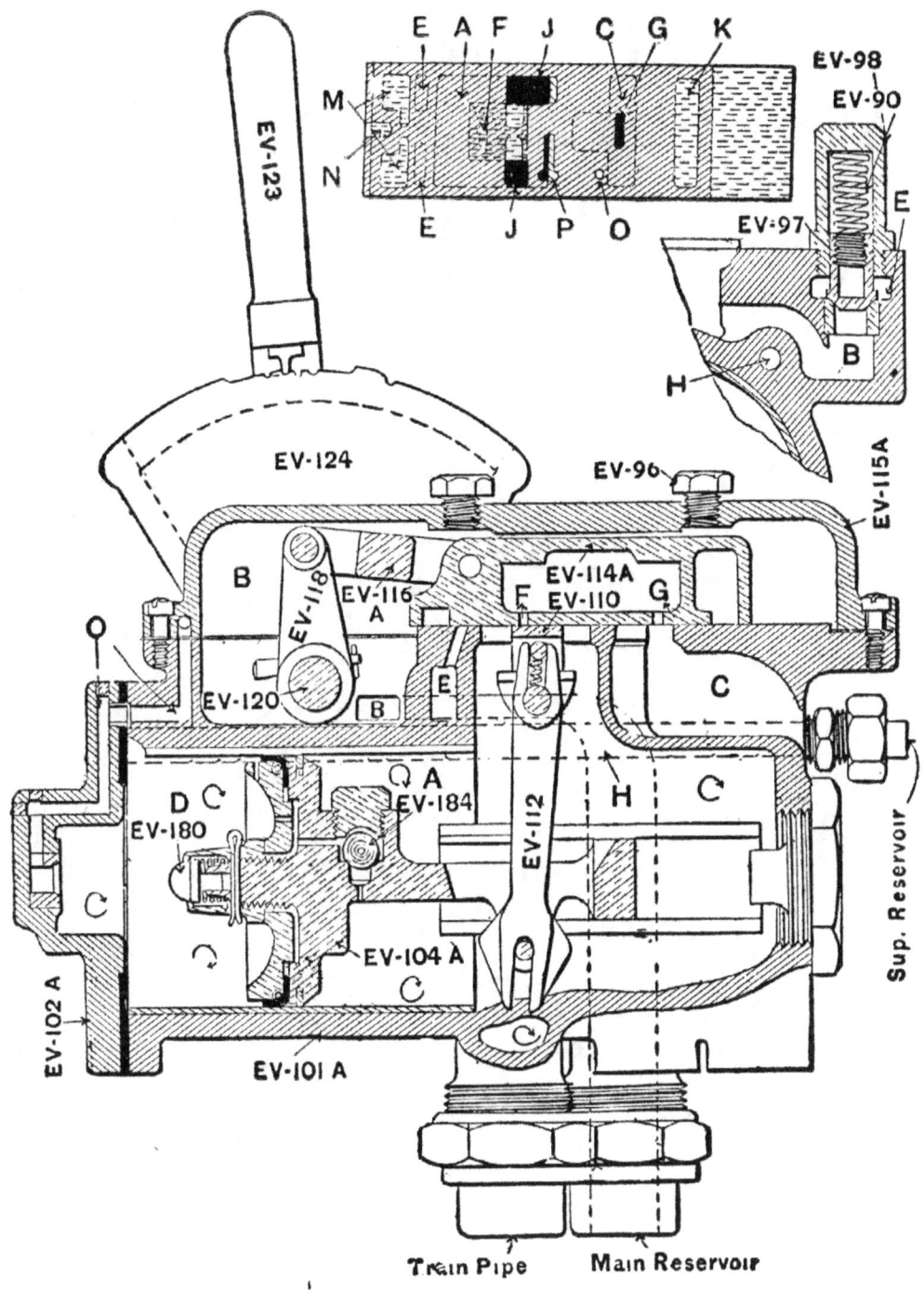

CHART 37—FIG. 6. SERVICE POSITION.

Brake Valve, Style B. (The old style A is not being made any more, and it will not be necessary to speak of it until we have fully described the present standard, which is Style B, or B 1.)

MODERN AIR-BRAKE PRACTICE

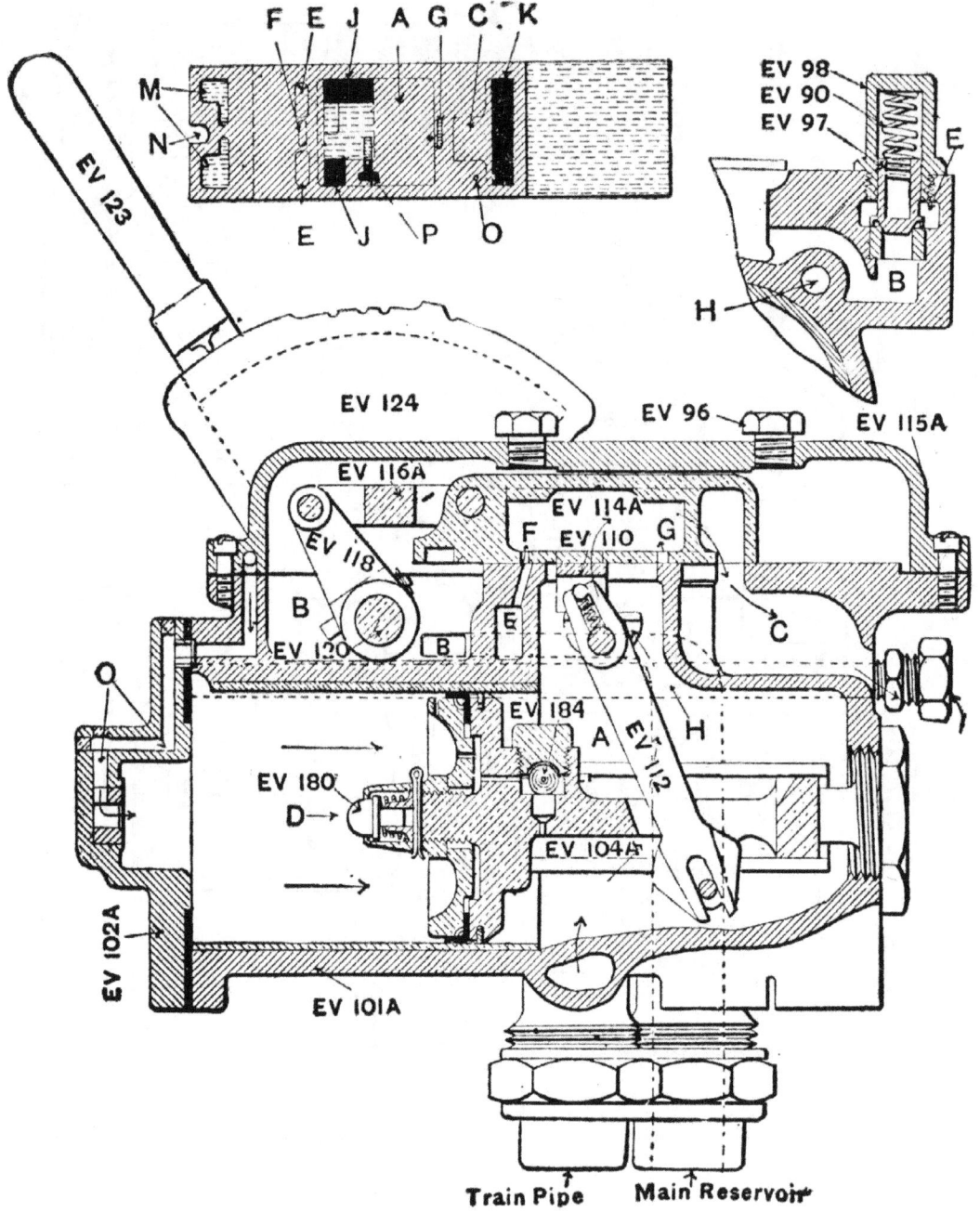

CHART 37—FIG. 7. EMERGENCY POSITION, NEW YORK BRAKE VALVE.

The essential parts to the New York Brake Valve, as shown in Chart 37, are (aside from the body and cover which contain the parts) a main

ITS USE AND ABUSE

slide-valve, which is connected by a link to a shaft operated by a handle, in which there is a lock bolt for the purpose of engaging the notches on the quadrant. Under the main slide-valve there is a small cut-off slide-valve which is controlled by an arm connected to a graduating piston. The graduating piston contains a small ball valve for the purpose of admitting air into chamber D or supplementary reservoir, and a ball-faced vent valve fastened to the end of the equalizing piston for the purpose of closing port O.

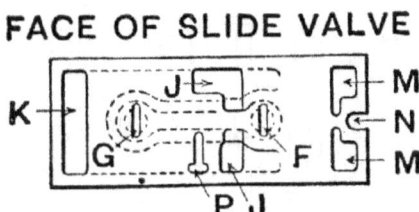

CHART 37—FIG. 8. FACE OF SLIDE VALVE.

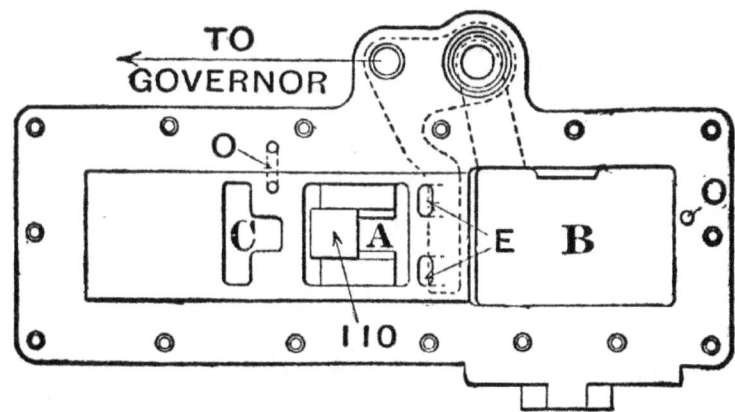

CHART 37—FIG. 9. SHOWING PORT "O" IN MAIN SLIDE VALVE SEAT.

MODERN AIR-BRAKE PRACTICE

The diagram of the brake valve in which the duplex gauge is shown (Chart 37, Fig. 1), also illustrates how the single governor, supplementary reservoir, main reservoir, and trainpipe are connected to the brake-valve, making in all six pipe connections.

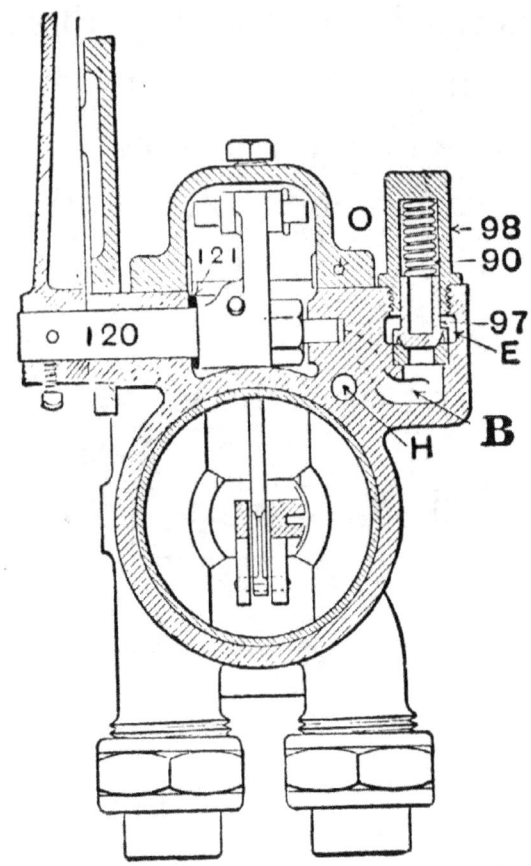

CHART 37—FIG. 10. CROSS SECTION OF NEW YORK BRAKE VALVE, SHOWING PASSAGE "H" IN BODY AND PASSAGE "O" IN THE VALVE COVER.

Fig. 2, Chart 37, shows the handle in full release position. In studying this illustration you will notice on the top right-hand corner a sectional view of the excess pressure valve. The location

IT'S USE AND ABUSE

of this valve on the brake valve is shown in Fig. 10, Chart 37.

In the left top corner of Fig. 2 you will notice a view of the main slide-valve and valve seat. In studying this view of the face of the main slide-valve you must remember that you see it as though you were looking directly through the top of the valve, and that the slanting lines represent the face of the main slide-valve while the dotted horizontal lines represent the slide-valve seat. You will notice that port K in the main slide valve extends across nearly the whole width of the valve, as does also the cavity marked M, and that the ports F and G are directly in the center, while port J and cavity P are on the side, therefore when looking at the main slide-valve sectionalized you must remember that you are looking at it as though it were cut half in two. This view shows plainly cavities M, F-G, K, and also Ports F and G, but you cannot see in the sectional view ports O, J, or cavity P. The port marked A in the valve seat is the opening that leads from chamber B by the end of the main slide-valve into chamber A, and this port A is controlled by the face of the main slide-valve.

In Fig. 2 you will notice that port F is closed by the main valve seat, whereas in the diagram of running position port F is closed by the small cut-off valve. The position of the cut-off valve is indicated by dotted lines in the view illustrating the face of the main valve. The large exhaust-cavity C is also indicated by dotted lines in the plan view of the main slide-valve seat.

The main slide-valve has four cavities, which are designated as M, F-G, P, and K. The ports in the main slide-valve are F, G, J, K, and N. The ports in the main valve seat are designated as E, A, C, and O.

As this great number of ports is likely to somewhat confuse the student, we will try to simplify the matter by saying that when the main slide-valve is moved to full release position main reservoir air from chamber B passes by the end of the slide-valve directly through the large port A into chamber A and straight into the trainpipe.

When the handle of the brake valve is in running position, main reservoir pressure passes through port E in the slide valve seat and cavity M in the slide-valve into port A and the trainpipe. While the air is passing from chamber B through cavity M it is also passing through chamber E into the pump-governor pipe.

In lap positions ports E and A in the slide-valve seat are closed by the main slide-valve, and

ITS USE AND ABUSE

exhaust port F is kept closed by the small cut-off valve, as shown in Fig. 4, Chart 37.

In service graduating position (Fig. 5), ports E and A are closed by the main slide-valve, but port F is moved back of the cut-off valve, so that while main reservoir pressure is shut off, trainpipe pressure can pass up through port F in the main slide-valve and out through port G into the main exhaust cavity C.

The handle being in service graduating position, when trainpipe pressure has exhausted below the pressure in the supplementary reservoir or chamber D, the equalizing piston is then forced forward by chamber D pressure, which causes the cut-off valve to move over and close exhaust port F. With the handle in service graduating position the main slide-valve closes the top end of port O, for, if it did not, when the equalizing piston moved forward the unseating of the ball-faced check valve would permit all the air from the supplementary reservoir to escape, and thereby prevent the automatic lapping of the brake valve.

Should the handle be moved to another service graduating notch, just as soon as the trainpipe pressure had exhausted below what was left in the supplementary reservoir the equalizing piston would again move forward and cause the cut-off valve to again lap exhaust port F. This action would continue in each of the graduating notches, but when the handle is moved to emergency position the valve does not automatically lap itself, for the reason that the equalizing piston has then made its full stroke.

When the handle is in emergency position, Fig. 7, the large port in the main slide-valve, marked J, is connected to the large exhaust port marked K, which causes the trainpipe pressure to pass out through exhaust passage C and be reduced suddenly, thereby causing all the triple valves in the train to go to emergency position.

When the handle is thrown from emergency, service or lap position back to full release, the raising of the trainpipe pressure drives the equalizing piston back, which causes the vent-valve (180) in the end of the piston (104 A) to close the bottom end of passage O, for in full release, running or positive lap position the top end of port O is open to exhaust cavity C by way of cavity P in the main slide-valve.

In order to firmly fix in the student's mind the purpose of the several ports and cavities we will run over them again as follows:

ITS USE AND ABUSE

Port E in the slide-valve seat and cavity M in the main slide-valve are primarily used for the purpose of directing the main reservoir pressure through the excess pressure valve into the train-pipe.

Chamber E supplies trainpipe pressure to the pump governor.

Ports F and G in the main slide-valve are trainpipe exhaust ports for the purpose of making a service application of the brakes.

Ports J and K are primarily used for the purpose of making an emergency application in connection with exhaust port C.

Passage C is the main exhaust port of the brake valve.

Cavity P is for the purpose of connecting port O in the main valve seat with the exhaust passage C.

Port N in the main valve is for the purpose of increasing the area of port A when the handle is in full release position, thereby allowing a full and free passage of main reservoir air into the trainpipe when releasing the brakes.

Passage O begins in the cap of the valve body (102 A), and passes through the wall of the cover (115 A) of the brake valve when it sinks into the valve body (101 A) and ends up in the main slide-valve seat, under the main slide-valve.

Passage H, which leads from chamber D, passes through the body of the brake valve to the pipe connection with the supplementary reservoir.

A small ball-valve (184) in the equalizing piston is for the purpose of supplying air to the supplementary reservoir so that the trainpipe and chamber D pressures may equalize when the brake valve is in either running or release position.

The vent-valve in the end of the equalizing piston is for the purpose of controlling the bottom end of passage O.

The purpose of passage and port O is to allow the pressure in chamber D to escape to the atmosphere when the equalizing piston is forced back to its normal position.

The purpose of the excess pressure valve (97) is to maintain a given pressure in the trainpipe when the handle of the brake valve is in running position.

The purpose of the several notches on the quadrant are as follows: When the handle is in the extreme forward position main reservoir pressure is fed direct into the trainpipe; the first notch, which is also indicated by a small pin on

ITS USE AND ABUSE

the side of the quadrant, is running position, and in this position main reservoir pressure is fed into the trainpipe through an indirect passage, or by way of the excess pressure valve; the next notch on the quadrant is known as positive lap position, and when the handle is in this position all ports in the brake valve are closed between the main reservoir and the trainpipe and between the trainpipe and the atmosphere, and in this position the pressure from the trainpipe and pump governor is also shut off, and it is on account of this fact that a duplex pump governor is necessary with the New York Brake Valve. The next notch after positive lap is the first service graduating notch, and when the handle is in this position the brake valve will allow about five pounds of trainpipe pressure to exhaust when it will automatically lap itself; the second notch will allow about eight pounds of trainpipe pressure to exhaust when the valve will automatically lap itself; the next notch will cause an exhaust from the trainpipe of eleven pounds; the fourth notch causes a sixteen pound reduction, and the fifth notch is the twenty-three pound or full service position.

When the handle is placed in any of the service graduating positions the brake valve will automatically lap itself, but when the handle is thrown to emergency position then the automatic lap feature does not operate. When the valve automatically laps itself the equalizing piston moves the cut-off valve so that it covers exhaust port F, but when the handle of the valve is moved to positive lap position the main slide-valve places port F over the cut-off valve.

Should the handle be placed in the five-pound notch and while it was in this position trainpipe leakages should cause the trainpipe pressure to be reduced to sixty pounds or less (when working with a seventy pound standard) then when the handle was moved to the eight pound notch there would be no exhaust from the trainpipe, for the reason that the trainpipe leakages would cause the pressure in chamber D to push the equalizing piston forward and cause the cut-off valve to keep exhaust port F closed. This is a splendid feature of the New York Brake Valve.

As the automatic lap feature is dependent upon the proper movement of the equalizing piston, you will at once see that should there be any leakages from chamber D or the supplementary reservoir it would prevent the equalizing piston from moving forward and causing the cut-off valve to close exhaust port F. There are several other things besides direct leakage from chamber D to the atmosphere which will prevent the automatic lapping of the valve, and they are as follows:

A leak by the piston packing-leather of the equalizing piston will prevent the automatic lap.

Should the ball check-valve fail to seat it will prevent the automatic lap.

Should the face of the main slide-valve be scratched so that it did not seat properly on the cut-off valve it would prevent the automatic lap.

Should the seat of the cut-off valve be scratched so that it did not seat properly it would prevent the automatic lap.

Should the arm (112) connecting the cut-off valve to the equalizing piston become bent or disarranged it would prevent the automatic lap.

These several defects and all others pertaining to the New York Brake Valve will be treated under the head of Defects and Diseases and Their Remedies.

You will notice two cap screws in the cover of the brake valve; these are for the purpose of admitting oil to the main slide-valve seat.

To oil the slide-valve seat let off all main reservoir pressure, cut out the trainpipe from the brake valve, exhaust all air pressure and remove the cap screws from the valve cover, then throw the handle to the full release position and drop in just a little good oil onto the valve seat; then throw the handle to emergency position and drop a little oil on that end of the valve seat, and work the handle backward and forward several times in order to distribute the oil. Be careful not to use too much oil as it will gum up the valve.

While the air pressure is off unscrew the cap nut of the excess pressure valve and wipe that valve off with kerosene, and be sure that it is wiped dry before you put it back.

In setting the regulating spring of the excess pressure valve, place the brake valve handle in running position, and let the pressure pump up until the red hand of the gauge shows twenty pounds before the black hand begins to move. Should the black hand begin to move before the red hand reaches twenty pounds it indicates that the graduating spring needs to be tightened down, while, on the other hand, if the black hand of the gauge did not begin to move until the red hand had passed the twenty pound mark then the graduating spring should be loosened up.

With the New York Brake Valve handle in running position the excess pressure is accumulated *before* the trainpipe pressure begins to show,

ITS USE AND ABUSE

whereas with the Westinghouse brake valve it is just the opposite; for you do not get your excess pressure until *after* the trainpipe is fully charged.

The old style A New York Brake Valve differs from the present style B and B 1, in that it does not have the vent-valve in the end of the equalizing piston, neither does it have the ball check-valve nor port O in the valve seat, and as a consequence when you make a service application and go to full release position, and then move the handle direct from release to service graduating position the valve will not automatically lap for the reason that in order to get the automatic lap feature of either of the New York Brake Valves it is necessary to have the supplementary reservoir pressure equal to trainpipe pressure at the beginning of a service application, and with the old style A valve, which does not have the ball check-valve, the only way in which the supplementary reservoir can be charged up is by placing the handle in running position.

THE NEW YORK PLAIN TRIPLE VALVE

The New York Plain Triple Valve is so nearly like the Westinghouse plain triple that it requires no special description here, and the same

MODERN AIR-BRAKE PRACTICE

instructions regarding the Westinghouse plain triple will apply equally to the New York plain triple.

THE NEW YORK QUICK ACTION TRIPLE VALVE

In studying the diagrams of the New York Quick Action Triple Valve, Chart 38, you must remember that the real valve does not have the shape shown in Figs. 1, 2, 3 and 4, but that portion of the triple which shows passage H, port J, vent valve 137 and check valve 139 are shown in Figs. 5 and 6. The object in drawing the diagram in the way in which it is shown is to make plain to you how these ports, passages and valves are related to the rest of the triple.

The principal operative parts of the New York Quick Action Triple are the main triple piston (128), the exhaust slide valve (38), the graduating sliding valve (48), the vent piston (129), the rubber seated vent-valve (131), and spring (132), emergency piston (137), with rubber seated quick action valve (139), and spring (140), non-return brake cylinder check valve (117), and its spring (118).

You will notice that the vent piston (129) has a port F which leads through the center of it into chamber G of the main triple piston. This

ITS USE AND ABUSE

allows trainpipe pressure to get in between the pistons, forming a cushion which does away with

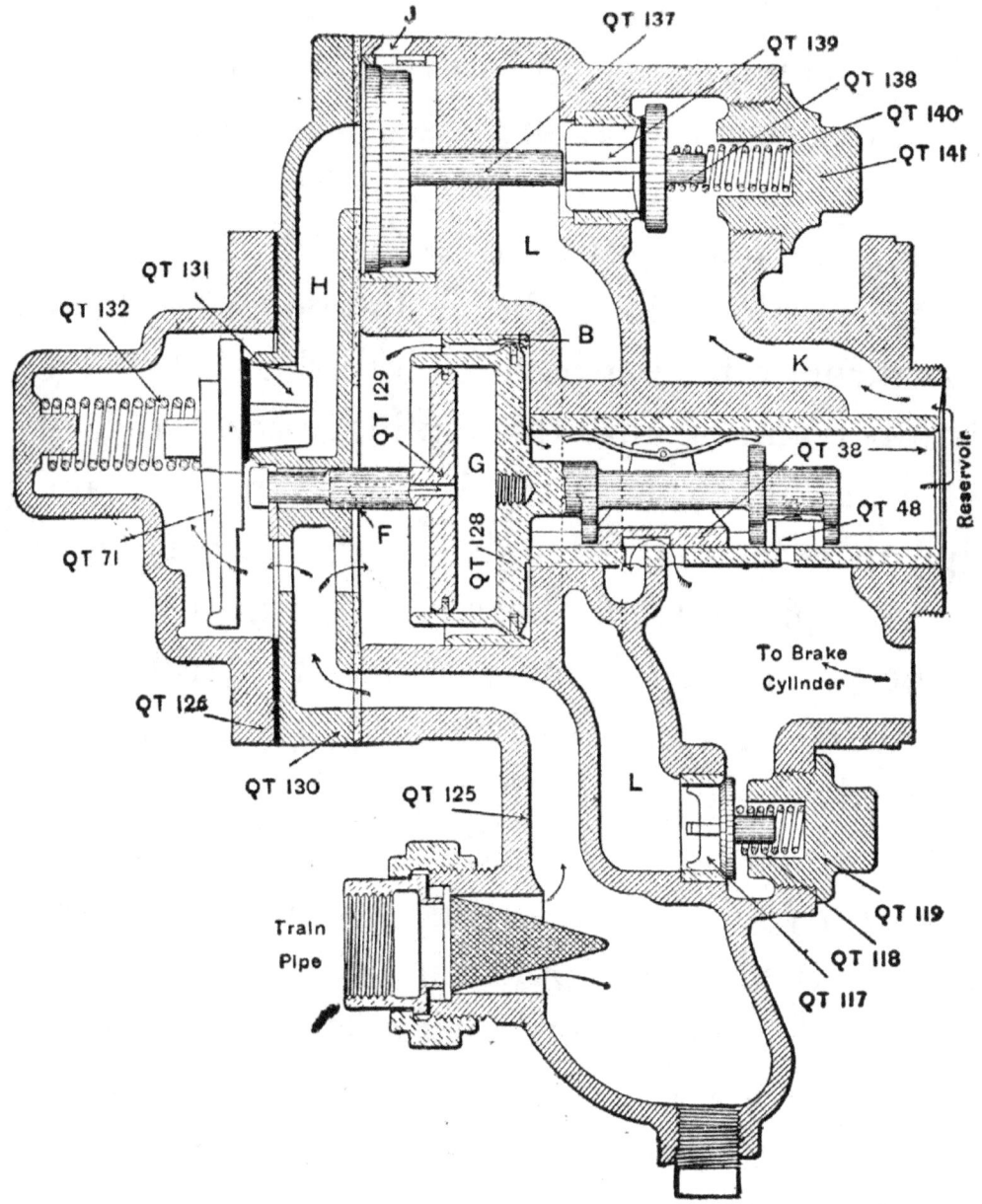

CHART 38—FIG. 1. NEW YORK QUICK ACTION TRIPLE VALVE, RELEASE POSITION.

the graduating spring as used in the Westinghouse triple.

The passage of the air through the New York

Quick Action Triple is as follows: referring to (Fig. 1, Chart 38) trainpipe pressure passes through the strainer, fills the cavity back of the rubber seated vent-valve (131), thereby holding that valve to its seat, and also passes through a large opening into the main piston chamber causing the main piston to be forced to charging position, which allows the trainpipe pressure to pass through feed groove B into the slide-valve chamber and on into the auxiliary reservoir; at the same time this action is taking place trainpipe air is feeding through port F in the stem of the vent-piston (129), thereby charging chamber G, between the pistons. When the trainpipe, chamber G, and auxiliary reservoir are all charged equally to seventy pounds we are then ready to make an application of the brake.

Now if you will notice Fig. 2, which shows the triple in the service application position, you will see that the main triple piston has moved back until it touches the vent piston (129), and that it has moved this vent piston back far enough so that port F is just closed; as the trainpipe pressure is reduced the pressure in chamber G is also reduced, but as it reduces slower than the trainpipe pressure it graduates the movement of the main triple piston, so that when the main

ITS USE AND ABUSE

piston has made its full stroke it has not disturbed the rubber seated vent-valve (131), but has moved the graduating slide valve (48) to a

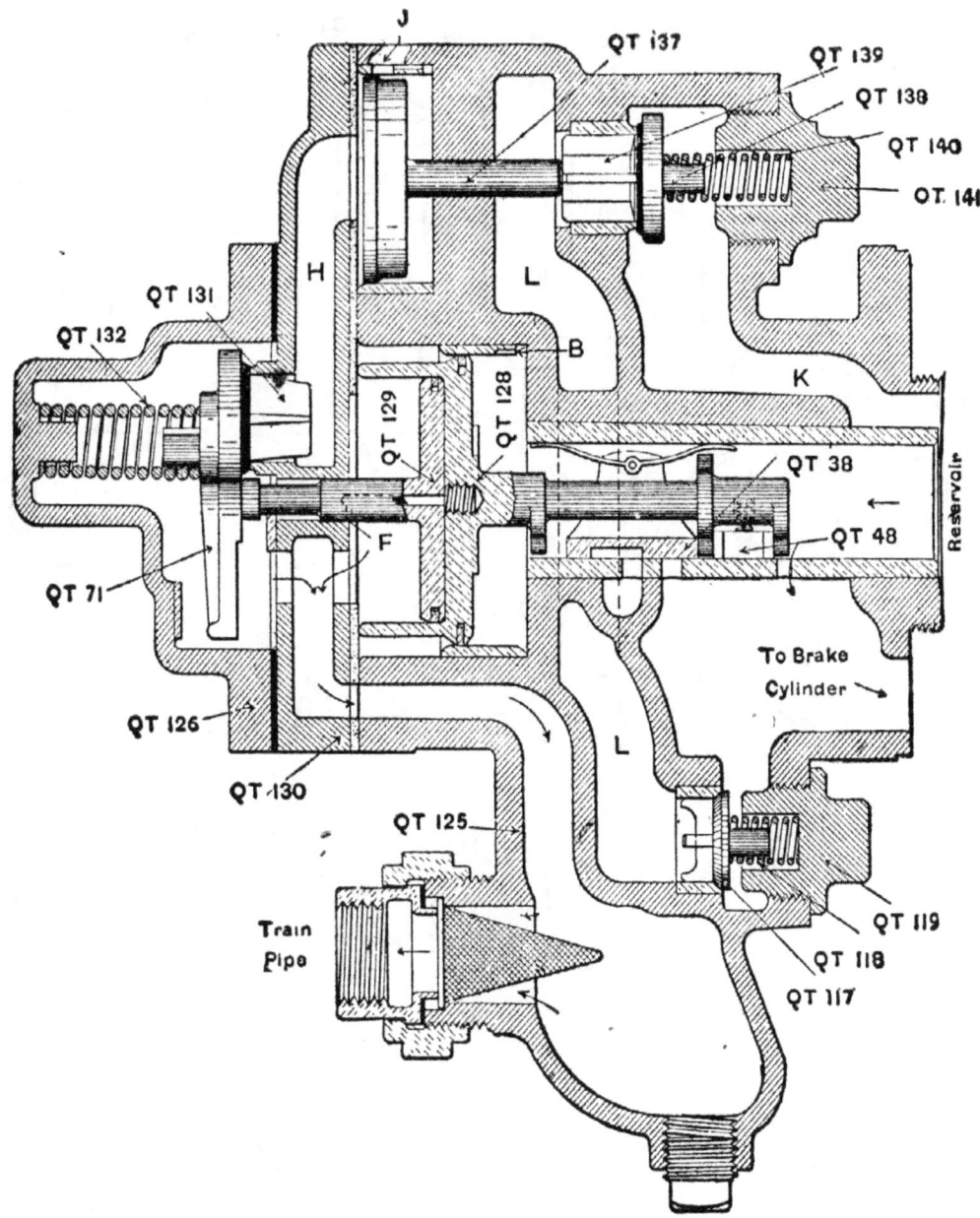

CHART 38—FIG. 2.   NEW YORK QUICK ACTION TRIPLE, SERVICE APPLICATION POSITION.

position which opens the supply port from the auxiliary reservoir to the brake cylinder, and at

MODERN AIR-BRAKE PRACTICE

the same time has moved the exhaust slide-valve (138) forward and closed the exhaust port from the brake cylinder to the atmosphere. When

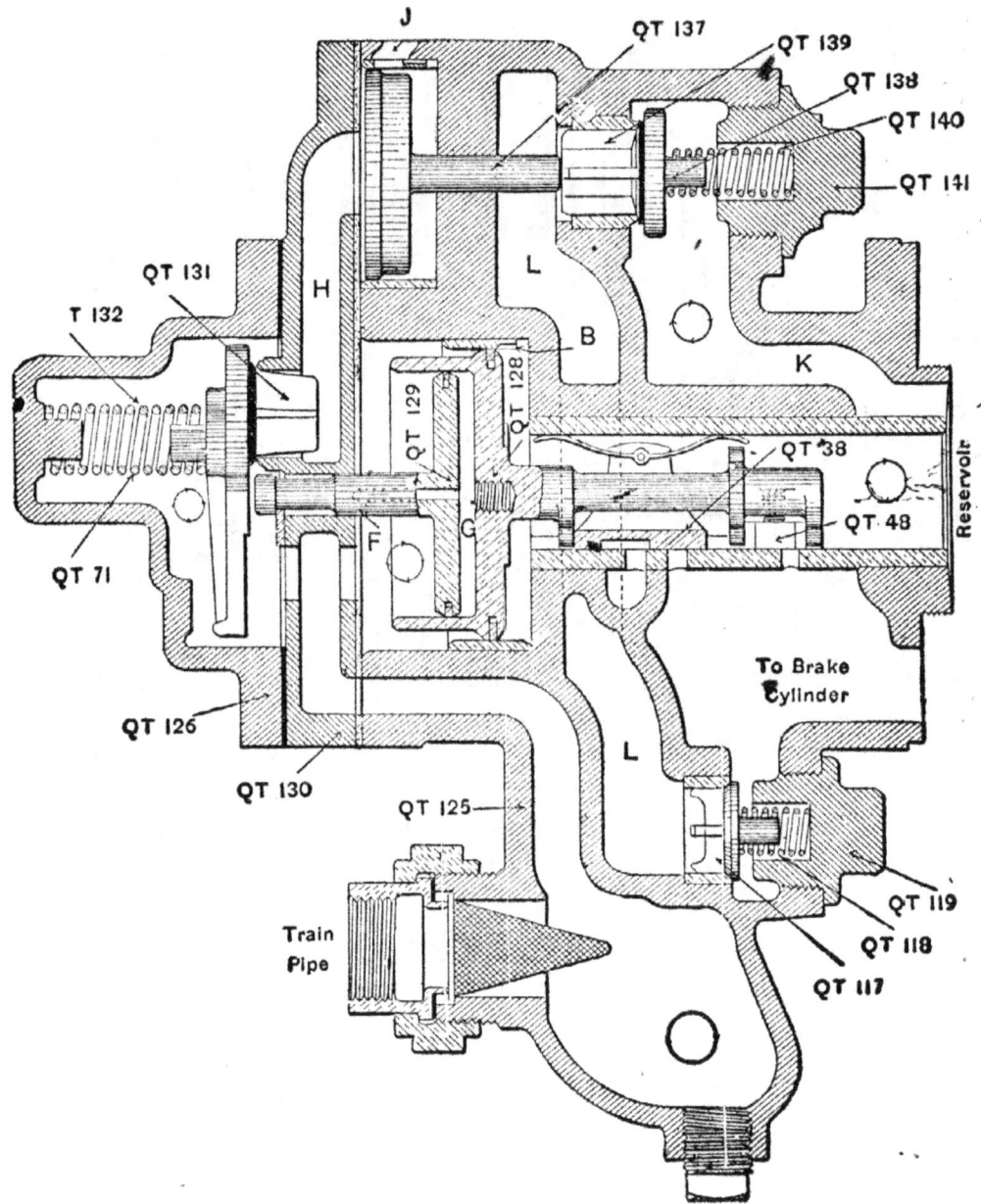

CHART 38—FIG. 3. NEW YORK QUICK ACTION TRIPLE VALVE, AUTOMATIC LAP POSITION.

the main piston moves forward it gradually closes port F before all the pressure from

ITS USE AND ABUSE

chamber G has exhausted, consequently when the auxiliary pressure has reduced to a degree

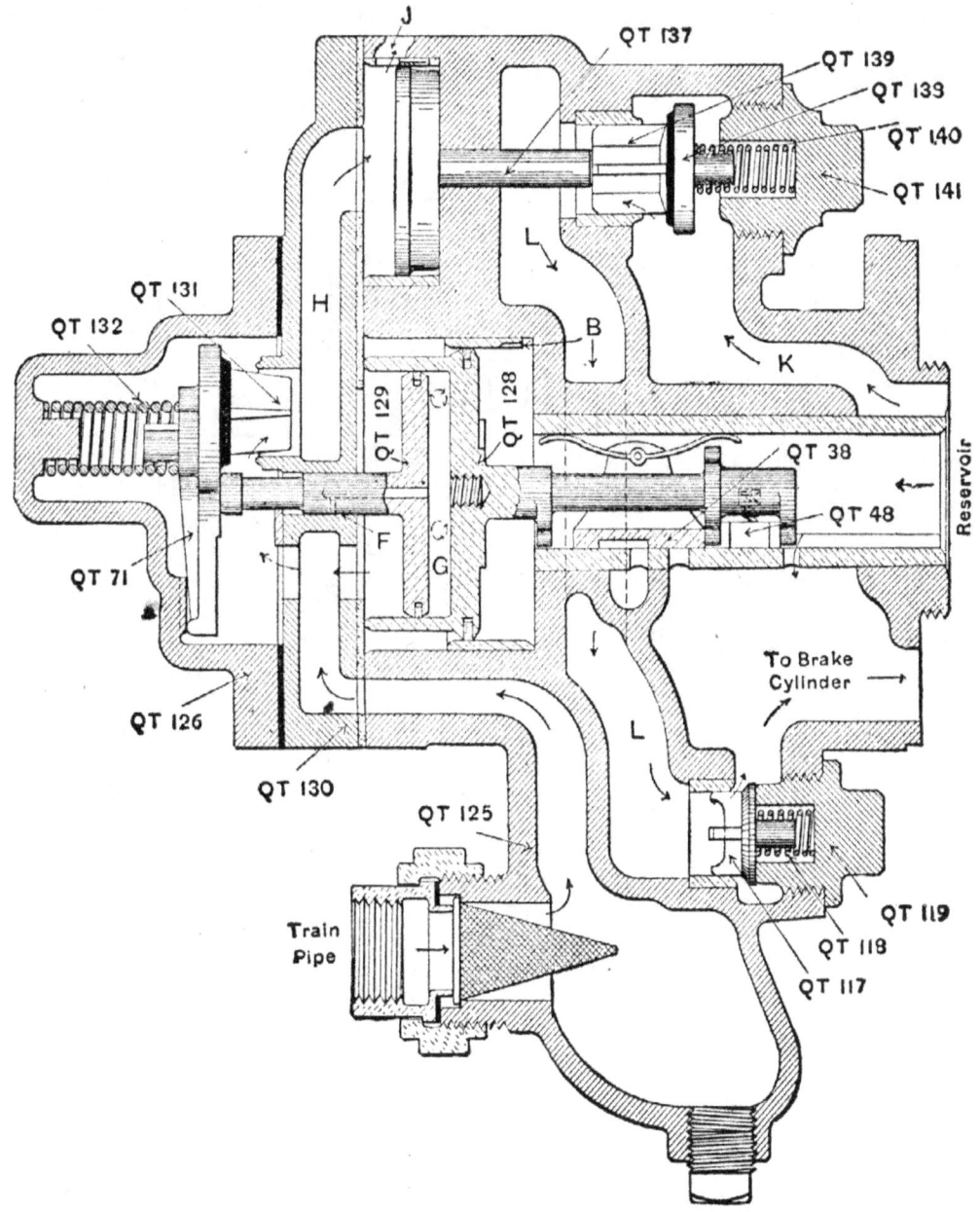

CHART 38—FIG. 4A. NEW YORK QUICK ACTION TRIPLE VALVE, EMERGENCY POSITION (SPECIAL VIEW FOR SHOWING EMERGENCY VALVE).

slightly less than trainpipe pressure the air which is confined in chamber G expands and

MODERN AIR-BRAKE PRACTICE

forces the main piston back just a little, which causes the graduating slide-valve to close the port from the auxiliary reservoir to the brake cylinder without disturbing the exhaust slide-

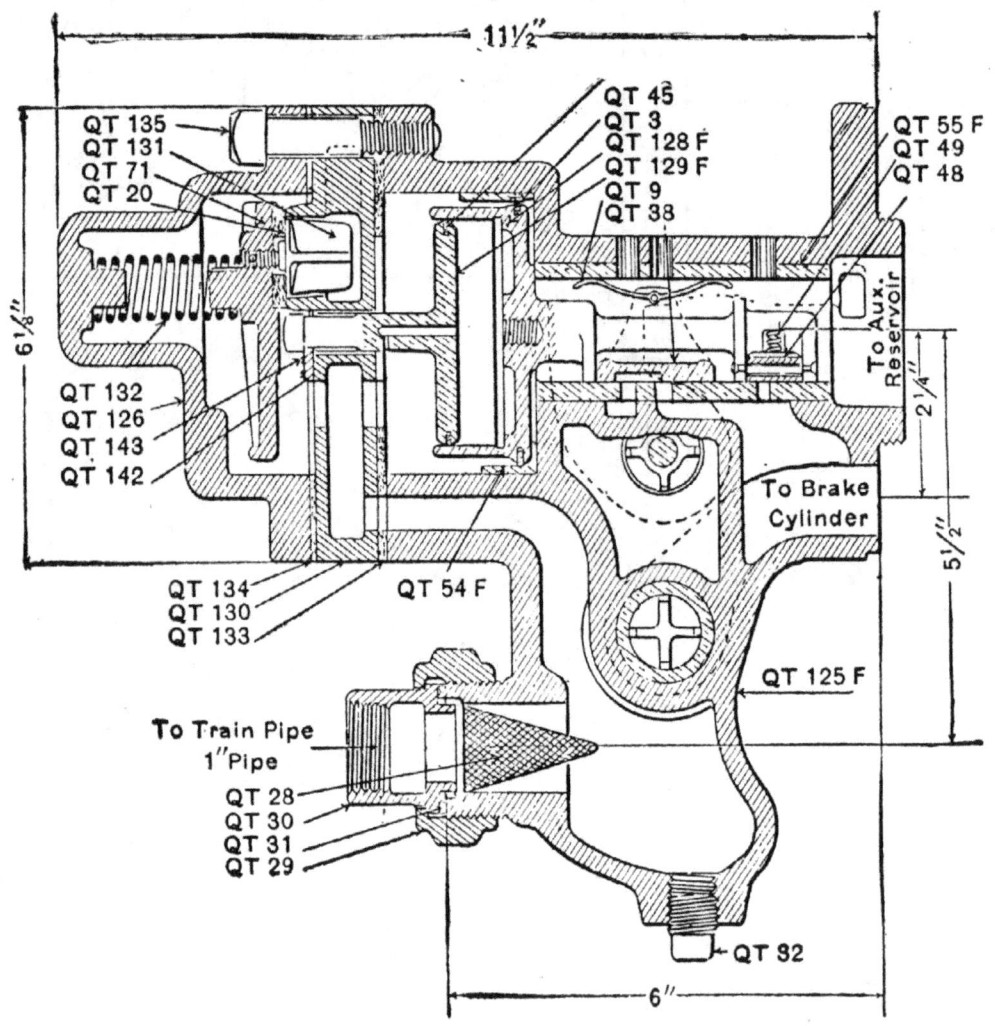

CHART 38—FIG. 4. NEW YORK QUICK ACTION TRIPLE VALVE.

valve that controls the exhaust port from the brake cylinder to the atmosphere. The triple is now in lap position as shown by Fig. 3.

ITS USE AND ABUSE

The emergency action of the triple (Fig. 4) is brought about as follows: The air cushion in chamber G cannot be reduced through port F as quickly as the trainpipe pressure is reduced,

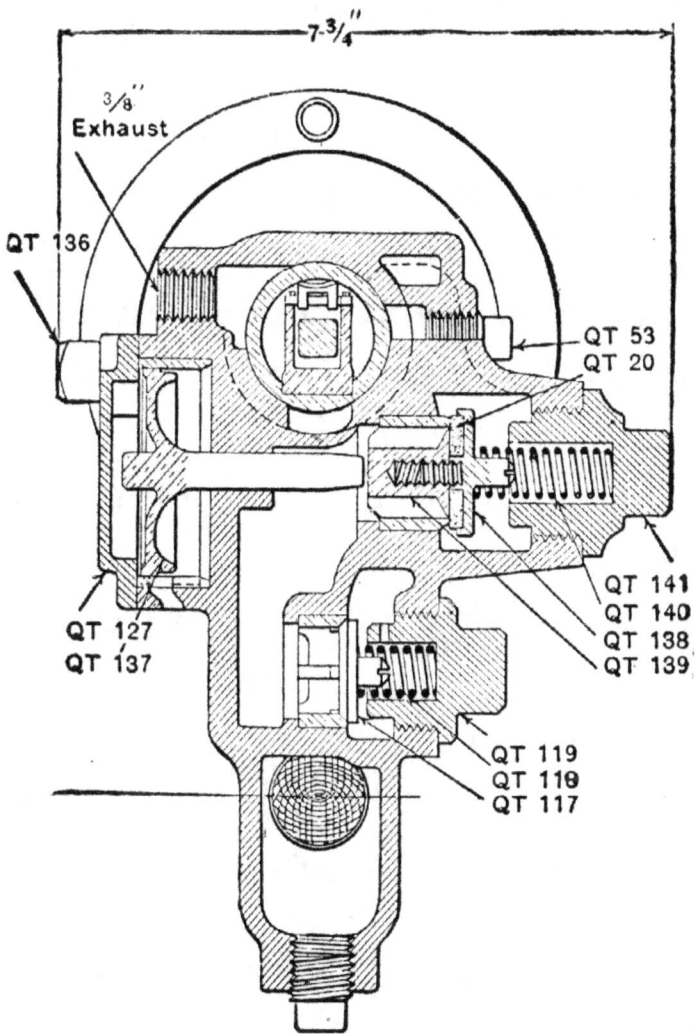

CHART 38—FIG. 6. NEW YORK QUICK ACTION TRIPLE VALVE.

consequently when a sudden reduction is made on the trainpipe pressure it causes the auxiliary pressure to drive the main piston back so quickly that port F is closed before chamber G can empty itself, and with an air cushion between the two pistons you will at once see that the stem of the vent-piston strikes the rubber seated vent valve and drives it from its seat, which allows trainpipe pressure to pass into passage H and thereby forces the emergency piston (137) forward, which action not only opens port J to the atmosphere for the purpose of still further reducing the trainpipe pressure, but it also unseats the rubber seated emergency valve (139) which allows the auxiliary pressure to flow from chamber K by the rubber seated valve into chamber L, and unseat the non-return check valve, thereby causing the auxiliary reservoir pressure to quickly equalize with the brake cylinder. When the trainpipe pressure has reduced less than the auxiliary reservoir pressure the emergency valve (139) is forced to its seat and the brake cylinder pressure equalizes with the pressure in chamber L causing the non-return check valve to go to its seat, and it is held there both by the brake cylinder pressure and the spring (118).

THE NEW YORK COMBINED AUTOMATIC AND STRAIGHT AIR BRAKE VALVE

The New York Straight Air Brake Valve performs the same functions as the Westing-

ITS USE AND ABUSE

house, that is it applies the engine and tender brakes independent of the triple valve when the triple is in release position.

The New York straight air equipment consists of a straight air brake valve, a reducing valve, a double check-valve, a brake cylinder guage and a safety valve on the brake cylinder, the same as is used in the Westinghouse system, but the New York straight air valve is modeled after their automatic engineer's brake valve, for you will see by the diagram illustrating the straight air brake valve that the essential parts to this valve (aside from the case) is a slide-valve operated by a handle working over a quadrant. There are two oil plugs for the purpose of oiling the slide-valve seat, the same as with the automatic brake valve. There are two pipe connections and one exhaust. One pipe connection admits main reservoir pressure into the brake valve and the other pipe connection allows the pressure to pass into the brake cylinder.

There are four positions on the New York straight air brake valve as follows: Release, Lap, Service, and Emergency.

By looking at Fig. 1, Chart 40, you will notice that the handle is in full release position, and brake cylinder pressure can pass under the slide-

MODERN AIR-BRAKE PRACTICE

valve and out at the exhaust cavity. Should the handle be moved to lap position you will notice that the slide valve will close the passage leading to the brake cylinder, thereby preventing main reservoir pressure from getting into the cylinder

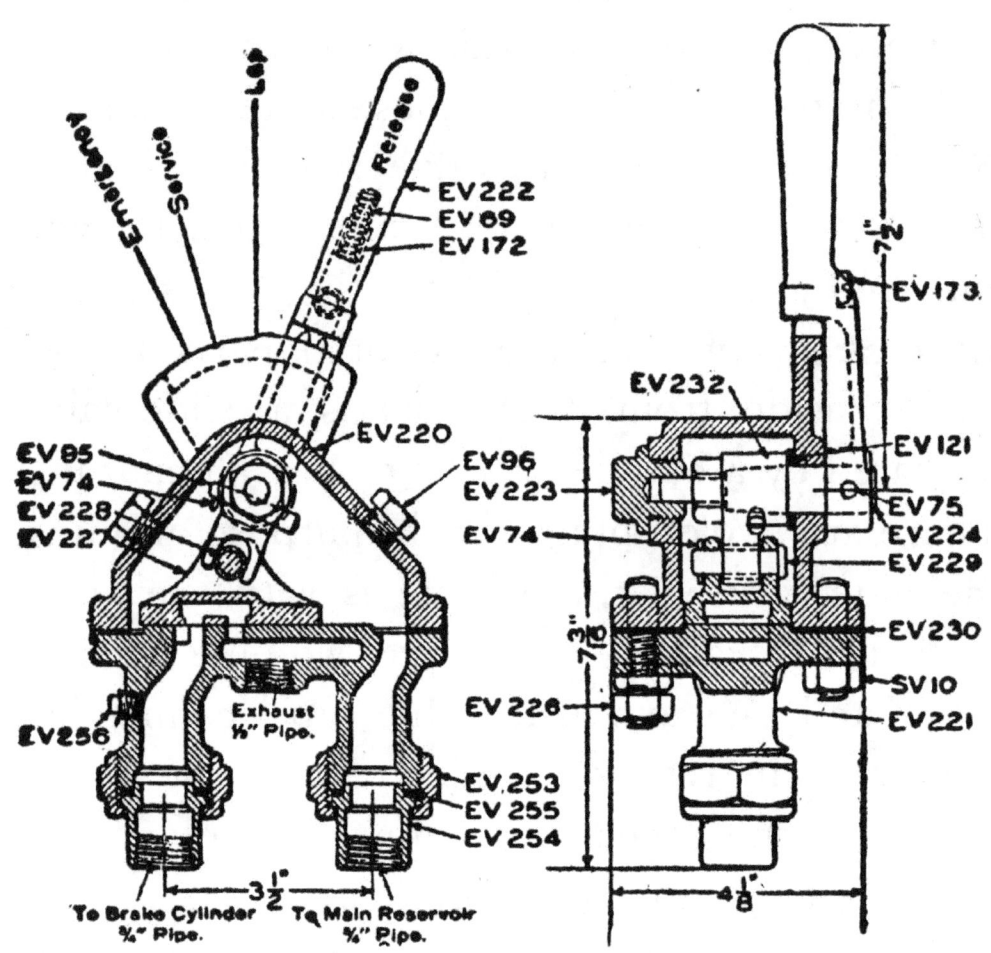

CHART 40—FIG. 1. NEW YORK STRAIGHT AIR BRAKE VALVE.

and also preventing the cylinder pressure from escaping to the atmosphere. Now should the handle be moved to the next notch, or service position, the slide-valve will be moved further

ITS USE AND ABUSE

back, thereby creating a small opening to the brake cylinder and allowing the engine brakes to be set gradually, but should the handle be thrown to emergency position the slide-valve will be moved still further back, so that the

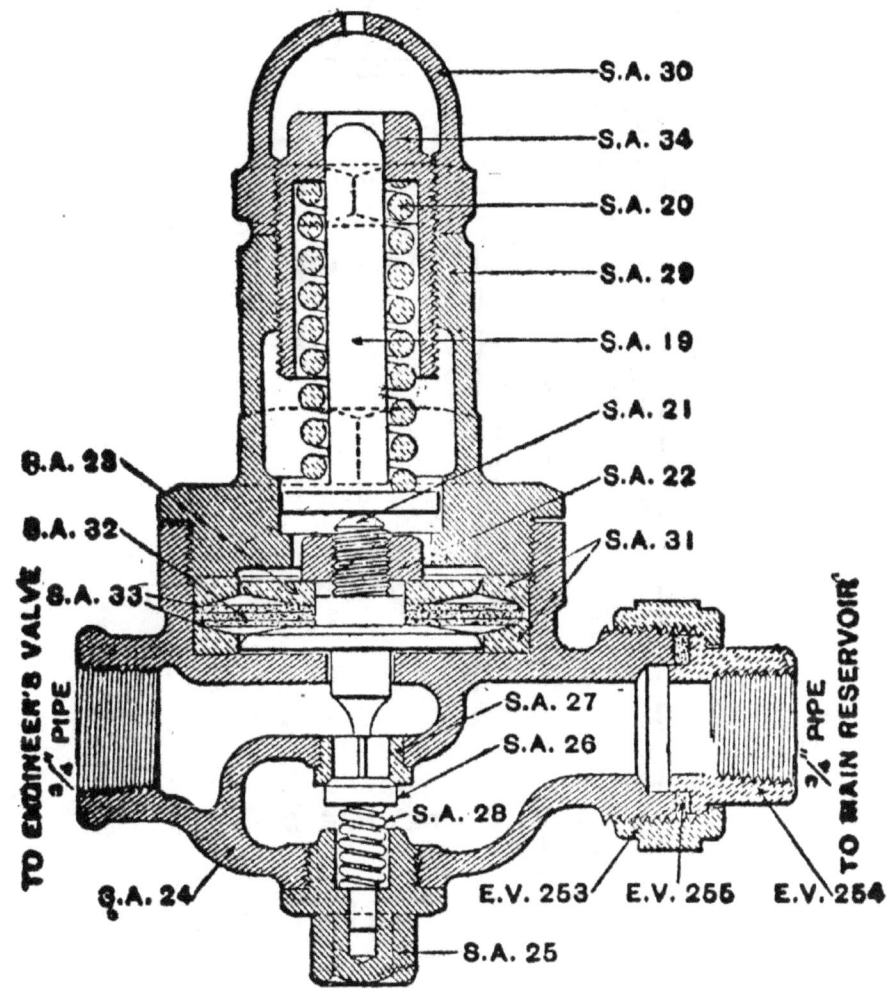

CHART 40—FIG. 2. NEW YORK STRAIGHT AIR PRESSURE REDUCING VALVE.

passage to the brake cylinder is wide open, which permits a quick rush of air into the cylinders. Between the main reservoir and the straight air brake valve there is a reducing valve, Fig. 2, Chart 40, for the purpose of keeping the main reservoir pressure down to a predetermined standard, which is usually forty-five pounds.

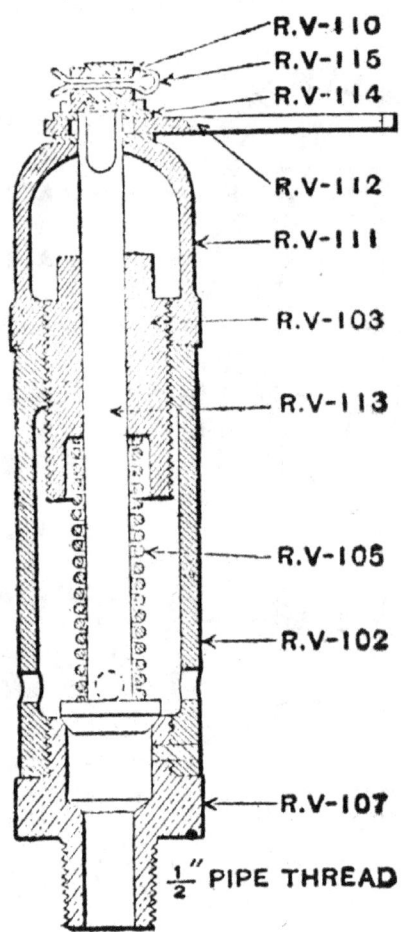

CHART 40—FIG. 3. NEW YORK SAFETY VALVE, WITH HAND RELEASE.

The straight air reducing valve, as shown in Chart 39, is connected at one end to the main reservoir and at the other end to the straight air brake valve, and as the regulating spring is supposed to be screwed down to forty-five

ITS USE AND ABUSE

pounds, you will readily see by Fig. 2, Chart 40, that the force of the graduating spring will drive

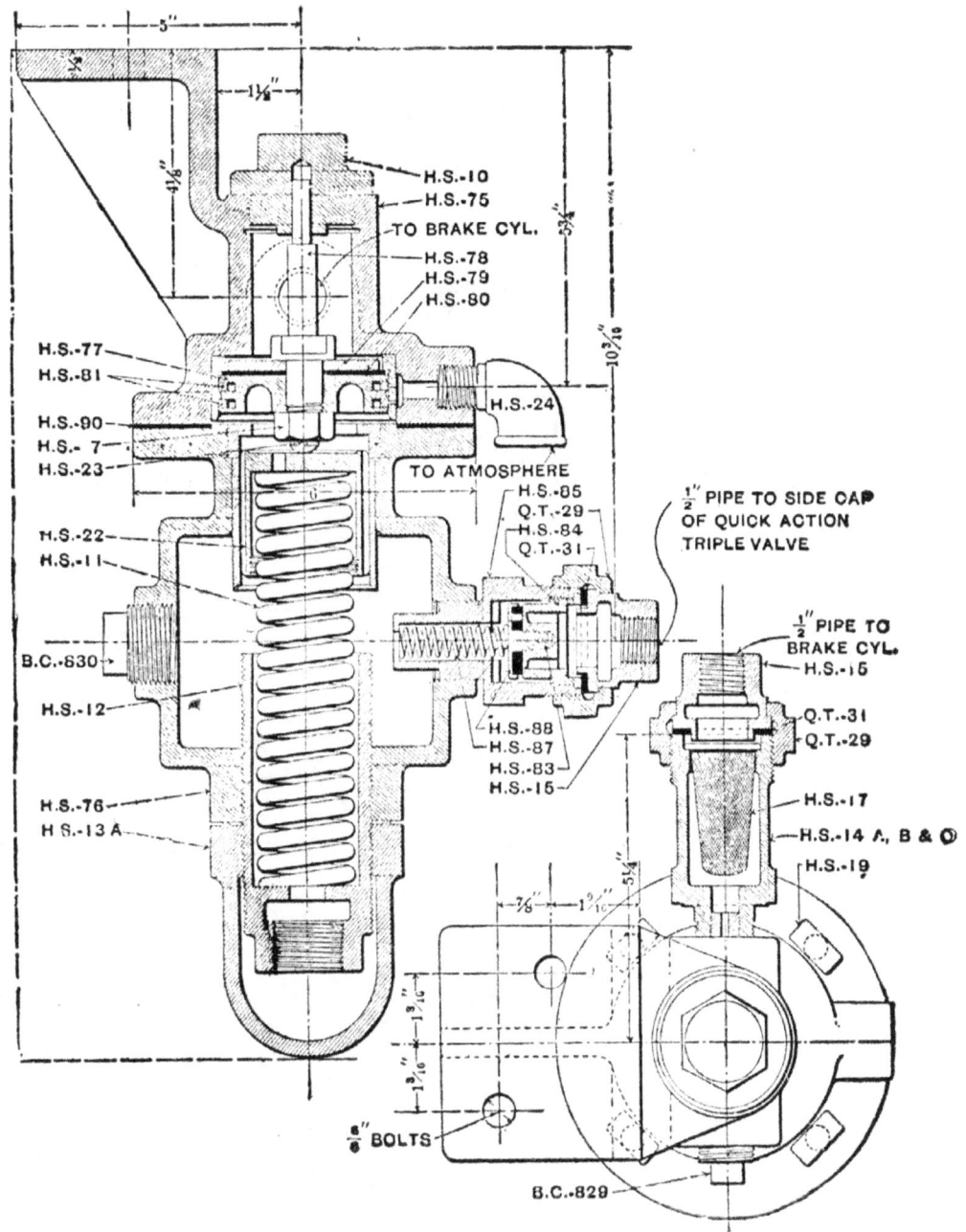

CHART 40—FIG. 4. STYLE A COMPENSATING VALVE, HIGH SPEED BRAKE.

the diaphragm down so that it unseats the check-valve (26), therefore when no air is in the brake

MODERN AIR-BRAKE PRACTICE

cylinder the main reservoir pressure can pass by the check-valve and flow through the pipe

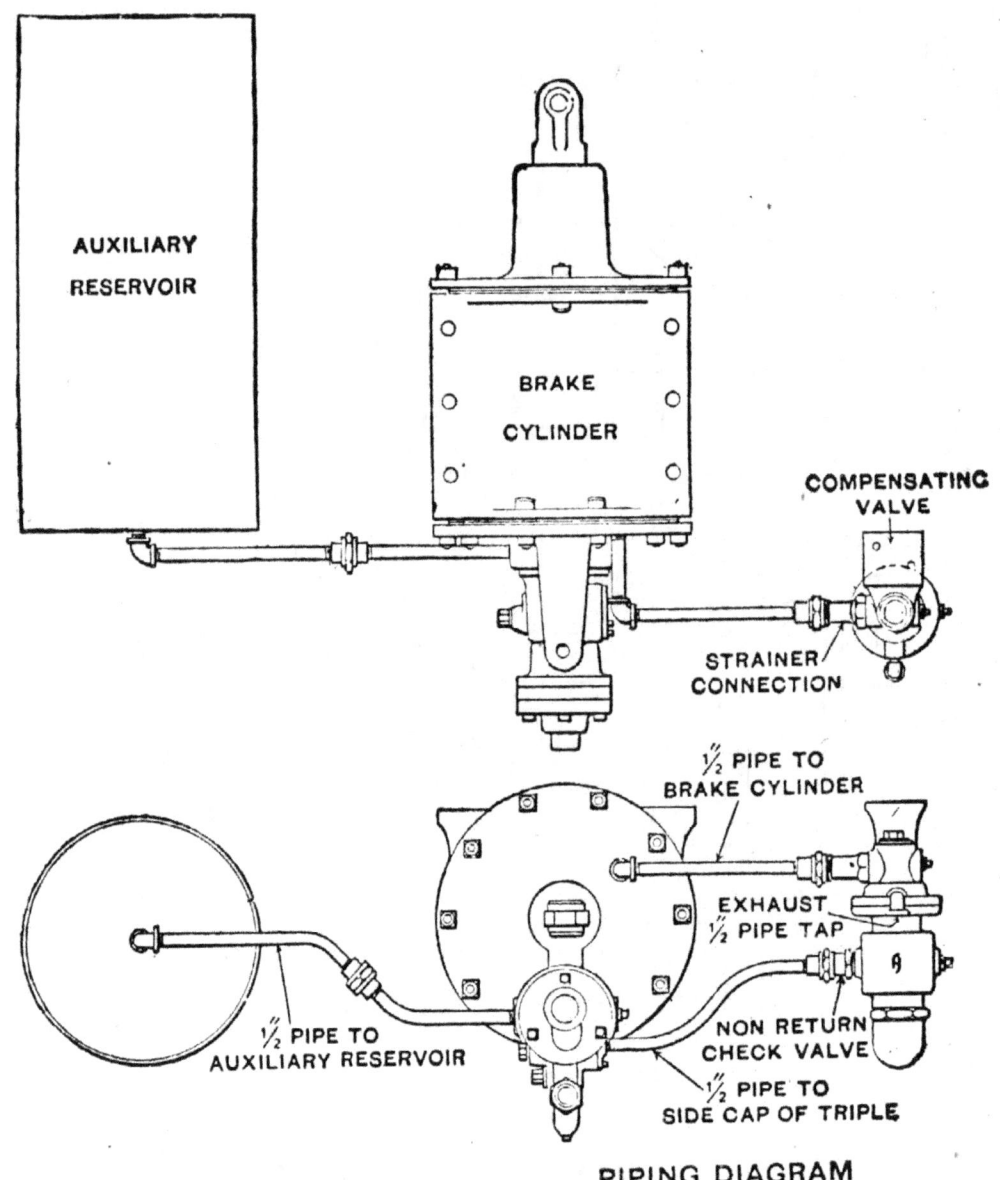

PIPING DIAGRAM
COMPENSATING VALVE, STYLE A.
Diagram 114
Diagram Showing Method of Piping Style A and A-1 Compensating Valve.
CHART 40—FIG. 5.

connection leading to the straight air brake valve, and when the pressure under the

ITS USE AND ABUSE

diaphragm becomes a fraction greater than what the regulating spring (20) is set at, the diaphragm will be moved up, thereby allowing the check-valve to reseat and shut off the main reservoir pressure, but should the brake cylinder leathers leak, as soon as the leakage brings the pressure down below the tension of the graduating spring the diaphragm will be forced down and again unseat the check-valve to admit main reservoir pressure.

This action you will see enables the engineer to place the straight air brake valve in service position and work under his engine with perfect safety, because he knows that as long as the pump works the straight air brake valve will automatically supply main reservoir pressure to the brake cylinders, and keep the engine from moving.

One of the greatest benefits that the straight air brake valve confers in road service is that it enables the engineer to set the engine brakes independently of the train brakes, so that in slowing down or in making a stop he can keep the train bunched, and thereby prevent a break-in-two.

The safety valve (Fig. 3, Chart 40), on the brake cylinders is for the purpose of taking care of any leakage in the reducing valve, for should the check-valve of the reducing valve leak the main reservoir pressure would, of course equalize with the brake cylinder pressure, and in order to prevent this the safety valve is placed on the brake cylinder to allow any extra pressure that might get into the cylinders to automatically blow down.

The diagram which illustrates the general arrangement and method of piping the New York Combined Automatic and Straight Air Brake Valve (Chart 39), will show the relative positions of the several parts.

You will notice in the pipe that leads to the brake cylinders that there is a safety valve, and you will also notice in this same pipe there is a double check-valve which is the same as used in the Westinghouse system, which see. You will notice that from the brake pipe there is shown in dotted lines another pipe on the end of which there is a cock. This cock is for the purpose of releasing the air from the brake cylinders when descending heavy grades or in case of a hose bursting, thereby saving the engine tires from being loosened or skidded.

By noticing this same diagram you will see that on the tender there is also a safety valve, a

ITS USE AND ABUSE

double check-valve, and the line of pipe in dotted lines has a release cock on the end of it, the same as on the engine. These release cocks are placed one in the engine cab and the other in the gangway on the tender. The descending

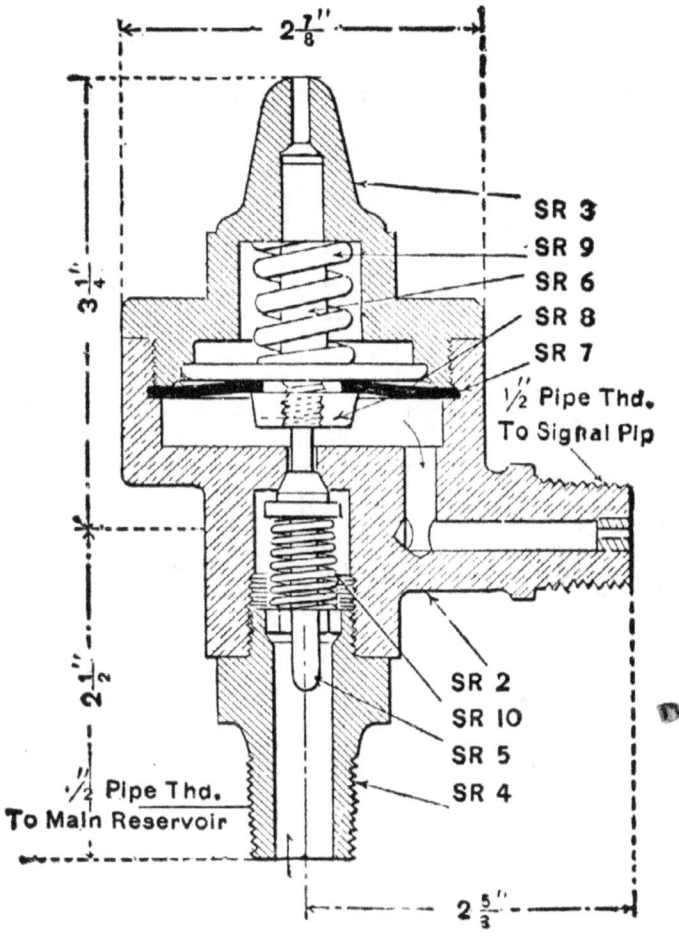

CHART 41—FIG. 1. NEW YORK AIR SIGNAL PRESSURE REDUCING VALVE.

of long heavy grades makes it absolutely essential to have some means by which the engine and tender brake cylinder pressure can be reduced without having to release the train brakes, and we all know how important it is to release the engine and tender brakes when a hose bursts, providing we want to save the engine tires from being loosened or flattened.

NEW YORK AIR SIGNAL EQUIPMENT

The number of parts constituting the equipment of the New York Signal is the same as the

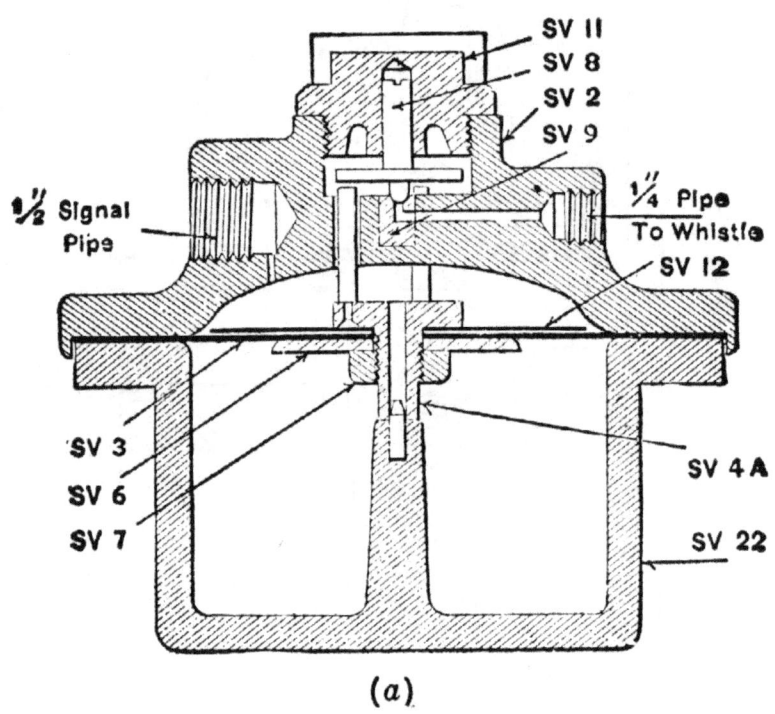

CHART 41—FIG. 2 STYLE B NEW YORK AIR SIGNAL VALVE.

Westinghouse, but the construction of the parts is somewhat different. The essential parts are a pressure reducing valve (Fig. 1, Chart 41); and air signal valve (Fig. 2), the car discharge valve (Fig. 3) and the signal whistle (Fig. 4).

The operation of the reducing valve is as fol-

ITS USE AND ABUSE

lows. When the regulating spring (6) is screwed down to forty pounds it causes the diaphragm (8) to force the check-valve (5) from its seat, which allows the main reservoir pressure to flow by the check-valve into the diaphragm chamber and out at the pipe connection leading to the signal pipe, and when the signal pipe is charged up to a fraction over forty pounds it causes the diaphragm to be forced away from the check-valve, and thereby enables the check-valve to reseat and shut off main reservoir pressure. As soon as the signal pipe pressure is again reduced below the tension of the graduating spring, the diaphragm again unseats the check-valve and allows the main reservoir pressure to again flow into the signal pipe.

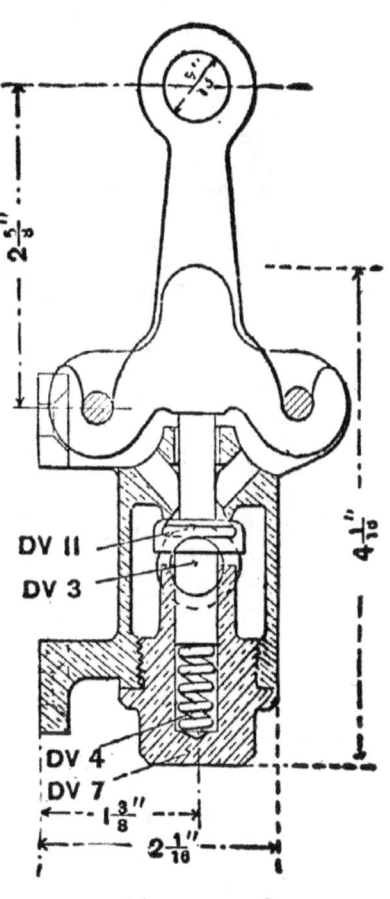

CHART 41—FIG. 3. CAR DISCHARGE VALVE.

The air signal valve (Fig. 2) is connected to the signal pipe and to the signal whistle, and when the pressure from the signal pipe enters the signal valve it passes down through the port in the diaphragm stem into the lower air chamber, and equalizes on both sides of the diaphragm, and at the same time passes up into the air valve chamber and equalizes on both sides of that valve. When a sudden reduction is made from the signal pipe the air on the top side of the diaphragm is also reduced so that the pressure in the lower air chamber lifts the diaphragm and causes the upright pins which are fastened to the diaphragm stem to force the air valve (8) from its seat, and thereby permit the air from the lower chamber to blow into the whistle. As soon as the signal pipe pressure has stopped exhausting, the diaphragm is forced back to its seat which allows the air valve to again drop to its seat. The same instructions regarding the manner of operating the air signal that governs the Westinghouse system apply equally as well to the New York Air Signal.

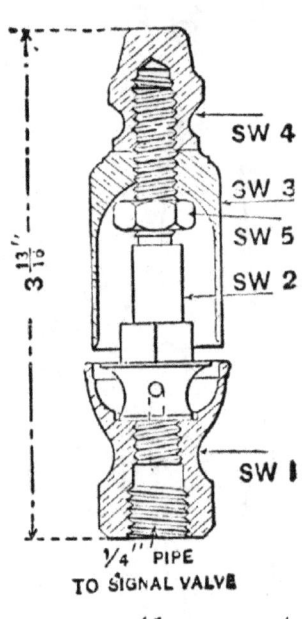

CHART 41—FIG. 4.
SIGNAL WHISTLE.

QUESTIONS AND ANSWERS TO SECTION 4

THE PARTS OF THE NEW YORK AUTOMATIC, HIGH SPEED AND STRAIGHT AIR BRAKE EQUIPMENT AND THEIR DUTIES

Note.— As many of the answers given in Section 1 will apply equally to the New York Equipment the same as they do to the Westinghouse, whenever such is the case I will indicate the answer by referring to the answer in Section 2.

62. When an engine is equipped with the New York Automatic Quick Action Brake what are the essential parts and what are their duties?

Ans. See Question 2, Section 2.

63. What additional apparatus is needed to change the quick action equipment into the high speed brake?

Ans. A triplex pump governor and an automatic reducing valve for the engine and tender brake cylinders. The triplex governor with the New York equipment performs the same functions as the extra slide-valve feed-valve does with the Westinghouse brake; that is, it controls the pressure.

64. What apparatus is required on a freight car with the New York equipment?

Ans. See Question 7, Section 2.

65. What are the parts which constitute the equipment on a passenger coach?

Ans. See Question 8, Section 2.

66. When a passenger car is equipped with a high speed brake what additional apparatus is needed?

Ans. The compensatory valve connected to the brake cylinder, for the purpose of automatically reducing the cylinder pressure as the train slows down. The compensating valve of the New York brake is somewhat different from the Westinghouse High Speed Reducing Valve for the reason that it does not permit the cylinder pressure to begin to exhaust as quickly as it does with the Westinghouse valve. The compensatory valve is fully illustrated in Fig. 4 of Chart 40, and aside from the fact that it retards the brake cylinder pressure for a longer period, the general operation of it otherwise is the same as the Westinghouse High Speed reducing valve. Where the Westinghouse valve uses a slide-valve to control the exhaust the New York compensating valve performs this function with a piston, as shown in Fig. 41, Chart 40.

ITS USE AND ABUSE

67. What is a main distinguishing feature between the New York and the Westinghouse pump?

Ans. The New York pump is duplex—that is, it has a double set of cylinders, two air cylinders and two steam cylinders. The steam cylinders are on the lower instead of on the top end, as on the Westinghouse.

68. Besides being a duplex pump, what other distinguishing feature is there to the New York pump, as compared with all the Westinghouse pumps excepting their new compound pump?

Ans. All of the New York pumps are not only duplex, but are also compound compressors, for the reason that the air cylinders have different diameters (the low pressure cylinder being equal to twice the size of the high pressure cylinder), and the air from the low pressure cylinder is passed into the high pressure cylinder and again compressed before entering the main reservoir. Another distinguishing feature between the New York and Westinghouse pumps is that the valve gear of the steam end of the New York pump consists merely of a small reversing valve for each steam cylinder, whereas the Westinghouse pump not only has a reversing valve but also has a main steam valve gear, consisting of three pistons in the eight-inch pump and two pistons and a slide valve in the 9½-inch, 11-inch and new compound pump.

69. This being true, what particular fact should an engineer always keep in mind when operating a New York pump?

Ans. He should be careful not to allow the lubricator to feed oil too rapidly into the steam end of the pump, for the reason that as there are only two small slide-valves to lubricate besides the main pistons, if too much oil were allowed to get into the reversing valve chamber it would have a tendency to force the reversing valves from their seats and thereby lower the efficiency of the pump.

70. Why is it said that with the New York pump one measure of steam will generate three measures of air?

Ans. Because of the fact that the low pressure cylinder has a volume capacity twice that of the high pressure cylinder, and as the high pressure cylinder receives a charge of air direct from the atmosphere every time it makes a stroke, and as the piston does not make a second stroke until the low pressure cylinder has discharged its air into the high pressure cylinder, it means that on the return stroke of the high pressure piston a volume of air equal to three times the volume of the high pressure cylinder is forced into the main reservoir.

71. What is the principal difference in the construction of the No. 1, No. 2, No. 6, and No. 5 New York pump?

Ans. The general construction of all of these pumps is the same, except that the No. 6 and No. 5 pumps have two air inlets (one for each air cylinder) and two sets of receiving valves in the air cylinders. The lift of the air valve of the No. 5 pump is 3-16 of an inch, in order to accommodate a large volume of air, and all air valves are interchangeable.

72. Why does one piston wait for the other before making its return stroke?

Ans. Because the reversing valve under one piston controls the action of the opposite one.

73. What is the difference between the reversing valve of the New York and Westinghouse pumps?

Ans. The operation of the valve is the same, as both have a reversing-valve rod with a button on one end and a shoulder on the other, and a reversing-valve plate to move the rod up or down, but the reversing valve in the Westinghouse pump is made to control three ports, whereas the New York reversing valve is just a common D slide valve for the purpose of controlling two ports.

74. How is the air end of the New York pump oiled?

Ans. By an automatic oil cup in the head of the air cylinders by which the flow of the oil can be regulated. Shown in Fig. 7, Chart 35.

75. Can you explain the operation of the steam end of the New York pump?

Ans. When the pistons are at rest, and steam enters the pump from the boiler, it fills both of the slide-valve chambers with steam, and from the left-hand chamber live steam passes through port B to the under side of the right-hand piston and through port C, which leads from the right-hand chamber to the top side of the left-hand piston. When the right-hand piston is forced up it shifts the position of the right-hand reversing valve so that port C is connected to the exhaust port F, and port A is opened from the reversing valve chamber to the under side of the left-hand piston, which causes that piston to be forced up. As the exhaust passage from the right-hand piston is controlled by the left-hand slide-valve, the right-hand piston will remain up until the left-hand piston has made its stroke and pulled the

ITS USE AND ABUSE

reversing valve up, which connects port B from the under-side of the right-hand piston with the exhaust passage F, and at the same time opens port D so that live steam can get on top of the right-hand piston and drives it down. When this piston has made its full down stroke it again changes the right-hand reversing valve so that port C exhausts the steam of the left-hand piston, and port A is again opened and admits steam from the right-hand slide-valve chamber to the under side of the left-hand piston, as shown in Fig. 3 of Chart 35.

76. Can you explain the operation of the air end of the pump?

Ans. With the No. 1 and No. 2 pumps the intermediate valves and receiving valves are located between the air cylinders, consequently when the high pressure cylinder receives a charge of atmospheric pressure the lifting of the receiving valve also lifts the intermediate valve until the pressure has been equalized inside and out of the air cylinders, whereas with the No. 6 and No. 5 pump each cylinder has its own separate set of receiving valves, so that the intermediate valves are not moved excepting when air is passed from the low pressure to the high pressure cylinder. The reason the air valves are lifted and again reseated in response to the action of the piston is explained in Question 16 of Section 2.

77. What is the stroke of the piston in the several New York pumps?

Ans. The No. 1 and No. 2 is nine inches, the No. 6 is ten inches, the No. 5 is twelve inches.

78. What are the diameters of the steam and air cylinders of the several pumps?

Ans. The steam cylinders of the No. 1 pump are five inches, the No. 2 and No. 6 are 7 inches, and in the No. 5 pump they are 8 inches. The high pressure air cylinders in all pumps are the same in diameter as the steam cylinders. The low pressure air cylinder of the No. 1 pump is seven inches; the No. 2 is 10 inches, the No. 6 is eleven inches, and the No. 5 is twelve inches. The No. 6 pump is made to take the place of the No. 2.

79. Why are the intermediate valves called by that name?

Ans. Because these valves are intermediate between the low and high pressure air cylinders.

80. What difference is there in the way in which the pump governor is connected up with the New York equipment as compared with the Westinghouse?

ITS USE AND ABUSE

Ans. As the New York Brake Valve has an excess pressure valve instead of a trainpipe governor, it follows that the pump governor must be connected to the trainpipe pressure, and as the action of the brake valve is such that communication from the trainpipe and pump governor is shut off when the handle is in any position except running and full release, it makes it necessary to use the Duplex governor with the New York Brake Valve, in order to prevent main reservoir pressure from getting too high.

81. What is the difference between an ordinary governor and a duplex governor?

Ans. The duplex governor is one in which there are two air portions, so that the regulating spring on one portion can be set at one pressure and the other at a different pressure.

82. With an ordinary automatic brake at what pressure is the duplex governor set?

Ans. The air portion of the governor which connects to the trainpipe is usually set at seventy pounds and the portion which is connected to the main reservoir is set at ninety pounds, when these pressures are used as standards.

83. What is a triplex governor?

Ans. A governor with one steam portion and three air portions.

MODERN AIR-BRAKE PRACTICE

84. For what purpose is the triplex governor used?

Ans. For the high pressure control or high speed brake. For with a triplex governor one trainpipe portion can be set at seventy pounds, the other trainpipe portion at ninety pounds and the main reservoir pressure at 110 pounds, or they can be set at whatever pressure is desired.

85. What object is there in having two degrees of trainpipe pressure?

Ans. By closing the cut-out cock on the governor pipe leading to the low pressure air portion it will enable the engineer to change the automatic brake into either the high pressure control or high speed system.

86. How many positions are there on the New York Engineer's Brake Valve?

Ans. Five, the same as on the Westinghouse, except that service graduating position is divided into five notches which represent trainpipe reductions of about 5, 8, 11, 16 and 23 pounds.

87. How is the excess pressure gotten with the New York Brake Valve?

Ans. By placing the handle in running position the excess pressure is accumulated *before* any pressure enters the trainpipe, for when the handle is in this position main reservoir pressure must

ITS USE AND ABUSE

be greater than the tension of the regulating spring of the excess pressure valve before it can lift that valve from its seat.

88. When the handle is in running position how does the air get from the main reservoir into the trainpipe?

Ans. It enters the brake valve through passage B into chamber B and passes through port E of the seat of the slide-valve through cavity M in the slide-valve and through port A in the slide-valve seat into chamber A, which is the same as trainpipe, as shown in Fig. 3 of Chart 37.

89. When the handle is in full release position how does the main reservoir pressure get into the trainpipe?

Ans. It passes direct from chamber B by the end of the main slide-valve through port A into the trainpipe.

90. In making a service application how does the air exhaust from the trainpipe?

Ans. By way of ports F and G in the main slide-valve and out through exhaust passage C, as shown in Fig. 5, Chart 37.

91. In making an emergency application how does the air exhaust from the trainpipe?

Ans. By way of the large ports J and K and exhaust passage C, as shown in Fig. 7, Chart 37.

MODERN AIR-BRAKE PRACTICE

92. What causes the New York Brake Valve to automatically lap itself?

Ans. When the handle is placed in any one of the service graduating notches it causes the main slide-valve to be moved so that port F is uncovered and the top end of port O in the slide-valve seat is closed, consequently when trainpipe pressure is reduced slightly less than the pressure in the supplementary reservoir, that pressure forces the equalizing piston back, which causes the cut-off valve to gradually close port F and stop the trainpipe pressure from further exhausting.

93. How does air get into the supplementary reservoir?

Ans. In either running or release position the air that feeds into the trainpipe lifts the ball check-valve in the equalizing piston and allows trainpipe pressure to flow into Chamber D on through passage H into the supplementary reservoir.

94. For what purpose is the vent-valve in the end of the equalizing piston?

Ans. When the handle of the brake valve is in either release or running position the top end of port O connects with cavity P and exhaust cavity C, so that if the bottom end of passage O were not closed it would not only drain the sup-

ITS USE AND ABUSE

plementary reservoir but would make a constant leak from the trainpipe.

95. In making a service application with the New York Brake Valve what should the engineer do if the automatic lap feature should fail to operate?

Ans. After he has exhausted the required amount of trainpipe pressure the handle should be placed on positive lap position.

96. With the New York Brake Valve, would the length of the train have anything to do with the way in which a service application should be made?

Ans. Yes, for with a train of four cars or less if the handle was not placed upon the first notch to start with, it is likely to cause an emergency application for the reason that the small trainpipe volume would rush out so rapidly as to cause a sudden reduction of trainpipe pressure, but with a train of five cars or more there is sufficient volume of trainpipe to overcome this difficulty.

97. What defects would destroy the automatic lap feature of the New York Valve?

Ans. There are several. Should the small cut-off valve become scratched, or should the connecting arm become bent so as to prevent the valve from seating, or should the packing of the piston leak so as to permit the pressures to equalize, or should there be any leak from the supplementary reservoir, any one of these would prevent the automatic lap.

98. Is there any material difference in the method of regulating the New York pump governor from that of the Westinghouse?

Ans. No, as both governors will shut off at a lower pressure by loosening the regulating nut, and will carry a higher pressure by screwing down on the regulating nut.

99. What is the essential difference in the manner of producing quick action with the New York triple as compared with the Westinghouse?

Ans. In either case a sudden reduction is necessary, but the difference is that with the Westinghouse triple a portion of the trainpipe pressure enters the brake cylinder, whereas with the New York quick action triple the trainpipe pressure is exhausted to the atmosphere when an emergency application is made.

100. What is the object of the piston which works in conjunction with the main slide valve piston of the New York quick action triple?

Ans. It takes the place of the graduating spring of the Westinghouse triple, for the reason

ITS USE AND ABUSE

that trainpipe pressure fills the space between these two pistons, and when a reduction is made on the trainpipe pressure the pressure between the two pistons is partially confined so that it acts as a cushion for the main slide valve piston; but should the trainpipe pressure be reduced suddenly these two pistons would be kept apart on account of the air not being able to get from between them quick enough, and consequently the stem of the smaller piston would strike against the emergency vent valve, marked 71, which would permit trainpipe pressure to enter passage H and cause piston 137 to unseat check valves 139 and 117, and thereby produce an emergency application of the brake.

101. How does the New York Quick Action Triple Valve operate?

Ans. When trainpipe pressure enters the triple at the strainer it fills the cavity back of the rubber seated vent valve and at the same time it passes through the feed groove into the auxiliary reservoir. It is fed through a port in the stem of the vent piston which allows air to charge up the chamber between the vent piston and the main triple piston, so that when trainpipe pressure is reduced, which causes the auxiliary pressure to force the main piston back, the air that is between the vent piston and the main piston does not entirely escape, and consequently when the trainpipe and auxiliary pressures equalize, the portion of air confined between the piston and the main piston expands and moves the main piston ahead, which closes the supply port between the auxiliary and the brake cylinder thereby lapping the triple valve.

102. What is the action of the triple in an emergency application?

Ans. When a sudden reduction is made on the trainpipe pressure the main piston is moved so quickly back that the port in the vent piston is kept closed, and as a consequence the air between the two pistons keeps them apart, which results in the stem of the vent piston striking against the vent valve, thereby unseating it and allowing trainpipe pressure to get into passage H, and drive the emergency piston against the rubber seated quick action valve, which action permits the auxiliary pressure to rush into chamber L and unseat the non-return brake cylinder check valve, thereby allowing the auxiliary and brake cylinder to quickly equalize. When the air in passage H forces the emergency piston forward it opens a small exhaust port which allows a further reduction in the trainpipe pres-

ITS USE AND ABUSE

sure, and as the reduction is at once felt by the next triple, it causes all other triples in the train to act quickly.

103. How is the brake released?

Ans. The same as with the Westinghouse triple, that is when the trainpipe pressure is raised higher than that in the auxiliary, the main triple piston is forced back so that the exhaust cavity in the slide-valve connects the brake cylinder with the atmosphere.

104. As there are two packing rings in the New York Triple Valve, does this fact have a tendency to cause the triple to fail to release properly?

Ans. No. But should the port in the vent piston become clogged after the auxiliary is charged it is likely to produce an emergency application when making a service application, for the reason that, if the air which is between the two pistons cannot be reduced about as fast as the trainpipe pressure, the stem of the vent piston will unseat the vent valve and cause an emergency application.

105. What care should be given to the New York triple?

Ans. Just the same as with the Westinghouse, or any other triple, for you cannot expect to keep machinery in proper working order if it is not looked after.

106. What is the principal difference in the construction of the New York Straight Air Brake Valve as compared with the Westinghouse?

Ans. The New York Straight Air Brake Valve contains a slide valve, and has four positions, whereas the Westinghouse Straight Air Brake Valve contains two lift valves and has only three positions. The general pipe arrangement is the same and both systems require double check valves, safety valves, two extra exhaust valves, gauges, etc.

107. How does the Straight Air Brake Valve operate?

Ans. There is a pipe from the main reservoir, on which there is a reducing valve set at 45 pounds, and the other end of this pipe is connected to the straight air brake valve, so that when the handle of the brake valve is moved so that the slide-valve uncovers the port leading into the brake cylinder, main reservoir pressure can then pass through the reducing valve, through the brake valve and into the brake cylinder until the cylinder pressure becomes slightly greater than the tension of the graduating spring in the reducing valve, when the flow of air is shut off

ITS USE AND ABUSE

from the main reservoir automatically. Should the handle of the brake valve be moved to service position for just a short time and then brought back to lap there would only be a partial application of the brakes, for the reason that the movement of the slide-valve would prevent air from getting into the cylinder regardless of the action of the reducing valve.

108. When an engine is equipped with the New York or Westinghouse Straight Air Brake Valve, and if the two additional exhaust valves should be omitted, could the brakes on the engine be released with the straight air brake valve when a hose bursts? Why?

Ans. No. For with the ordinary straight air brake valve equipment the double-check valve would prevent the passage of brake cylinder pressure in one direction and the slide valve in the triple would prevent it from exhausting in the opposite direction, and it is therefore on account of this fact that the two additional exhaust valves are necessary with the old style straight air equipment.

109. For what purpose is the double-check-valve?

Ans. It is to automatically close communication between the brake cylinder and the straight air brake valve when auxiliary pressure is flowing into the brake cylinder, and also to shut off communication between the brake cylinder and the triple when the straight air brake valve is being used.

110. Should an engine be equipped with the two additional exhaust valves besides the straight air valves, could they be used to release the engine brakes quick enough to prevent the shock of cars against the engine when the hose bursts? Why?

Ans. No. For the reason that these additional exhaust valves are connected merely to the brake cylinders, and should the brake cylinders on the engine and tender be of the larger style used in modern practice, the auxiliary reservoir pressure would continue to flow into the brake cylinders even though the brake cylinder exhaust valves were kept open, and thereby prevent the brakes from releasing quick enough to avoid the shock.

111. What additional equipment would therefore be needed in order to quickly release the brake cylinder pressure on engine and tender when a hose bursts in order to avoid the shock and minimize the possibility of buckling the train?

ITS USE AND ABUSE

Ans. It would be necessary to have additional exhaust valves connected to the auxiliary reservoirs on the engine and tender, that is in addition to the straight air brake valve there should be four additional valves in order to insure a quick release of engine and tender brakes.

112. With these additional valves could the straight air brake valve be used to retain the automatic application of the locomotive brakes without making a straight air application, that is without using the main reservoir pressure, while the train brakes are being released?

Ans. No. For the reason that in order to retain the automatic application of the engine brakes with the old style straight air equipment it is necessary to control the triple exhaust ports on both the engine and the tender, which would require retaining valves in addition to the valves already mentioned.

113. With the New York high speed brake is it necessary to have quick action triples on the tender?

Ans. No. But they may be used if desired.

114. Why is it that with the high speed brake a trainpipe pressure of 110 pounds can be used and still avoid sliding the wheels?

Ans. For the reason that with the high speed

brake all wheels on the engine, tender and cars are braked, and therefore the braking power is applied more uniformly than it used to be in former days. When the power is applied to all wheels alike the danger of sliding wheels is reduced to the lowest possible point, and on account of this fact some systems of air brakes are now using only one triple valve on the locomotive and tender in order to secure this result.

115. For what purpose is the compensating valve used in the high speed brake?

Ans. It is a safety valve for the purpose of automatically exhausting the brake cylinder pressure when it gets higher than it should.

116. At what pressure should the compensating valve exhaust?

Ans. When the pressure in the brake cylinder becomes greater than 60 pounds the compensating valve should operate to let off all pressure above that amount before the speed of the train is materially reduced, in other words, it would maintain the cylinder pressure at about 75 pounds for a few seconds but would close the exhaust when the cylinder pressure had dropped to 60 pounds.

117. Is there any material difference in the number of parts comprising the New York

ITS USE AND ABUSE

whistle signal equipment compared with the Westinghouse?

Ans. No. Both systems require a whistle, a signal valve, a reducing valve, whistle signal pipe and car discharge valve, and in order to blow the whistle with either system it is simply necessary to exhaust the pressure from the signal pipe, which action causes the signal valve to operate and allow the air to pass to the whistle.

118. Is there any material difference between the New York retaining valves and those of the Westinghouse?

Ans. None, so far as the operation and handling is concerned.

SECTION 5

CHAPTER V

THE DUKESMITH AIR BRAKE CONTROL SYSTEM—ITS PARTS AND THEIR DUTIES

As the tendency of modern railway practice is to heavy motive power and long trains, these conditions demand additional safe guards which have not heretofore been supplied by other Air Brake Companies, and in order to provide means to overcome the existing difficulties the Dukesmith Air Brake Company of Pittsburg, Penna., are now manufacturing what is known as the Dukesmith Air Brake Control System.

The Dukesmith Air Brake Control System consists of the following equipment:

An Engineers Automatic Brake Valve, known as style A, which performs the functions of the ordinary brake valve, or in other words by its use the brakes on the entire train can be applied and kept applied; released and kept released and the train-pipe pressure maintained at a lower point than that of the main reservoir

Style B of the Dukesmith Engineers Auto-

ITS USE AND ABUSE

matic Brake Valve performs all of the functions of style A, but in addition thereto it applies the engine and tender brakes with straight air when an emergency application is made.

Style C of the Dukesmith Engineers Auto-

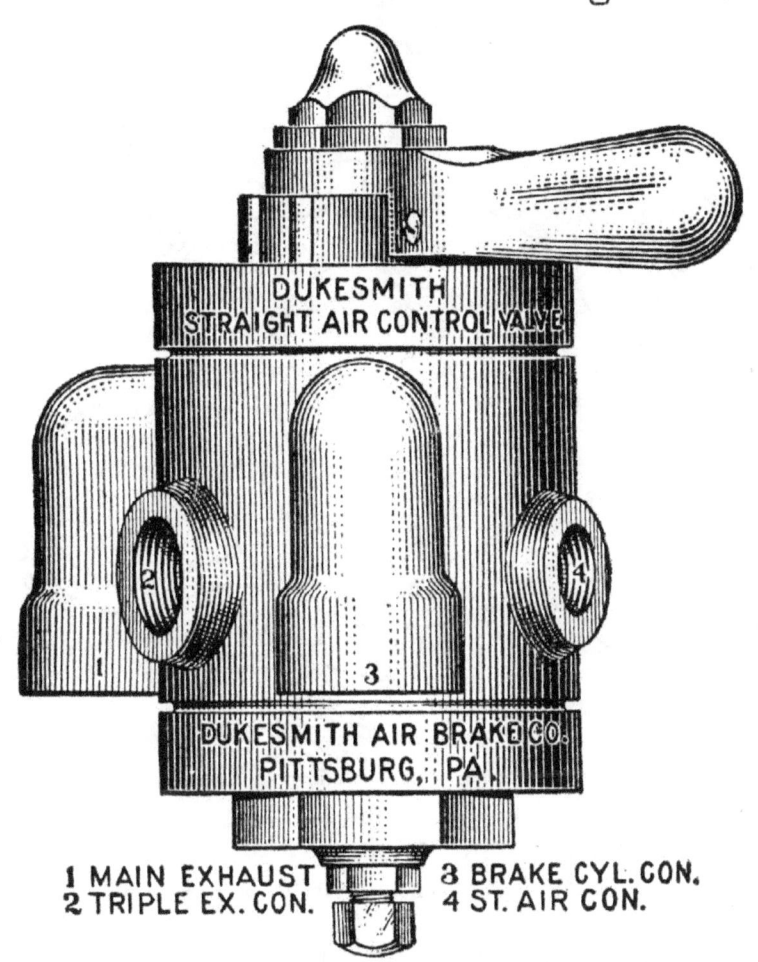

PLATE 72—THE DUKESMITH STRAIGHT AIR CONTROL VALVE, WITH AUXILIARY RELEASE.

matic Brake Valve performs all the functions of style A, but it also enables the engineer to apply and release the engine and tender brakes independently of the train brakes, and also enables him to hold the engine brake applied while the train brakes are being released, and also to release the engine and tender brakes when a hose bursts or when an emergency application has been made, which minimizes the possibility of buckling the train.

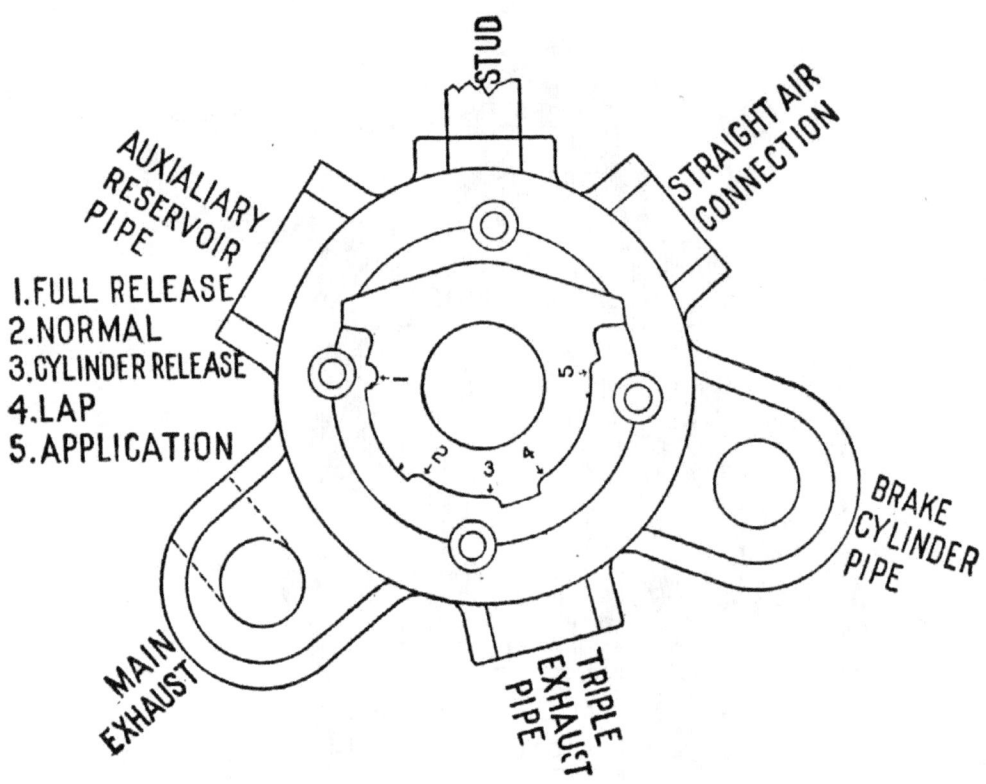

PLATE 73—DIAGRAM SHOWING POSITIONS ON QUADRANT AND PIPE CONNECTIONS OF DUKESMITH STRAIGHT AIR CONTROL VALVE.

The Dukesmith Driver Brake Control Valve style A, is used in connection with any automatic brake-valve and is for the purpose of combining in one valve an independent release and retaining valve for the locomotive.

The Dukesmith Straight Air Driver Brake Control Valve, style B, combines all the features

ITS USE AND ABUSE

of Control Valve A and in addition thereto it enables the engineer to apply the locomotive brakes with straight air independent of the automatic system, or to release the engine and tender brakes independent of the triple valve, and with

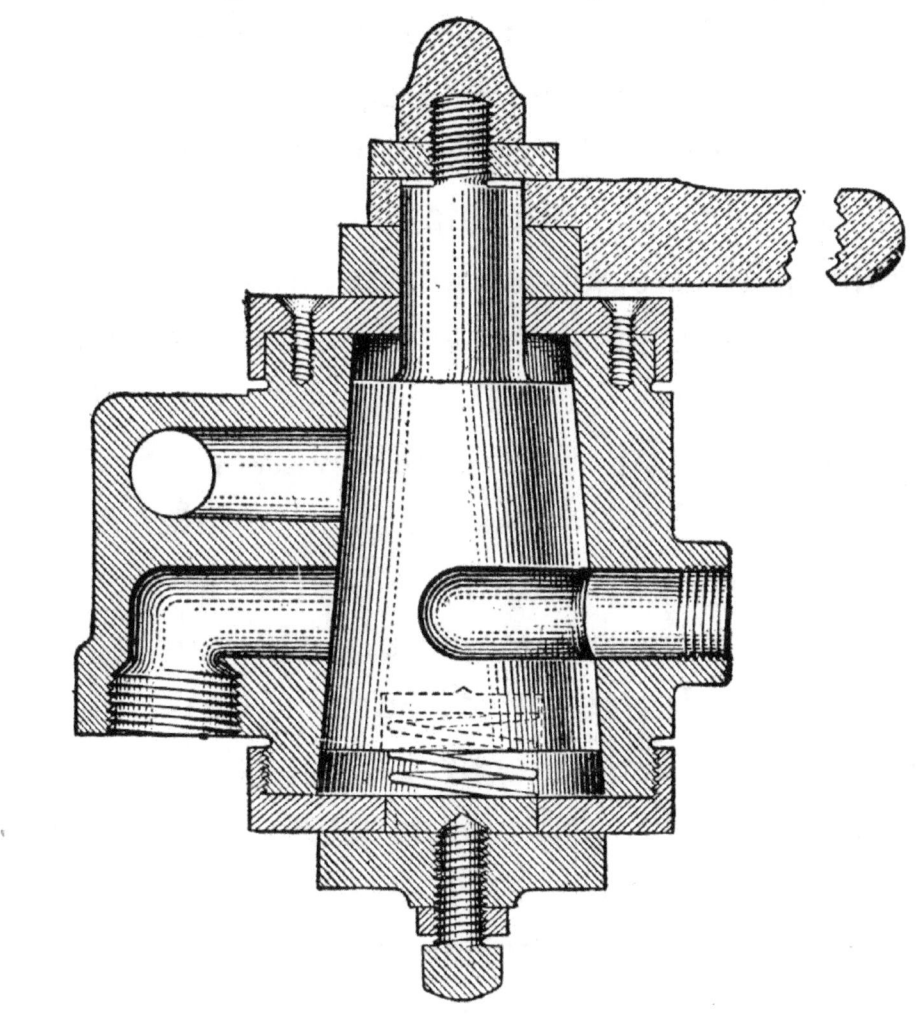

PLATE 74—VERTICAL SECTIONAL VIEW OF DUKESMITH STRAIGHT AIR CONTROL VALVE WITH HANDLE IN APPLICATION POSITION.

this valve the entire brakes on the locomotive can be let off in from 10 to 12 seconds when a hose bursts, by reason of the fact that this valve is not only connected direct to the brake cylinders on the engine and tender but is also connected to the auxiliary reservoirs on the locomotive, which permits of the quickest possible release.

The Automatic Release Signal is made in two styles; style A is for car service and style B is for engine service. Style A Release Signal carries a large metal signal in order that it may be seen at a distance, it automatically exhausts the cylinder pressure above a predetermined amount and in addition has an independent exhaust valve for the purpose of releasing a stuck brake independently of the triple valve. Style B Release Signal is contained in a circular casing, and automatically exhausts the brake cylinder pressure above a predetermined amount, and has a graduating device for the purpose of regulating the point at which the brake cylinder pressure should be automatically exhausted. Either style of release-signal indicates at all times the exact operation of the brake, as it tells whether the brake power is too great or too little; whether a brake is leaking off or releasing off; what the brake piston travel is or whether a brake is stuck or not.

The Dukesmith Car Control Valve performs four functions, as follows: First it can be used to

ITS USE AND ABUSE

apply the brakes on the entire train either in a

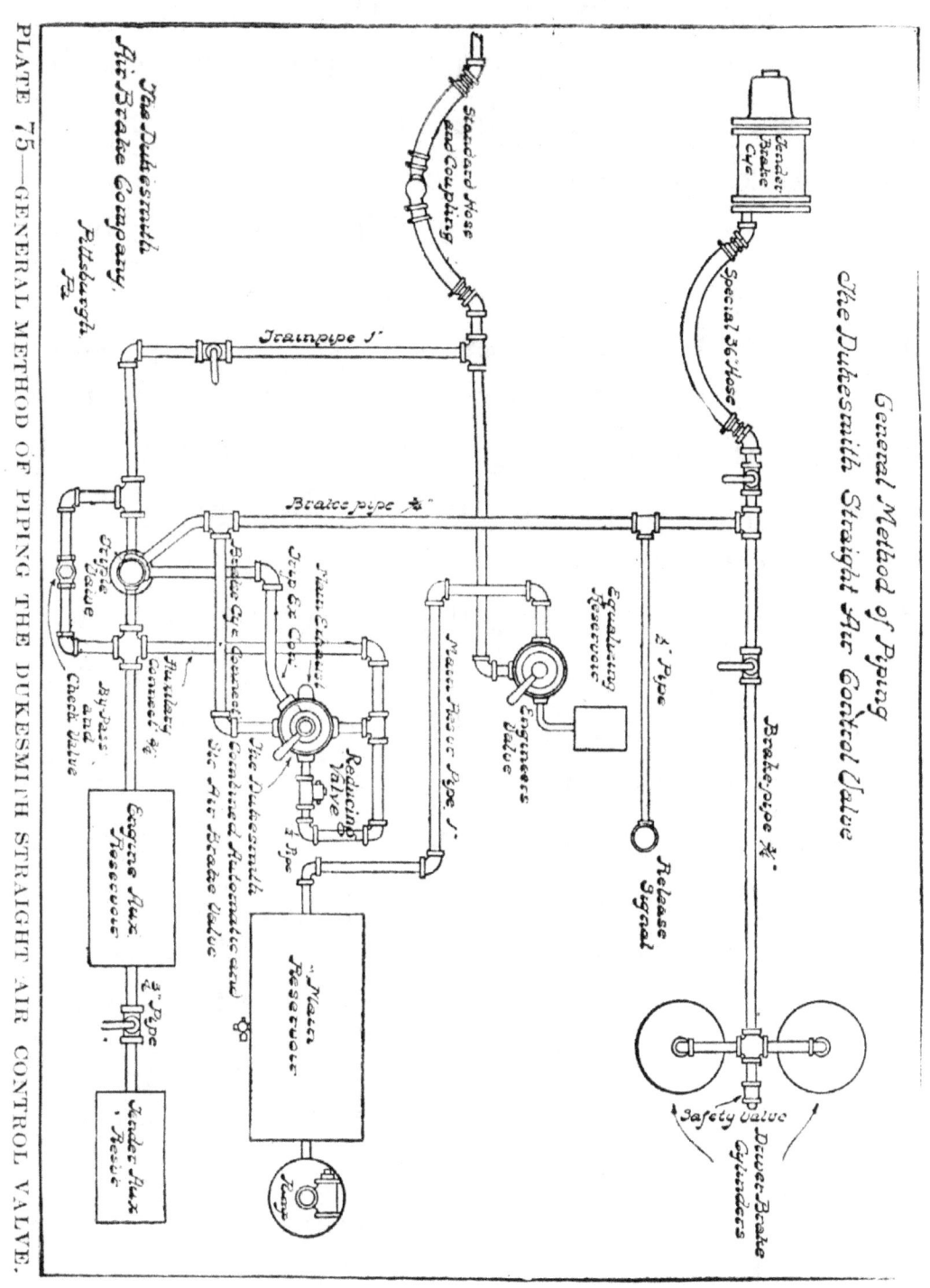

PLATE 75—GENERAL METHOD OF PIPING THE DUKESMITH STRAIGHT AIR CONTROL VALVE.

service or emergency application; Second it can be used to release the brake on the car to which

MODERN AIR-BRAKE PRACTICE

it is attached independently of the triple valve, and without having to stop the train to do so; Third it can be used to retain the brake on the car to which it is attached, and fourth it can be used to keep the brake cut out without having to stop the train to do so. There are four positions in which the handle of the Car Control Valve may be placed, as follows: Normal, Lap, Release and Application. It occupies the same position in passenger coaches as the old style conductor's valve, and it may be operated to apply the brakes in emergency either by turning the handle itself or by pulling a rope attached to the handle and extending through the car. There is very little possibility of the car control valve getting out of order, as the working parts consist merely of a tapered key working in a casing and a handle attached to the key.

The Dukesmith Emergency Cut-Out Cock is a device used for double-heading, and takes the place of the old style cut-out cock located in the trainpipe under the brake valve on the engine. The Emergency Cut-Out Cock when closed prevents the engineer on the second engine from accidently charging the trainpipe, which would release the brakes, and also prevents him from making a service application, but it does enable

ITS USE AND ABUSE

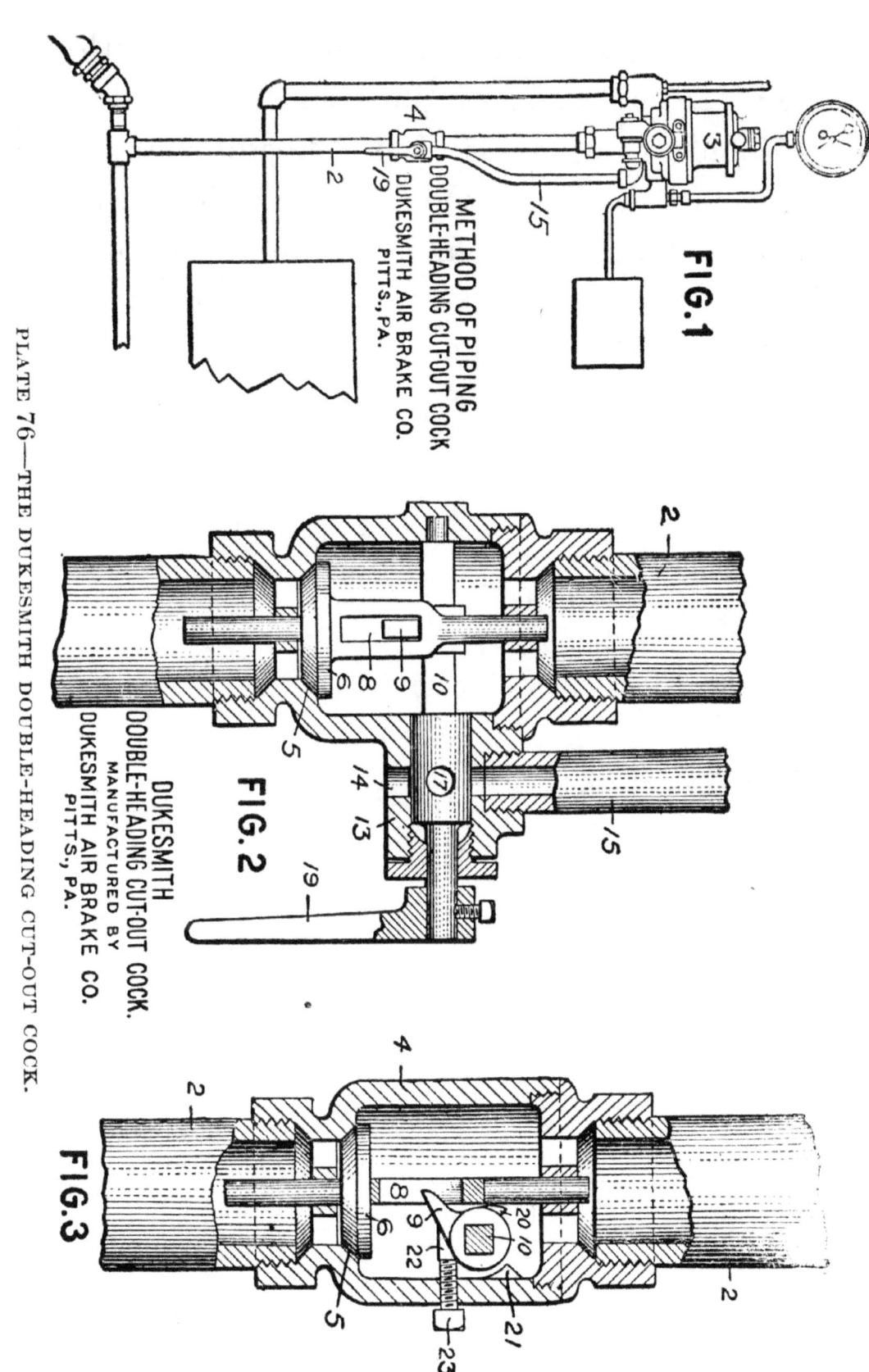

PLATE 76—THE DUKESMITH DOUBLE-HEADING CUT-OUT COCK.

him to make an emergency application without having to cut in the cut-out cock.

When an engine is single heading and the emergency cock is open it is no different from any other cut-out cock, as in this position the check valve is locked in an open position and cannot seat of its own accord.

SECTION 6

CHAPTER VI

OPERATION, HANDLING AND MAINTENANCE OF THE DUKESMITH AIR BRAKE CONTROL SYSTEM

QUESTIONS AND ANSWERS TO SECTION 5

119. What is the construction of style B Dukesmith Engineer's Automatic Brake Valve?

Ans. There is a case in which is contained a tapered key or plug, and through the top of the key there is an angular port, near the bottom of the key there is an annular groove and near the top of the key there is another small groove.

120. For what purpose is the angular port near the top of the key?

Ans. For controlling the passage of the air direct from the main reservoir to the trainpipe.

121. For what purpose is the annular groove near the bottom of the key?

Ans. It is for three purposes, as in running position it permits the passage of main reservoir pressure into the trainpipe governor, in service position it permits trainpipe pressure to exhaust

MODERN AIR-BRAKE PRACTICE

to the atmosphere through the trainpipe exhaust port, and in emergency position it permits trainpipe pressure to exhaust to the atmosphere through both the service and emergency exhaust ports.

122. For what purpose is the small groove near the top of the key?

Ans. For the purpose of admitting main reservoir pressure direct into the brake cylinders on the locomotive when the handle of the valve is placed in emergency position. This feature is to provide against leaky cylinder leathers.

123. What is the construction of style A Dukesmith Engineer's Automatic Brake Valve?

Ans. The same as style B, excepting that it does not contain the small groove near the top of the key, as this valve performs the same functions of any of the old style brake valves.

124. What is the construction of style C Dukesmith Engineer's Automatic Brake Valve?

Ans. The same as style A excepting that near the top of the key there is a small groove for the purpose of connecting the main trainpipe with the engine trainpipe when the valve is in running position, and a small vertical groove near the top of the key for the purpose of con-

ITS USE AND ABUSE

necting the engine trainpipe with the atmosphere when the handle of the valve is in emergency position.

125. With style C brake valve are there any additional parts required aside from the ordinary trainpipe governor or feed valve?

Ans. Yes, with style C brake valve there is a device known as the automatic exhaust valve, which is located in the supplementary trainpipe that connects the brake valve with the triple valve on the engine, and this exhaust valve is for the purpose of enabling the engineer to apply the engine brakes without applying the train brakes. As there is a continuous exhaust from the exhaust valve it follows that when the handle of the brake valve is placed in either lap or holding position the supply of air is cut off from the supplementary trainpipe, and, as a consequence, the pressure in the supplementary trainpipe forces the piston in the exhaust valve up, and thereby opens a port in the exhaust valve which allows supplementary trainpipe pressure to equalize with the exhaust chamber of the exhaust valve, so that a reduction is thus made which causes the engine triple to move and apply the brakes on the engine and tender. When the handle of the brake valve is again brought to running position it establishes communication between the main trainpipe and the supplementary trainpipe, which causes the piston in the exhaust valve to be moved down, and thereby exhausts the air from the exhaust chamber and at the same time allows the pressure from the main trainpipe to recharge the supplementary trainpipe and release the brakes on the engine.

126. In how many positions can the handle of style C Brake Valve be placed, and for what purpose?

Ans. Full release, holding, running, lap, service, emergency and emergency-release. In full release position the main reservoir pressure passes directly into the main trainpipe, but is cut off from entering the supplementary trainpipe, so that in this position the engine brakes are kept set while releasing the train brakes; in holding position main reservoir pressure passes through the bottom groove in the key to the trainpipe governor but does not enter the supplementary trainpipe, so that in this position the main trainpipe can be kept charged while the engine brakes are still applied; in running position main reservoir pressure passes through the bottom groove

ITS USE AND ABUSE

through the trainpipe governor into the main trainpipe and from thence through the small groove in the top of the key into the supplementary trainpipe so that the pressure in both trainpipes is maintained at the same degree, and this is the only position in which the handle can be placed to release the brakes on the locomotive, excepting emergency-release position. Lap position closes all ports in the brake valve, so that it naturally follows that in this position the supplementary trainpipe will automatically exhaust its pressure down to whatever the exhaust valve is set at, which is usually from five to seven pounds, consequently if a service application of five or seven pounds is made without first placing the handle of the brake valve in lap position the result would be just the same as if any other brake valve were used; in service position the main trainpipe is in communication with the supplementary trainpipe by way of the small groove in the top of the key, and is in communication to the atmosphere through the service exhaust port by way of the bottom groove in the key; in emergency position the main trainpipe is in communication with the atmosphere through both the service and emergency exhaust ports by way of the bottom groove in the key, and the supplementary trainpipe is in communication with the atmosphere by way of the small vertical groove in the top of the key; emergency-release position is the last position on the valve, and in order to get the handle to that position it is necessary to press against the locking device in the handle in order to allow the bolt to pass over the raised part of the quadrant. This position is for the purpose of quickly releasing the engine brakes in case of a hose bursting (or after an emergency application has been made), for the reason that by releasing the engine brakes it reduces to a minimum the possibility of buckling the train when the surge of the cars rush against the engine. The reason that this position releases the brakes on the engine is due to the fact that in this position the bottom groove in the key places the auxiliary reservoir on the engine in direct communication to the atmosphere through the emergency exhaust port which allows the auxiliary to be quickly emptied, so that the pressure in the brake cylinders can lift the slide valve of the engine triple and pass out to the atmosphere through the emergency exhaust.

ITS USE AND ABUSE

127. What is there to get out of order with any of the Dukesmith Engineer's Brake Valves?

Ans. Nothing excepting the natural wear and tear of a tapered key working in a tapered casing, and if the key is kept properly lubricated there is nothing more to do except to see that the key is properly seated by keeping the tension spring just tight enough to hold it to its seat without causing it to bind.

128. What particular construction of this valve is it that prevents it from binding the same as ordinary plug valves would bind?

Ans. In the bottom of the valve casing there are two anti-friction metal disks, one at the top and one at the bottom of the tension spring, and these disks have a small tapered point so that the weight of the valve is carried by these points, which reduces the friction to almost nothing; another reason why the key of this valve cannot bind is because of the fact that the handle rests on a shoulder on the stem of the key, which prevents it from binding against the top of the case.

129. Does the valve operate any different so far as friction is concerned when the pressure is pumped up or when it is not?

Ans. Yes, when there is no pressure in the

valve it works harder than when the pressure is pumped up, for the reason that when the pressure fills the ports of the valve it balances it perfectly and overcomes the tension of the spring, and it is for this reason that the tension spring should be regulated while the pressure is on. The spring should be regulated so that the valve will just seat, but should not be made any tighter.

130. What is the construction of Driver Brake Control Valve style A?

Ans. It is a disk or rotary valve having three ports, one of which is connected directly to the brake cylinders on the engine, one to the triple exhaust port of the engine triple and one to the atmosphere.

131. How many positions are there on this valve?

Ans. Three; full release, normal (or running), and lap (or retaining).

132. What is the construction of the Dukesmith Straight Air Control Valve style B?

Ans. The same as the Dukesmith Engineer's Automatic Brake Valve style A.

133. What then is the difference between the two valves?

ITS USE AND ABUSE

Ans. The manner in which the pipe connections are made.

134. How many positions are there on the Straight Air Control Valve?

Ans. Five; full release, normal (or running), cylinder release, lap and application position.

135. What pipe connections has this valve different from control valve style A?

Ans. It has an auxiliary connection for straight air and also an exhaust connection from the engine auxiliary reservoir, besides the connection to the exhaust of the engine triple and the connection to the brake cylinders on the locomotive.

136. What is the construction of the Dukesmith Car Control Valve?

Ans. About the same as the Straight Air Driver Brake Control Valve, excepting that it is smaller and, in addition to the annular groove in the bottom of the key, there is also an angular port. The working parts consist of a tapered key working in a tapered case, operated by a handle, and there are four positions in which the handle may be placed, normal (or running), lap, cylinder-release and application position.

137. For what purpose is the Car Control Valve?

MODERN AIR-BRAKE PRACTICE

Ans. It takes the place of the old style conductor's valve used in passenger coaches, and is also a retaining valve and an independent release valve, and can be used to cut out a brake if it should become defective.

138. When the handle is in normal position what ports are open?

Ans. The port leading from the triple exhaust port to the atmosphere.

139. When the handle is in lap position what ports are open?

Ans. None, as in lap position all ports are closed.

140. In cylinder-release position what ports are open?

A. The port leading direct from the brake cylinder to the atmoshere.

141. In application position what ports are open?

Ans. The port leading direct from the train-pipe to the atmosphere.

142. Should this valve leak, how would you overcome the leak?

Ans. If it is properly lubricated I would simply tighten up the tension of the spring, but would not make it any tighter than just enough to seat the valve.

ITS USE AND ABUSE

143. What amount of brake cylinder pressure is retained when the handle of the car control valve is placed in lap position?

Ans. That depends entirely on the weight of the lift valve in the retaining valve located on the pipe leading from the triple exhaust to the Car Control Valve, which may be either 15, 25, or 50 pounds according to the requirements of the service.

144. What is the construction of the Automatic Release Signal?

Ans. The Release Signal consists of a cylinder which moves up and down over a piston connected to a hollow piston rod, between the bottom of the cylinder and the under side of the piston there is a graduated spring, the tension of which can be regulated by a sleeve which enters the bottom of the cylinder, and the top of this sleeve has a flange which contacts with the stem of a small vent valve in the piston, and there are a number of small vent ports in the bottom of the cylinder so that when the brake cylinder pressure passes up through the hollow piston rod to the top of the piston it strikes on the under side of the top of the cylinder and forces the cylinder up in proportion to the pressure. Should the pressure be great enough it would lift the release signal cylinder until the flange on the sleeve strikes against the stem of the vent valve and would keep it unseated until the pressure had dropped slightly below the tension of the spring when the cylinder would be moved down by the spring and permit the vent valve to seat. The remaining pressure would continue to hold the cylinder up until the pressure was exhausted either in the regular way or by the independent exhaust valve connected to the release signal when the cylinder would be forced to its normal position by the spring.

Style C Engine Release Signal has the safety valve screwed in the top cap of the signal cylinder, and is regulated the same as any standard safety valve.

145. What is the construction and operation of the Dukesmith Emergency Cut-Out Cock?

Ans. It consists of a casing having pipe connections at top and bottom, and having a rod extending horizontally through its center, one end of which is connected to a handle, and this rod also controls the exhaust of the pipe which is connected to the service exhaust port of the brake valve, and this same rod, which runs horizontally through the cut-out cock, controls a check-valve in the cut-out cock, for when the

ITS USE AND ABUSE

handle is in normal position the check-valve is locked open by a projection on the rod. When the handle of the cut-out cock is closed the service exhaust port of the brake valve is closed, but the check-valve is free, so that when an emergency application is made with the brake valve the trainpipe pressure is free to exhaust, but should the brake valve handle be placed in either running or full release position the check-valve prevents main reservoir pressure from getting into the trainpipe.

146. As the method of piping the Dukesmith Straight Air Control Valve only requires one triple valve on the locomotive, what means are provided for quickly recharging the engine and tender auxiliary reservoir?

Ans. There is a by-pass from the trainpipe to the auxiliary reservoir pipe in which there is a check-valve, and between the check-valve and the trainpipe there is a reducing stud or diaphragm, which makes the aperture equal to an ordinary feed groove in a triple.

147. Should the brake rigging on the tender become defective what should be done?

Ans. The cut-out cock on the brake cylinder pipe between engine and tender should be closed, and also the cut-out cock on the auxiliary pipe leading to the tender auxiliary.

148. In operating the Dukesmith Straight Air Control Valve what point should an engineer keep particularly in view in regard to bunching the train?

Ans. He should be careful to make only a light application with the straight air control valve before making an automatic application, and he should also be careful to allow the cars plenty of time to bunch against the engine before making the automatic application.

149. Should the engineer desire to lower the brake cylinder pressure on the engine and tender without releasing it entirely, or without losing any auxiliary reservoir pressure what should he do?

Ans. Place the handle in cylinder-release position long enough to let off whatever pressure he desires, and then return the handle to lap position.

150. How can an engineer keep the slack of a train by using the Dukesmith Straight Air Control Valve?

Ans. By leaving the handle of the control valve on lap position while releasing the train brakes, and the handle should not be brought to

ITS USE AND ABUSE

normal position until the train has come to a full stop.

151. Does it require any particular training in order to successfully handle the Straight Air Control Valve?

Ans. It only requires such knowledge as every engineer ought to possess, backed by good judgment, the same as is required in handling any brake valve.

SECTION 7

CHAPTER VII

THE PHILOSOPHY OF AIR-BRAKE HANDLING—RULES AND TABLES FOR COMPUTING BRAKE POWER—BRAKE LEVERAGE—EQUALIZATION OF PRESSURE, ETC.—SIZES OF CYLINDERS AND RESERVOIRS—TESTING AND INSPECTION OF AIR BRAKES—ETC.

After an engineer has learned the name and duty of every part of the air-brake equipment, his knowledge is of but little use either to himself or his employers unless he also learns the philosophy of air brake handling.

What is meant by the philosophy of air-brake handling is a clear and definite understanding of the effects produced by different volumes and pressures when the varying conditions of the brake equipment, track, load, grade and speed are taken into account.

One of the first things an engineer should learn is the value of maintaining correct standards of pressures in the different parts of the equipment.

For example, the cars in a freight train are braked to only seventy per cent of their light weight, which means that with Westinghouse triples the leverage is arranged with the understanding that 8-inch cylinders must contain a pressure of sixty pounds to the square inch in order to produce a brake power of seventy per cent. This means that an emergency application is required to be made, and the piston-travel not over eight inches, if the full seventy per cent is to be gotten.

Therefore, if a train of fifty cars, with everything in first class condition, was running forty miles an hour and the brakes were thrown on with an emergency application, it would have to run about 675 feet, or an eighth of a mile, before coming to a stop. This is because the stopping power is only equal to seventy per cent of the weight to be stopped. If these same fifty cars were all loaded to their capacity of 60,000 pounds each, the per cent of brake power to the weight to be stopped would be entirely changed, for with fifty cars of 30,000 pounds weight each, the total weight to be stopped would be 1,500,000 pounds, and if the brakes were properly adjusted there would be an available stopping power of 1,050,000 pounds, but when the cars are loaded the weight to be stopped is 4,500,000 pounds, and with an emergency application you have only got a stopping power equal to twenty-three and a third per cent of the weight, and as it requires a greater force to check the momentum of a heavy weight than it does for a light one, the loaded cars will run a considerable distance further than the empty cars would before stopping.

The reason the percentage of stopping distance is not greater in proportion to the decreased brake power is because when once the momentum is checked the force of gravity causes the heavy weight to settle quicker than a light weight. It is on this account that a train running twenty miles an hour can be stopped in a much shorter distance than one running forty miles an hour. At twenty miles an hour a fifty-car train can be stopped in less than 200 feet.

When making a service application the pressure in the brake cylinder is only fifty pounds to the square inch, and the brake power is thus reduced one-sixth, consequently there is only a fraction over fifty-eight per cent available stopping power on a light car, and only about nine-

ITS USE AND ABUSE

teen per cent on a loaded one, but if the brake shoes are hung from the body of the car the piston-travel will be increased from one to three inches when the car is loaded, as the shoes strike the wheels lower down when loaded than empty. This means that if such a car was braking to seventy per cent light it would only be braking to a fraction over fourteen per cent loaded, and if the piston-travel was over eight inches when the car was empty the brake power would be still further reduced.

If the piston was allowed to travel its full stroke, there would be no brake power exerted against the wheels, as all the force would be against the cylinder head.

While these facts should be self-evident to all enginemen and trainmen, a great number of them, however, seem to think if the leaks in the trainpipe are stopped that the brakes are all right.

It should never be lost sight of that whenever you change the piston-travel or load you also change the per cent of brake power.

An example of the stopping distance required for a heavy car, as compared with a lighter one, was recently given when a Pullman car weighing about 100,000 pounds was "kicked" off while running thirty miles an hour, and it stopped in 416 feet, while a coach weighing about 60,000 pounds stopped in 202 feet after being kicked off at thirty miles an hour. In both cases the braking power was ninety per cent of the weight, but there was a difference of forty per cent in the weight of the cars.

Train Handling. While different conditions require different handling of trains, there are, however, two distinct points to be remembered as regards the difference between stopping a freight train and a passenger train.

In stopping a passenger train running twenty-five miles an hour or over, *two applications* should be made, and the final release made just before the train comes to a stop. If the stop is made on a grade, reapply the brakes to prevent drifting. (See Straight Air Control.)

In stopping a freight train but *one application* should be made, and never release until the train comes to a standstill, or you are very liable to pull your train in two. (See Str. Air Control.)

An "application" is from the time the first reduction is made until the brakes are released, several "reductions" can be made during one application.

Why two applications should be made with a

ITS USE AND ABUSE

passenger train is, first, because the speed has to be reduced before the stop can be made, and, second, the train should be absolutely under control in approaching a station, as something or some one may be on the track, and if the engineer was making a "one application stop" the auxiliaries and cylinders would have equalized some distance back of the actual stopping place, and the train would drift to the usual place in spite of anything the engineer could do.

In making a two application stop the first reduction should be about ten pounds, followed in a few seconds by about five pounds, and again, in a few seconds, by five more, which will equalize the pressures. By this time the train is only running about fifteen or eighteen miles an hour, and you are nearly to the station, so you must now place the handle in *full release* just long enough to be sure that all brakes are off, and bring it to lap. This prevents the trainpipe pressure from becoming higher than that in the auxiliaries, and when you begin to make the actual stop a reduction of seven or eight pounds will cause the triples to move at once, and again set the brakes. When you feel that this reduction has produced the desired effect, make

MODERN AIR-BRAKE PRACTICE

another of four or five pounds and let the train drift to the usual place, and release just before it stops, which allows the trucks to right themselves, and no one is jerked off his feet in the coaches. (See Straight Air Control.)

By making two applications you get two shots out of each auxiliary, and, besides, after releasing the first time, you have a chance to get the added twenty per cent of brake power by using the emergency, if you have to; whereas if you were making the stop with one application you could never get more than a full service application after a ten-pound reduction, even if you used the emergency, which, of course, you should always do in case of danger.

With the high-speed brake a train can be stopped in about thirty per cent less distance than it can with a quick-action brake. For instance, a train running forty-five miles an hour can be stopped in 560 feet with the high-speed brake as against 710 feet with the quick-action brake. Consequently a train running sixty miles an hour can be stopped in 1,060 feet with the high-speed brake, making a net gain of 300 feet over a stop made with the quick-action brake, which requires 1,360 feet within which to stop a train running sixty miles an hour.

ITS USE AND ABUSE

Two applications should always be made with the high-speed brake in making a stop, but the initial reduction can be fifteen pounds instead of ten pounds, as would be proper when using the automatic brake. In making a ten or fifteen-pound reduction with the high-speed brake from 110-pound trainpipe pressure, the same cylinder pressure is produced as there would be if the same reduction were made with the automatic brake from seventy-pound trainpipe pressure, provided the piston-travel is the same. But, after the brake cylinder pressure has been raised with the high-speed brake to the point at which equalization would take place with the automatic brake, then any further reduction of the trainpipe pressure with the high-speed brake would raise the brake cylinder pressure accordingly. For example, an emergency application of the automatic brake with an 8-inch cylinder would produce a cylinder pressure of sixty pounds, but with the high-speed brake an emergency application will produce a brake cylinder pressure of about eighty-eight pounds. As this pressure is equivalent to a brake power of 130 per cent of the weight of the car, you will understand why it is necessary to have an automatic reducing valve to let the high pressure escape, as the

MODERN AIR-BRAKE PRACTICE

train slows down, in order to prevent wheel sliding. (See Automatic Release Signal.)

If the auxiliary and trainpipe pressure (after making a reduction) equalizes at any point above sixty pounds, just as soon as the auxiliary pressure gets a fraction lower than the trainpipe pressure the triple will automatically lap itself, so that while the brake cylinder, owing to the reducing valve, may only have sixty pounds in it there might still be seventy-five pounds, or more, in the auxiliary reservoir and trainpipe. Therefore, with the high-speed brake equipment an engineer can make two full service reductions of twenty pounds and release his brakes and still have seventy pounds pressure left in the auxiliaries with which to stop, if necessary, without having to recharge.

Owing to the high pressure contained in the auxiliary reservoir with the high-speed brake the air is forced into the brake cylinder more quickly than it is with the automatic brake, and naturally takes hold quicker. But, as previously explained, there is no greater pressure per square inch in the brake cylinder from a ten or fifteen-pound reduction with the high-speed brake than there would be if made with the automatic brake.

ITS USE AND ABUSE

Handling a freight train is very different from handling a passenger train, and when handling a freight train the following points should always be kept in mind:

Good driver and tender brakes on the heavy class of freight engines are equal to the brake power furnished by six or seven 30,000-pound cars.

Always listen to the blow from the trainpipe exhaust when making a service application, as by the length of the blow you can tell the length of your trainpipe. This little item may save your life, as there are many ways for an angle-cock to become closed.

Always insist on having your train brakes carefully tested, and their condition and number reported to you before leaving a terminal, or where any change has been made in the train.

Always lap your brake valve if the brakes apply suddenly without any apparent cause, as a hose may have bursted or a conductor's valve opened, and you will need all your main reservoir pressure to release and recharge.

Always close the steam-throttle in case of a break-in-two, for with a partially equipped train the non-air cars will only hit the head end that

MODERN AIR-BRAKE PRACTICE

much harder if you try to pull away, as the air brakes will stop you anyway. Should the engine, however, be equipped with either the Dukesmith Release and Retaining Valve or the Dukesmith Straight Air Control Valve, you can release the driver and tender brakes within ten seconds by using this valve, which will cause the shock of the cars against the engine to be greatly modifed, and prevent a further break-in-two. (See Dukesmith Driver Brake Control System.)

Never reverse an engine after applying brakes if your engine brakes are any good, as it will flatten the tires if you reverse with the brakes set.

In using sand, be sure to get it upon the rail before the speed of the train has been materially reduced, or it will slide the wheels, and if sand is used while the wheels are sliding it is certain to put bad spots on them.

In making a service application you must be governed by circumstances, as regards speed, load and grade, but never make less than a five-pound reduction to start with, as less than that will not push the brake piston out past the leakage groove. Ordinarily from five to seven pounds will be right, but you must always wait a few seconds between the first and second reductions to allow the slack to run out.

In handling loaded trains on heavy grades, it is always best to make about a ten-pound reduction to start with. (See "Train handling on heavy grades.")

Never make over a twenty-five pound reduc-

ITS USE AND ABUSE

tion in service applications, for with correct piston-travel a twenty-pound reduction will equalize the pressures, and any further reduction is a waste of air.

Always make a running test with a passenger train, and also with a freight train where track conditions will permit it. Some hobo may turn an angle-cock on you.

In all cases of emergency throw the handle to emergency position and leave it there until the train comes to a standstill. But with a passenger train the brakes may be released while running if the danger has been removed.

Releasing the Brakes. Never try to release brakes in running position, with the old style automatic brake valves, for it is just this kind of foolishness that causes many flat and broken wheels. When the brake valve is in running position the trainpipe pressure raises comparatively slow, and if there should be any leaky triple piston packing rings the trainpipe and auxiliary pressures will equalize without moving the slidevalve, and consequently the brakes on all such cars will stick, and on poor rail the wheels on such cars may catch and slide while going slow, or if a brake sticks for any considerable time it will overheat the wheel and cause it to burst and wreck the train.

The amount of money paid out annually by railroad companies on account of "brakes sticking" is something enormous. The money paid out on account of doubling hills from brakes sticking would make a nice fortune.

To release brakes, always use full release position, no matter how long the train is.

Never open the throttle just after releasing brakes on a freight train, but allow the slack to adjust itself first. If you don't, you are almost sure to pull out a draw head and part your train.

To insure a prompt release, when coupling onto an empty or partially charged train, always make about a fifteen-pound reduction and then release and lap the valve until the trainman has made the coupling and opened both angle-cocks. Some engineers are always complaining about their tender brake sticking, when the probability is they have allowed the auxiliary to charge up to seventy pounds, and when the trainman opens the angle-cock between the tender and train it naturally reduces the trainpipe pressure and sets the brake on the tender, and as the volume of space in the trainpipe prevents a quick raising of the pressure, and as a very slight leak by the triple piston packing ring will allow the pressures to equalize, it is easy to understand why the tender brake sticks.

Failure to release brakes is commonly caused by not carrying sufficient excess pressure, for unless the trainpipe pressure is raised *suddenly* the slight leaks by the packing rings in the triples will cause the brakes to stick, for as the head triples are moved first, the feed grooves in the triples allow the trainpipe pressure to become lower every time a brake is released, so that on a long train the pressure would become so low as it neared the rear end that it would not be strong enough to force the triples to release position. Sometimes a brake can be released by making another heavy reduction, which changes the relation between the trainpipe and main reservoir pressures so that the excess thus created will give the triple a hammer blow and drive it to release position when the handle is thrown to full release. But, of course, if the brake is sticking on account of a leaky packing ring it would have to be bled off either by the auxiliary bleed cock or the release-signal valve.

In taking water, with a freight train, it is

always best to stop short of the water plug, cut off, and run up with the engine alone.

In setting out cars, always apply the brakes before the train is cut, because there can be no danger then of pulling out with an angle-cock closed against you.

Train Handling on Heavy Grades. Trains are frequently stalled on heavy grades because the engineer keeps throwing the handle of the brake valve to full release and then bringing it back to running position. By doing this he soon gets the trainpipe charged higher than what the feed valve is set for, and then, in running position, the brakes are sure to creep on, for the trainpipe pressure must be reduced before the feed valve will open to admit main reservoir pressure.

A heavy initial reduction is proper with loaded trains on heavy grades, because a certain amount of the brake power is necessary to overcome the "drop" or downward movement caused by gravity, which materially reduces the amount left for holding the train at a certain speed.

As soon as the train passes a summit the brakes should be applied, in order to know for a certainty what they are capable of doing. By waiting until the train is well under way before applying the brakes is very liable to cause a

ITS USE AND ABUSE

runaway, as the trainmen have a poor chance of stopping it by hand brakes, should the occasion arise. The man who is not afraid to call for hand brakes when he thinks there are not enough good air brakes to hold the train, is much safer for the railroad company than the fellow who is afraid to do so because the train crew will think "he has lost his nerve."

In descending a grade, always try to keep the trainpipe pressure as near standard as possible, by recharging as often as may be required, for in case a stop has to be made you will need all the power you can get. (See Str. Air Control.)

Always recharge in full release position. If the trainpipe pressure shows up on the gauge to be above standard, bring the handle to running position for a few seconds to allow it to equalize, and then place it on full release just for a second to kick off any forward brakes that may have set, owing to the auxiliaries on the forward cars charging up faster than the others.

Comparatively slight trainpipe leaks are more dangerous on a heavy grade than leaks which are readily noticed, for after a light application, unless the gauge is watched very close, the slight leaks will cause the brakes to continue to set until the pressures are equalized, when it would be impossible to apply them any harder should a stop have to be made.

Hostlers should remember that there are more than two positions (emergency and full release) on the brake valve. An emergency application is only intended to be used when the full brake power is required. A full service application is only necessary when running at a high rate of speed, therefore when handling an engine through the yards, make light applications of about five pounds to start with, and gradually increase the reduction as occasion demands. If you are running slow, don't try to use a high speed application, as you are very liable to slide the wheels. Never use the emergency on the turn table.

TESTING AND INSPECTION OF AIR BRAKES

No train should ever leave a terminal until the brakes have been thoroughly tested and put in good order.

In testing a train, begin at the rear end and close the angle-cock, and, if it is a freight train, couple the hose between the caboose and the first car, after knocking the hose-couplings together to jar out any dirt that may be lodged in them, then turn the angle-cocks straight with

ITS USE AND ABUSE

the pipe; next, see that the brake is cut in at the cross-over pipe; examine the retainer to see that the handle is turned down, and notice if the hand brake is released. Treat every car in the train alike, and when you reach the head end, before coupling the tender hose, always blow it out by opening the angle-cock.

While the train is being charged up, which will take about fifteen minutes, if it is a thirty or forty-car train, go over the train and stop all the leaks. If a bad blow is found at a triple gasket which can't be stopped by tightening the nuts, cut the brake out, bleed it and report it on a defect card. If the blow is at the hose-coupling, and a new gasket does not stop it, drive a small sliver of wood, or a match, between the lugs, which will force the heads together. Never use paper or a nail.

When the train is charged up and the brakes have been set, begin at the front end and examine the piston-travel on each car. If a piston is found to travel nine inches, or over, mark the car so that you will know whether to take up or let out the travel, after the engineer has released the brakes.

Should you come to a car where the brake is cut in and the auxiliary charged, but the piston is not out, have the engineer make a further reduction to ascertain if the brake "leaked off" or "released." If it releases you can hear it blow out of the retainer, and if it leaks off the air is escaping around the packing leather in the cylinder, which usually cannot be heard. In either case cut out the brake and report it correctly, for if you say it leaked off, the car repairer will go after the cylinder leather, and if you say it released, he would go after the triple.

Upon reaching the rear end of the train, signal the engineer to release, and then see if every car releases properly.

If a brake has failed to release, examine the retainer, and if it is found with the handle turned down, and the brake rigging is not caught, cut the brake out, bleed it and report it.

When you come again to any car which you had previously marked for changing the piston travel, take up or let out the slack by moving the truck dead lever forward or back, as the case may be, but be sure to take it up at both ends of the car alike.

Having finished inspecting the train, report to the engineer the number and condition of the brakes in working order.

When a train is equipped with the release sig-

ITS USE AND ABUSE

nal you can tell by the action of the signal just what the brake is doing and if the piston travel is too great.

BRAKE LEVERAGE

The subject of brake leverage is a very interesting one, but as all foundation brakes are supposed to be carefully figured out by competent experts when the car is built, an absolute knowledge of leverage is not required of enginemen or trainmen. I shall, however, explain the different kinds of levers and the manner of figuring them so that any one can, by a few simple calculations, tell if a car or engine is getting its proper braking power, and also lay out the proper leverage when building new work.

In order to tell the proper proportion of brake levers, or to ascertain what force is being exerted at any of the pins, it is necessary to take into account two forces and two distances.

The two forces represent the power applied at one pin and the weight lifted by the other pin, between which is the fulcrum; the two distances are figured from the fulcrum to the applied power and from the fulcrum to the weight.

In every case the applied power multiplied by the distance it is from the fulcrum divided by the distance from the fulcrum to the weight, will tell you what the weight is that is being lifted by the applied power.

The point that most bothers the new student is to tell where the fulcrum is, but this will come all right with a little practice.

Remember that in figuring leverage you must take "proportion" into account. If the applied power is proportionately one-third nearer the fulcrum than the weight is to the fulcrum, the power can only lift a weight equal to one-third of its force, and if the opposite is true then the power can lift a weight equal to three times its force.

For example, if a lever is forty inches long from the centers of the outside holes, and another hole is placed ten inches from either end, it would be called a one to three lever, for when you divide forty into two parts of ten and thirty, the result is that one portion is three times greater than the other, so that if you applied the power at either end and the weight at the other, then the fulcrum would be the ten-inch hole, and if the power of, say, 100 pounds, was nearest the fulcrum, you would multiply 100 by 10, which would equal 1,000, and when you divide 1,000 by 30 the result would be 33⅓, or one-third

ITS USE AND ABUSE

of the applied power. This would be called a lever of the first kind.

Now suppose that the power was at the long end of the lever, and the weight at the other, then the fulcrum would be at the thirty-inch hole. So that 100 multiplied by 30 would equal 3,000, which divided by 10 would equal 300 or three times the applied power. This is also a lever of the first kind.

Again, suppose the power was nearest the ten-inch hole, and the weight was *at* the ten-inch hole, then the fulcrum would be forty inches away from the applied power. In order to tell how much weight could now be lifted by the 100 pounds, you would multiply it by 40, which would equal 4,000, divided by 30 would equal the weight, 133⅓ pounds, for the reason that the fulcrum is three-thirds, or one whole number, away from the power, which gives 100 pounds lift, and the weight being one-third the distance from the power, gives a lift of one-third of the applied power, and the two combined equal one and one-third the force of the applied power. This is called a lever of the second kind, as the delivered force or weight is between the fulcrum and the applied power.

The third kind of lever is designated by having the applied force between the fulcrum and the delivered force, and is explained as follows: The applied power is now at the ten-inch hole, and the weight is at the end nearest the power, which would make the fulcrum at the opposite end, or thirty inches from the power. Multiply the 100 by 30, and you have 3,000 pounds, which divided by 40 (the distance the weight is from the fulcrum), and you have a lifting force of seventy-five pounds; for the reason that the applied power is located three-fourths the distance from the fulcrum to the weight. If you change the weight so that it would be at the thirty-inch hole the lifting force at the weight end would only be twenty-five pounds, because the applied power would then be located at a point equal to one-fourth the distance that the fulcrum is from the weight.

Therefore, with a lever of the third kind the lifting force is always increased in proportion to the distance that the applied power is from the fulcrum as the fulcrum is from the total length of the lever; in other words, by moving the applied power toward the weight increases the lifting force and moving the applied power away from the weight toward the fulcrum decreases the lifting force.

ITS USE AND ABUSE

Always remember that the applied power and the weight added together equal the strain at the fulcrum.

Should you wish to design a cylinder lever and wanted to know where to place the middle, or fulcrum pin, you would proceed as follows: Multiply the weight to be moved by the total length of the lever, between the two centers of the outside holes, and divide it by the applied force and weight combined; the result would be the distance in inches from the cylinder pin hole to the fulcrum. To prove it, multiply the applied force by the length, and divide by the force and weight combined, which should equal the number of inches from the fulcrum to the weight pin hole.

For example, suppose you had an eight-inch cylinder, with a quick-action triple, the applied power would be 3,000 pounds; now suppose you wanted a force of 1,500 pounds on the floating lever end of the cylinder lever, which is 33 inches long, you would multiply 3,000 by 33, which equals 99,000, now divide this by the required force (1,500) and the applied power (3,000) combined (or 4,500), and you have as a result 22, which is the number of inches the hole should be from the weight end of the lever which would

MODERN AIR-BRAKE PRACTICE

make the fulcrum eleven inches from the cylinder end of the lever. Prove this by multiplying

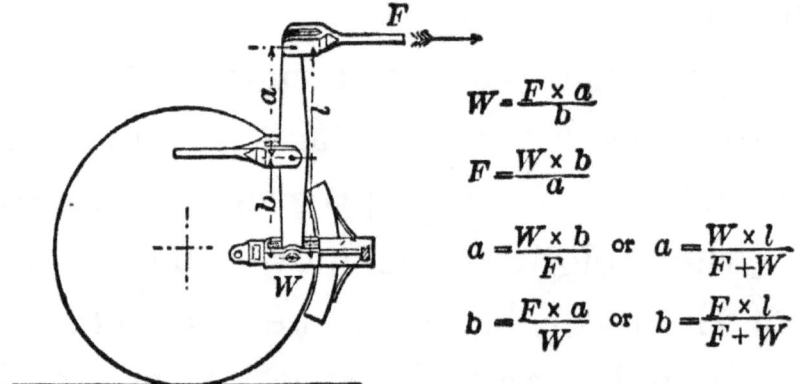

$$W = \frac{F \times a}{b}$$
$$F = \frac{W \times b}{a}$$
$$a = \frac{W \times b}{F} \text{ or } a = \frac{W \times l}{F+W}$$
$$b = \frac{F \times a}{W} \text{ or } b = \frac{F \times l}{F+W}$$

FULCRUM BETWEEN APPLIED AND DELIVERED FORCES.

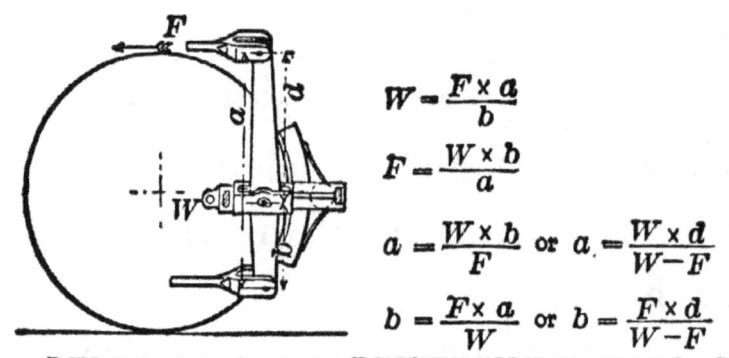

$$W = \frac{F \times a}{b}$$
$$F = \frac{W \times b}{a}$$
$$a = \frac{W \times b}{F} \text{ or } a = \frac{W \times d}{W-F}$$
$$b = \frac{F \times a}{W} \text{ or } b = \frac{F \times d}{W-F}$$

DELIVERED FORCE BETWEEN FULCRUM AND APPLIED FORCE.

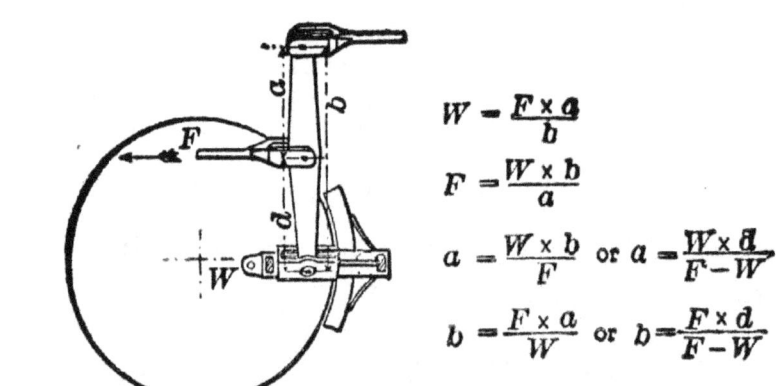

$$W = \frac{F \times a}{b}$$
$$F = \frac{W \times b}{a}$$
$$a = \frac{W \times b}{F} \text{ or } a = \frac{W \times d}{F-W}$$
$$b = \frac{F \times a}{W} \text{ or } b = \frac{F \times d}{F-W}$$

APPLIED FORCE BETWEEN FULCRUM AND DELIVERED FORCE.

PLATE NO. 77.—BRAKE LEVERS.

ITS USE AND ABUSE

3,000 by 11, and dividing by 22, and see if you don't get 1,500 as a result.

Plate 77 illustrates the formula for calculating the different kinds of levers. The first kind is where the fulcrum is in the middle; the second has the weight in the middle, and the third has the applied power in the middle.

The first formula translated into straight English would read as follows: The weight (W) is equal to the applied power (F) multiplied by the distance (a) from the power to the fulcrum, divided by the distance (b) from the fulcrum to the weight. From this you can read the others.

Plate 78 illustrates the two systems of brake levers used on passenger cars, and also the tender levers. The Hodge system is especially indicated as having a floating lever, which the Stevens system has not.

Plate 79 shows a freight equipment of levers with the brake shoes attached below the bottom rod. The plate shows the result of an **emergency** and a service application.

RULES FOR CALCULATING BRAKE POWER

The force exerted upon the piston depends upon the size of the cylinder and the air pressure in the cylinder.

MODERN AIR-BRAKE PRACTICE

To get the number of pounds push at the piston, multiply the number of square inches on the piston by the number of pounds pressure per square inch on the cylinder. For example, an

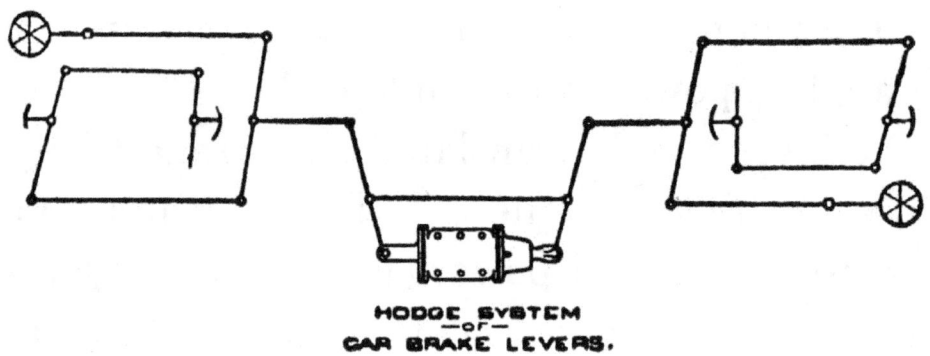

FIG. 1.

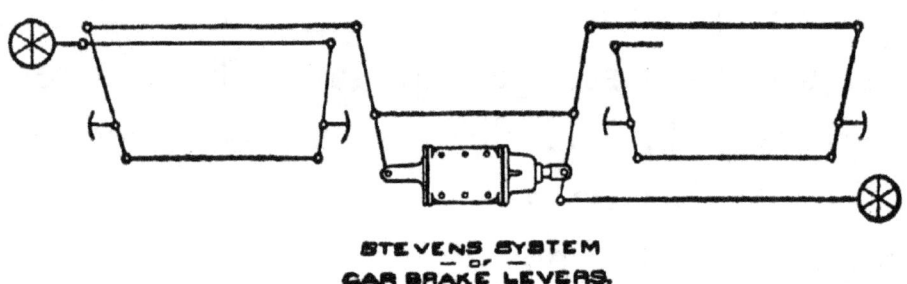

FIG. 2.

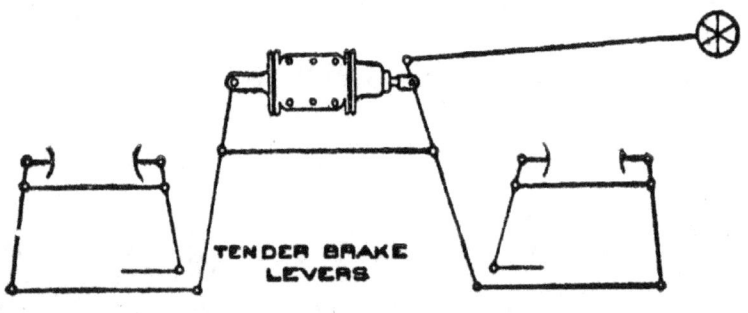

FIG. 3.

PLATE NO. 78.—CAR AND TENDER TRUCK BRAKE LEVERS.

eight-inch piston contains fifty square inches, which multiplied by fifty, the cylinder pressure, would give a push of 2,500 pounds at the end of the piston-rod.

To find the number of square inches on a piston, multiply the diameter by itself, and by the number thus obtained multiply .7854, and cut off the last four figures from the result, and the remainder will be the number of square inches. For example, 8 times 8 is 64, and .7854 multiplied by 64 equals 50.2656, or 50 inches and 2,656 ten thousandths of an inch, so you just cut off the ten-thousandths, unless they are equal to a half number or better, when you count them a half, as for instance a ten-inch cylinder would be counted as having 78½ square inches.

A short method is to multiply the diameter by itself, and the result by 11 and divide by 14.

To find at what pressure the auxiliary would equalize with the cylinder, find the number of cubic inches contained in the auxiliary by multiplying the number of square inches contained in its diameter by its length (minus the concavity in the heads), and then multiply the cubic inches by the pressure with fifteen pounds added, and divide by the combined cubic inch contents of the auxiliary and cylinder, and deduct the fifteen

MODERN AIR-BRAKE PRACTICE

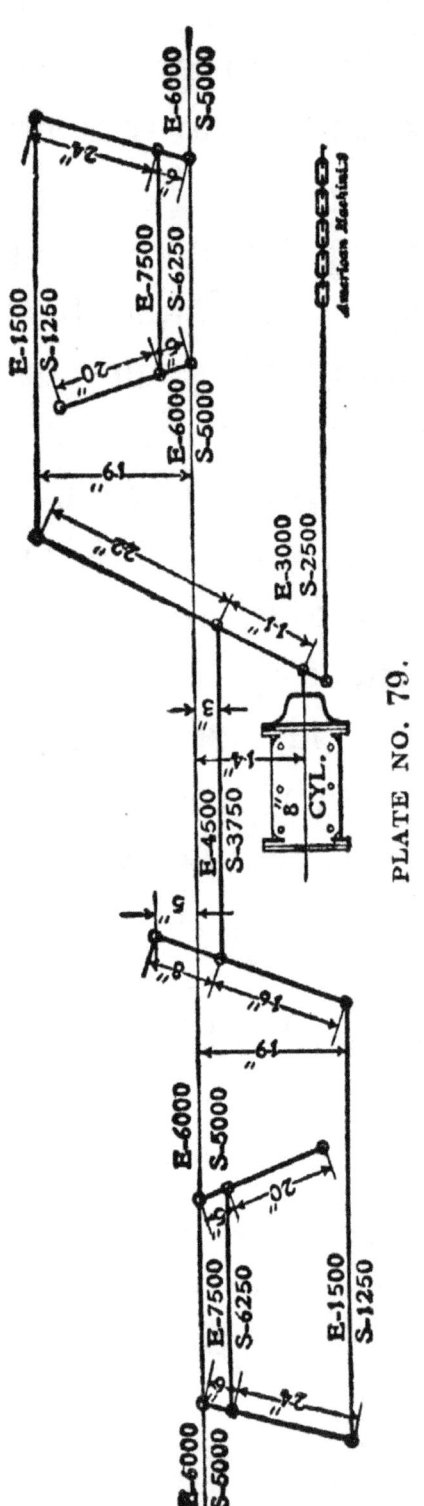

PLATE NO. 79.

Weight of car, 34,286 pounds; braking power, 24,000 pounds, or 70 per cent.

With the quick-action triple valve, E indicates forces in emergency applications and S indicates forces in full service applications; with the plain triple valve, S indicates the forces in either Full Service or Emergency Applications.

All distances of rods from center line of car body are those that should occur when the levers stand at right angles to it.

ITS USE AND ABUSE

pounds which you added, and the result will show the point of equalization.

For example, a freight auxiliary contains about 1,620 cubic inches, and the standard pressure is 70 pounds; to this add 15, which makes 85, now multiply 1,620 by 85 and you get 137,700. An eight-inch cylinder, with eight-inch piston travel, contains about 450 cubic inches. The cylinder and auxiliary together hold 2,070; now divide 137,700 by 2,070 and you get 66½, from which deduct 15, and the result is 51½, or the pounds pressure at which they equalize.

The following table gives the force exerted upon the pistons of the different sized cylinders with pressures of fifty and sixty pounds per square inch:

Size of cylinder,	6″	8″	10″	12″	14″	16″
50 lbs. pressure,	1,000	2,500	4,000	5,650	7,700	10,059
60 lbs. pressure,	1,700	3,000	4,700	6,700	9,200	12,050

SIZES OF AUXILIARY RESERVOIRS WHICH SHOULD BE USED WITH DIFFERENT SIZED CYLINDERS, WITH THE CUBIC-INCH CAPACITY OF EACH, WITH EIGHT-INCH PISTON TRAVEL

Eight-inch tender and truck cylinders, with 10×24 auxiliary: Cubic inches of cylinder, 450. Cubic inches of auxiliary, 1,491.

MODERN AIR-BRAKE PRACTICE

Eight-inch driver brake cylinders, with 10×33 auxiliary: Cubic inches in auxiliary, 2,050.

Ten-inch cylinders of all kinds, with 12×33 auxiliary: Cubic inches in cylinder, 628. Cubic inches in auxiliary, 3,030.

Twelve-inch cylinders of all kinds, with 14×33 auxiliary: Cubic inches in cylinder, 904. Cubic inches in auxiliary, 4,120.

Fourteen-inch cylinders of all kinds, with 16×33 auxiliary: Cubic inches in cylinder, 1,232. Cubic inches in auxiliary, 5,450.

Sixteen-inch cylinders of all kinds, with 16×42 auxiliary: Cubic inches in cylinder, 1,600. Cubic inches in auxiliary, 7,163.

CYLINDER PISTONS AND AUXILIARY DIAMETERS

The following tables show the number of square inches on the different sized pistons, and inside diameter of auxiliary resevoirs:

8-inch cylinder piston contains 50 square inches.
10 " " " " 78½ " "
12 " " " " 113 " "
14 " " " " 154 " "
16 " " " " 201 " "

10-inch auxiliary contains 71 square inches.
12 " " 103½ " "
14 " " 143 " "
16 " " 188½ " "

ITS USE AND ABUSE

PERCENTAGE OF BRAKING POWER REQUIRED

The following table shows the percentage of braking power required for engines, tenders, passenger and freight cars:

When plain triples are used the cylinder pressure is figured at fifty, but with quick-action triples it should be sixty pounds per square inch.

Engines, 75 per cent of weight on drivers.

Tenders, 100 per cent of light weight.

Passenger cars, 90 per cent of light weight.

Freight cars, 70 per cent of light weight.

When six-wheel trucks are used on passenger cars and have only four pairs of wheels braked, the braking power should be figured as 90 per cent of eight-twelfths of the total weight. A chair car weighing 90,000 pounds with only four pairs of wheels braked should only have a brake power of 54,000 pounds.

TO DESIGN LEVERS FOR A CAR

When designing the levers for a car you must begin by taking the total weight, and where four pairs of wheels are to be braked on an eight-wheel passenger car, take ninety per cent of the weight and divide it by four, which will give you the amount of power required for each brake beam. To find what the pull should be at the top of the live lever, measure the height of the truck, in order to know how long the live lever must be. Having found the length of the lever, and knowing what force there must be on the brake beam, you proceed as previously explained under "Brake Leverage," remembering that your live lever is of the second kind, as shown in plate 34.

THE OUTSIDE EQUALIZED DRIVER BRAKE

Plate 80 illustrates the regular outside equalized driver brake, which is now almost universally used on engines.

In order that you may better understand it, I will run through the figures for you. There is a fourteen-inch cylinder on each side, with a force at the piston of 7,650 pounds. The weight on drivers is 81,600, and 75 per cent of this is 61,200, which divided by 6 means that each wheel must have a brake power of 10,200. The length of the long arm of the cylinder lever is 24 inches, and the short arm is 6. So that 7,650 multiplied by 24 equals 183,600, which divided by 6 leaves 30,600 pounds at the bottom end of the short lever. This is carried to the first equalized lever, which is 4×8, or a one to two lever, and as the applied power is two-thirds the distance

ITS USE AND ABUSE

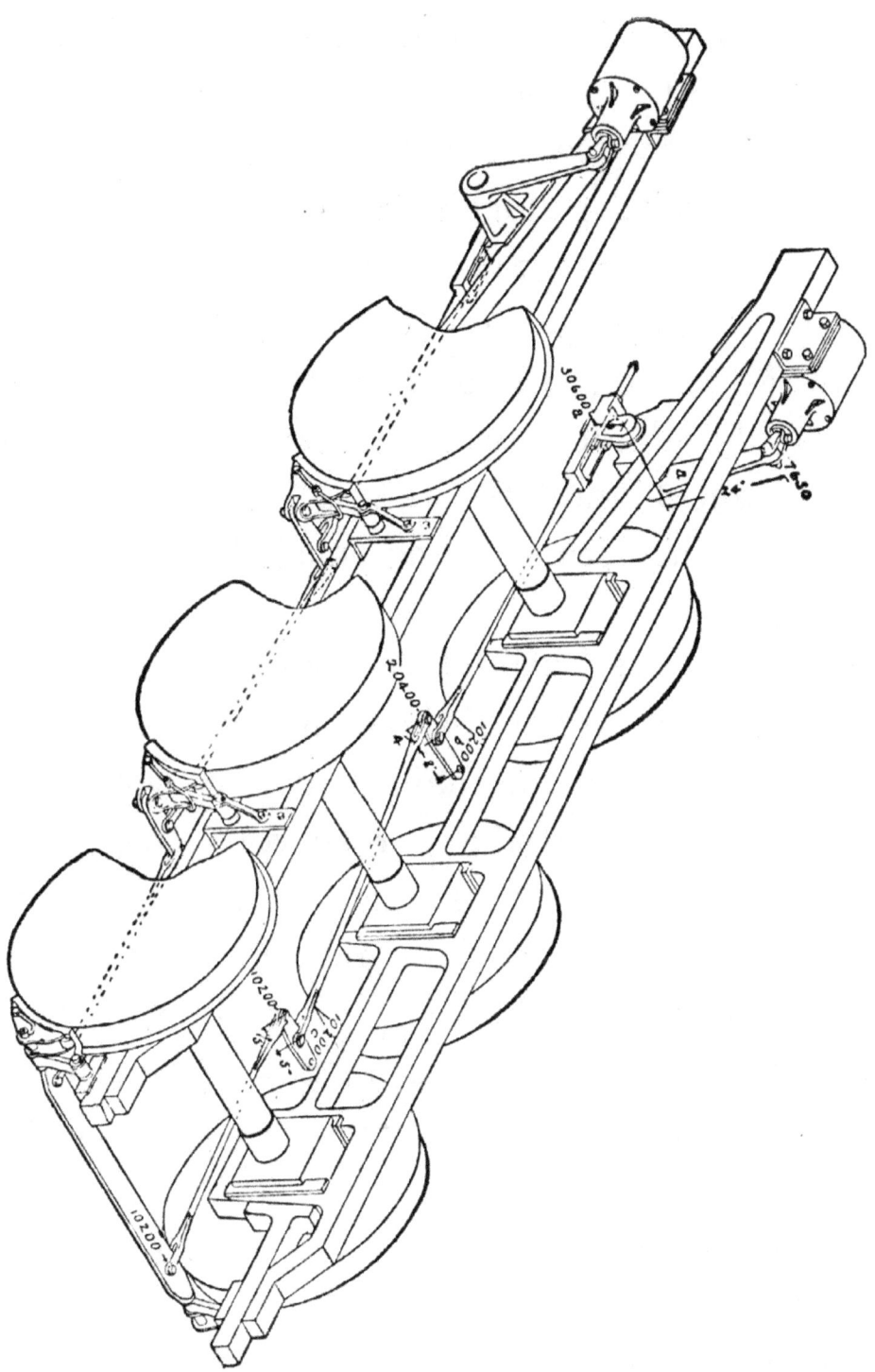

PLATE NO. 80—THE OUTSIDE EQUALIZED DRIVER BRAKE

from the fulcrum as the fulcrum is from the weight, the short end of the lever has a force of two-thirds of the applied power, or 20,400, which leaves the other one-third as the weight against the first wheel, which is the fulcrum. The applied power at the second equalized lever being 20,400 pounds, and as the applied power is midway between the fulcrum and the weight, it is equally divided, which leaves 10,200 at the second wheel, and the other 10,200 is carried to the third wheel, thus giving each wheel the same brake power.

THE CAM DRIVER BRAKE

The chief features of the Cam brake requiring consideration are the maintenance of such a piston-travel that auxiliary reservoir and brake cylinder pressures shall equalize at fifty pounds when the brakes are fully applied, and of such an adjustment of the cams that their point of contact shall be in line with the piston-rod; otherwise a bending influence will be exerted upon the piston-rod.

To adjust the cams, in order to shorten or lengthen the piston-travel or to secure a central point of contact, the check nut should be slacked off and the screw turned outward to shorten the piston-travel, or inward to lengthen it.

ITS USE AND ABUSE

To calculate the braking power, apply the brake and measure the piston-travel; then release the brake, insert pieces of one-quarter-inch steel wire crosswise between the tire and the shoe at the upper and lower ends, and again apply the brake; divide the difference of the piston-travel by the thickness of the steel, and multiply the result by the total force acting upon the piston. The result is the pressure of one shoe, which, multiplied by four, gives the total braking power. Divide this total by the total weight upon drivers to obtain the percentage of braking power.

EXAMPLE

Weight on drivers, 53,330 pounds.

Piston-travel, without inserting wires, three inches.

Piston-travel, with one-quarter-inch wires inserted, two inches.

Total force on piston (eight-inch cylinder brake, fully applied), 2,500 pounds.

1 divided ¼ equals 4. 4 multiplied by 2,500 equals 10,000 pounds.

10,000 pounds multiplied by 4 equals 40,000 pounds—the total braking power.

40,000 divided by 53,330 equals 75 per cent.

MODERN AIR-BRAKE PRACTICE

THE LOCOMOTIVE TRUCK BRAKE

Plate 81 illustrates the American Equalized Locomotive Truck Brake with Automatic Slack Adjuster. Inasmuch as a considerable proportion of the weight of certain types of locomotives is carried upon the truck, the importance of a well-designed brake upon that part of the equipment is self-evident, especially as the weight upon this truck frequently equals (and often exceeds) the weight of a large capacity car. This brake should be maintained in a high state of efficiency, which is readily accomplished by the aid of the automatic slack adjuster.

What has been said with reference to the maintenance and care of the driver brake applies with equal force to the truck brake.

In concluding this volume the author would earnestly impress upon the mind of the reader that he should carry the book with him in his daily work, as the time is already here when all railroad men are expected to have a thorough knowledge of the air brake, and unless you are already well posted you cannot expect to absorb the knowledge if your instruction book is left at home.

You should always keep in mind the fact that the human brain is naturally inclined to throw

ITS USE AND ABUSE

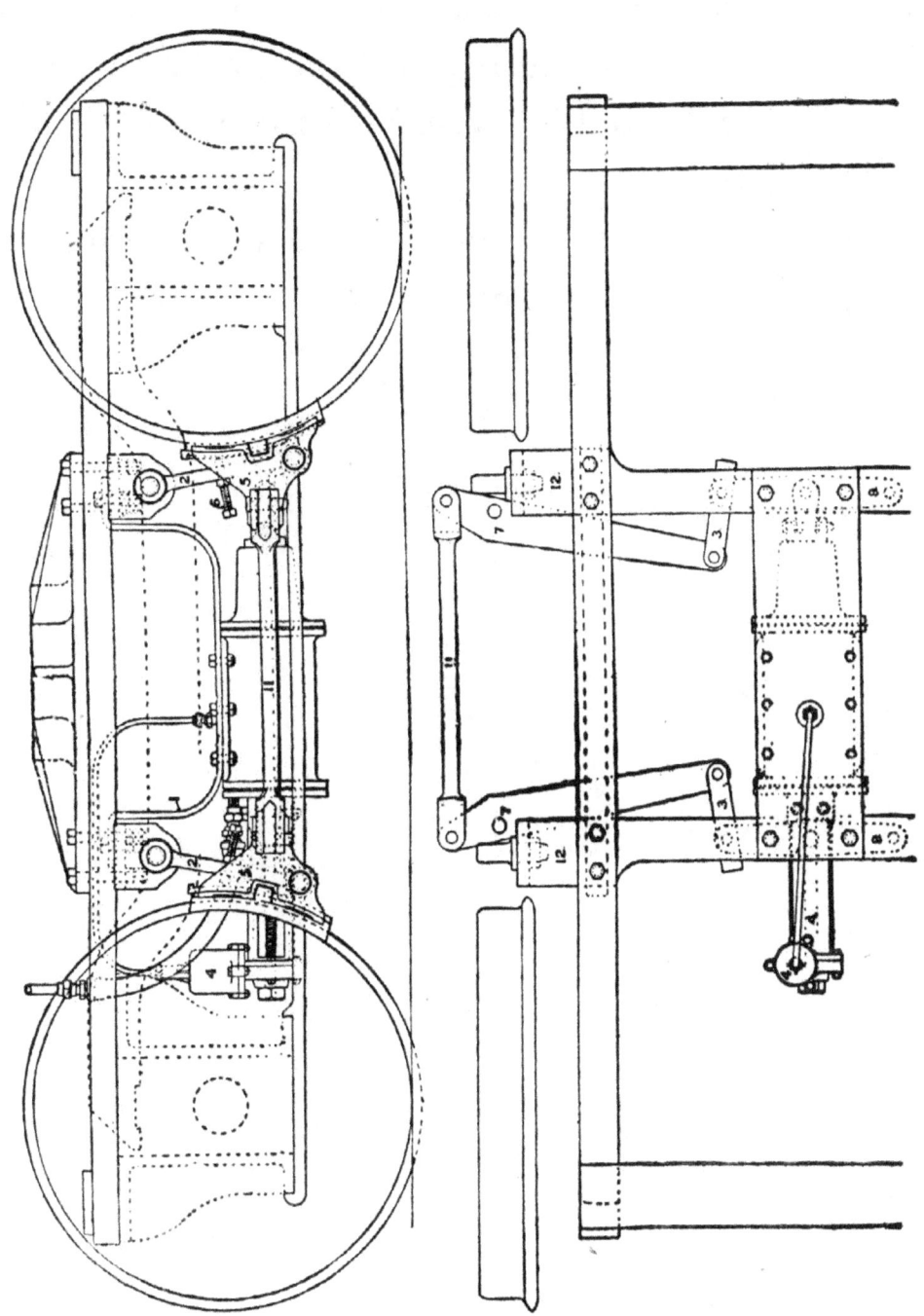

PLATE NO. 81—ENGINE TRUCK BRAKE.

off everything that tends to trouble it in any way, and as your success as a railroad man will depend upon your ability to acquire and retain knowledge, I would advise you to remember this little motto:

By reviewing what you think you know, you learn to know what you know you know.

QUESTIONS AND ANSWERS TO SECTION 7

THE PHILOSOPHY OF AIR-BRAKE HANDLING, BRAKE POWER LEVERAGE, ETC.

152. What is meant by the philosophy of air-brake handling?

Ans.—It is a clear and definite understanding of the effect produced by different volumes and pressures of air in relation to the varying conditions of track, load, grade and speed.

153. What is one of the first things an engineer should learn in handling trains?

Ans.—The value of maintaining correct standards of pressure in the different parts of the air-brake equipment.

154. What is the percentage of brake power allowed on freight cars, passenger coaches, engines and tenders?

Ans.—A tender is braked to 100 per cent of its light weight; an engine at 75 per cent of weight on drivers; a freight car at 70 per cent of its light weight, and a passenger coach at 90 per cent of its light weight. When a coach has six-wheel trucks and only four wheels braked, the braking power should be figured as 90 per cent of $\frac{8}{18}$ of the total weight.

155. What is the correct piston-travel for locomotives?

Ans.—As there are two cylinders on a locomotive both pistons should travel the same, and in order to determine what that travel should be, a test gauge should be applied and the piston-travel set so that the pressure equalizes with a full service or emergency application at fifty pounds. The smaller brake cylinders usually require about three inch piston-travel, and the larger cylinders four inches, and as the piston-travel on a locomotive is multiplied by two, a three-inch travel would mean six inches and the four-inch travel eight inches when both sides are counted.

156. What should be the piston-travel on cars?

Ans.—The piston-travel should be eight inches running along, and as the equipment while running allows a further piston-travel than can be obtained while the car is standing, the piston-travel should be set at about six inches. The rule is, never less than five, nor more than seven while the car is standing.

157. When a car is equipped with a slack

ITS USE AND ABUSE

adjuster how should the piston-travel be regulated?

Ans.—When a car is equipped with new shoes the piston-travel should be set at from 6 to 6½ inches by taking up the slack at the dead levers.

158. Is there anything that will cause the piston-travel to be too short on a car equipped with the automatic slack adjuster?

Ans.—Yes. If some of the slack has been taken up on the hand brake, or the position of the dead levers has been changed.

159. When a car is equipped with the slack adjuster, what may cause the piston-travel to become too long?

Ans.—Some obstruction may get into pipe B, or the pipe may leak, or the slack adjuster cylinder and packing leather may leak. If a car has been running with the hand brakes partly set it naturally takes up the slack, consequently when the brake is entirely released it will take the slack adjuster some time to readjust the piston-travel, as the cross head is only moved $\frac{1}{32}$ of an inch at each operation of the adjuster.

160. In applying new shoes to a car, what is necessary to be done in order to increase the shoe clearance?

Ans.—Turn the ratchet nut to the left and

after the shoes are applied the piston-travel may be shortened by turning the adjuster nut to the right.

161. What is the danger of operating trains with uneven piston-travel?

Ans.—Uneven piston-travel causes some of the brakes to be released sooner than others, and consequently where there is not sufficient excess pressure carried to promptly release all brakes, it is very liable to either pull the train in two or slide the wheels, or cause the wheels to become heated from the pressure of the brake shoe so that the wheels are broken and the train wrecked.

162. Is there any difference in the way in which a passenger or freight train should be stopped?

Ans.—Yes. All the difference in the world. A passenger train should be stopped by making two applications and the brake valve thrown to release position just before the train comes to a dead stop in order to avoid the shock. Whereas, with a freight train but one application should be made, and the train allowed to come to a full stop before releasing, unless the engine is equipped with a Straight Air Control Valve. A further difference is, that in making a passenger stop where the train is running twenty-five miles an hour or over, the initial reduction should not

ITS USE — AND ABUSE

be less than ten pounds, whereas, with a freight train the initial reduction should never be over seven pounds, excepting with a heavy train on descending grades, when a ten-pound application would be right and proper.

163. In coupling a locomotive on to a train, what should the engineer do?

Ans.—He should make at least a ten-pound reduction and hold his valve on lap until the train is fully coupled up in order to prevent the tender brake from sticking.

164. Why are the tender brakes liable to stick if the engineer fails to make a reduction when coupling on to a train?

Ans.—For the reason that when more trainpipe is connected to the tender the air flowing into the added trainpipe will cause a reduction at the tender triple and set the brake, and as the triple piston packing ring on the tender, more than any other, is liable to be slightly gummed, it follows that in charging up the train while the tender brakes are set with a high auxiliary pressure, the auxiliary and trainpipe pressure on the tender is liable to become equalized or nearly so, which prevents a sure release of the brake.

165. In using the high-speed brake, why is it that the brakes take hold quicker with the same

amount of reduction than they do when using the automatic brake?

Ans.—It is because the auxiliary is charged to 110 pounds instead of seventy, so that the air passes quicker into the brake cylinder.

166. Would you get a higher brake power with a ten or fifteen-pound reduction in using the high-speed brake than you would if using the automatic brake?

Ans.—No. Because the same pressure per square inch would show in the brake cylinder with a ten or fifteen-pound reduction no matter which equipment was being used. But should a reduction be made with a high-speed brake of twenty-two pounds or more, there would be a corresponding increase in the brake cylinder pressure.

167. Why is this?

Ans.—With the automatic brake a reduction of twenty pounds will cause the auxiliary and brake cylinder pressures to equalize so that a further reduction would be simply a waste of air; whereas, with the high-speed brake the auxiliary and brake cylinder pressures do not equalize until a reduction of about twenty-six pounds has been made, but when the auxiliary and brake cylinder equalizes with the automatic

ITS USE AND ABUSE

brake equipment in service application, there is only a pressure of fifty pounds in the cylinder if the piston-travel is correct; whereas, with the high-speed brake equipment the auxiliary and brake cylinder will equalize with a service application at about sixty-eight pounds, and in an emergency application at about eighty-eight pounds where 110-pound trainpipe pressure is used, with Westinghouse triples.

168. In making a two-application stop with a passenger train, how should the brakes be released after the first application and why?

Ans.—In making a two-application stop the initial reduction should be from ten to twelve pounds according to the speed of the train, and as soon as the momentum of the train has been checked, one or two more reductions should be made until the auxiliary and brake cylinder pressures have equalized, when the brake valve handle should then be thrown to full release position just long enough to insure that all brakes are released, when the handle should be brought to lap position, until it is time to make the second application. The reason for this is because the auxiliaries cannot be recharged between the first and second application, and by holding the valve on lap it allows the trainpipe and auxiliary pressures to equalize so that on the first reduction, when the second application is begun, the triples will move promptly and pass the air from the auxiliary into the brake cylinder.

169. In handling a freight train, why should the engineer always listen to the trainpipe exhaust in making a service application?

Ans.—Because by the length of the blow at the trainpipe exhaust he can tell whether he has a long or short trainpipe.

170. In case of a break-in-two, what should the engineer always do?

Ans.—Promptly close the throttle and lap the brake valve, in order to stop as quickly as possible so that the rear end of the train will not hit the forward end so hard when they come together. By using the Dukesmith Driver Brake Control Valve this shock is avoided.

171. How should sand be used on a rail?

Ans.—Sand should always be used before the speed of the train has been materially reduced, as otherwise it is almost sure to spot the wheels.

172. Why should the throttle never be opened immediately after releasing the brakes on a freight train?

ITS USE AND ABUSE

Ans.—Because the slack must be allowed to adjust itself first in order to prevent pulling the train in two.

173. What is a common cause for stalling trains on heavy grades?

Ans.—The continual throwing of the brake valve handle to full release and back to running position. By doing this the engineer gets the trainpipe charged higher than what the feed valve is set for and when the trainpipe pressure has been raised above seventy pounds, if the handle is placed in running position, no air can pass through the feed valve attachment until the trainpipe leaks have reduced the pressure below what the feed valve is set at, and as a consequence the brakes get to dragging and thereby stall the train.

174. What is the danger of bleeding off a stuck brake and allowing it to run without cutting it out?

Ans.—A brake that is inclined to stick and has to be bled off is almost sure to stick again at a time when the trainmen cannot get to it to bleed it unless there is a release signal on the car, and as a consequence the wheels are liable to become heated and either slide them or break them and ditch the train.

175. What number of cars in a train should be air braked?

Ans.—All of them, if possible, but a recent law passed by the National Congress requires that fifty per cent of all cars in a train shall be equipped with air brakes in good condition. and another new ruling requires seventy-five per cent of all cars in a train to be air braked by August, 1907.

SECTION 8

CHAPTER VIII

THE STRAIGHT AIR BRAKE AS USED ON ELECTRIC TRACTION CARS*

Motormen and conductors of electric cars should possess something more than a mere knowledge of how to apply and release brakes. They should understand the mechanical principles represented in the brake, and should know how to detect, remedy and report any and all kinds of defects which may arise in the straight air-brake equipment.

The changed conditions of the past few years have materially raised the personnel of electric railway employés. There was a time when almost any kind of a man could find employment on street railways, but to-day a man must possess a certain amount of ability before even his application will be considered. These changed conditions are mainly due to the advent of interurban railway traffic.

The cars operated in interurban service are not only very heavy, but the speed at which

---

*A special index for this chapter has been prepared and immediately follows the index of the Automatic Equipment.

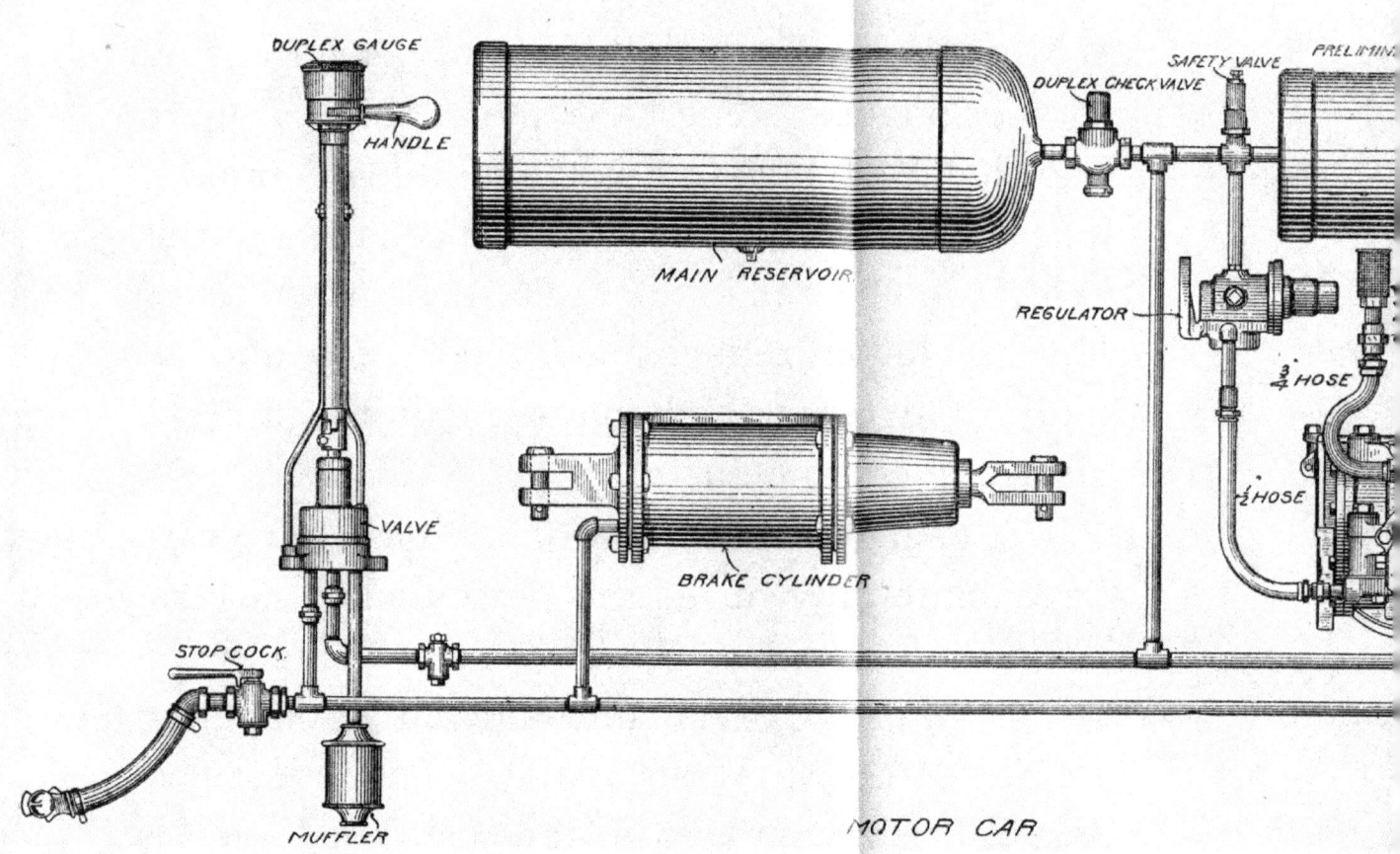

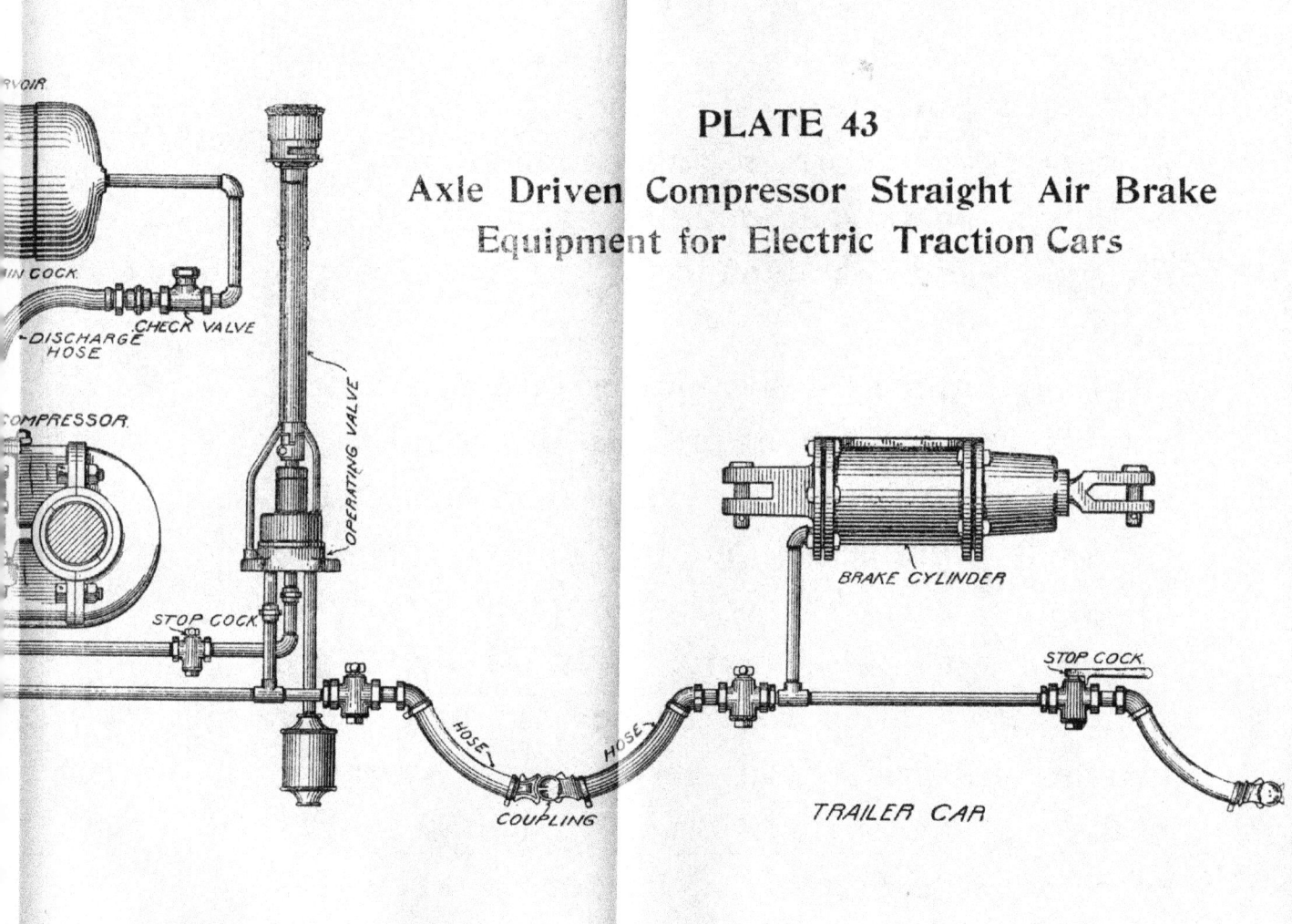

PLATE 43

Axle Driven Compressor Straight Air Brake Equipment for Electric Traction Cars

they travel is in many cases faster than ordinary steam railroad cars. Many steam railroads feel that they are doing very well if their freight schedules average twenty-five miles an hour, whereas nearly all interurban electric railways make an average schedule of at least thirty miles an hour. Such being the case, it is highly important that motormen and conductors master as fully as possible all detailed knowledge of the operation and maintenance of the brake apparatus.

The question of brake power on electric railways is quite a different proposition from that of steam railways, for the reason that with steam railroads there is but one kind of power brake in general use, which is the automatic air brake, whereas with the electric railways there are many kinds of power brakes in use. For instance, there is what is known as the hydraulic brake; and the magnetic brake; and of air brakes there are two systems in use, one is known as the automatic and the other the straight air brake.

Each kind of brake, of course, has its special advantages, but when all things are taken into account there is no brake so good and reliable for electric railways as an air brake, and as the

ITS USE AND ABUSE

majority of electric roads operate their cars singly, it naturally follows that the straight air brake is the best all-around brake that can be used.

While there are different systems of straight air brakes in use, some of which use what is known as the motor compressor, there is, however, a system now being universally installed, which is undoubtedly destined to supersede all others and is known as the *Standard Traction Brake Company's Axle Compressor System*, manufactured by the Westinghouse Air-Brake Company.

As the Westinghouse Air-Brake Company were the first to invent and operate the straight air brake, it naturally follows that they are in better position to bring the brake to a state of perfection than almost any one else, as in their immense plants in Wilmerding, Pa., they have every means at their command for doing so.

A student of straight air brakes as used on electric railways, should carefully read Section III of this book, as it will give him a general idea of brake handling and enable him to more readily grasp the points in which he is directly interested.

In order to understand the action of the

straight air brake, it is necessary to begin by studying the brake levers and cylinder under the body of the car.

By referring to Plate 82 you will see how the cylinder and levers are connected up. This diagram shows you how either the hand-brake staff or the air cylinder may be used to apply the brake. Where the hand-brake staff is used the levers are *pulled* forward, whereas when the brake is set by air the levers are *pushed* forward by the piston in the brake cylinder.

As the purpose of this book is to treat of compressed air only, I will describe the geared axle-driven compressor instead of the motor compressor, for the reason that with the axle-driven compressor no electricity is required to operate the compressor, or control the governor.

The parts necessary to complete the straight air-brake equipment on electric traction cars is as follows:

An air compressor, which is geared to and driven by the car axle, for the purpose of compressing the air used in the brake equipment.

An automatic regulator or governor, for the purpose of controlling the working of the compressor and regulating the amount of air to be compressed.

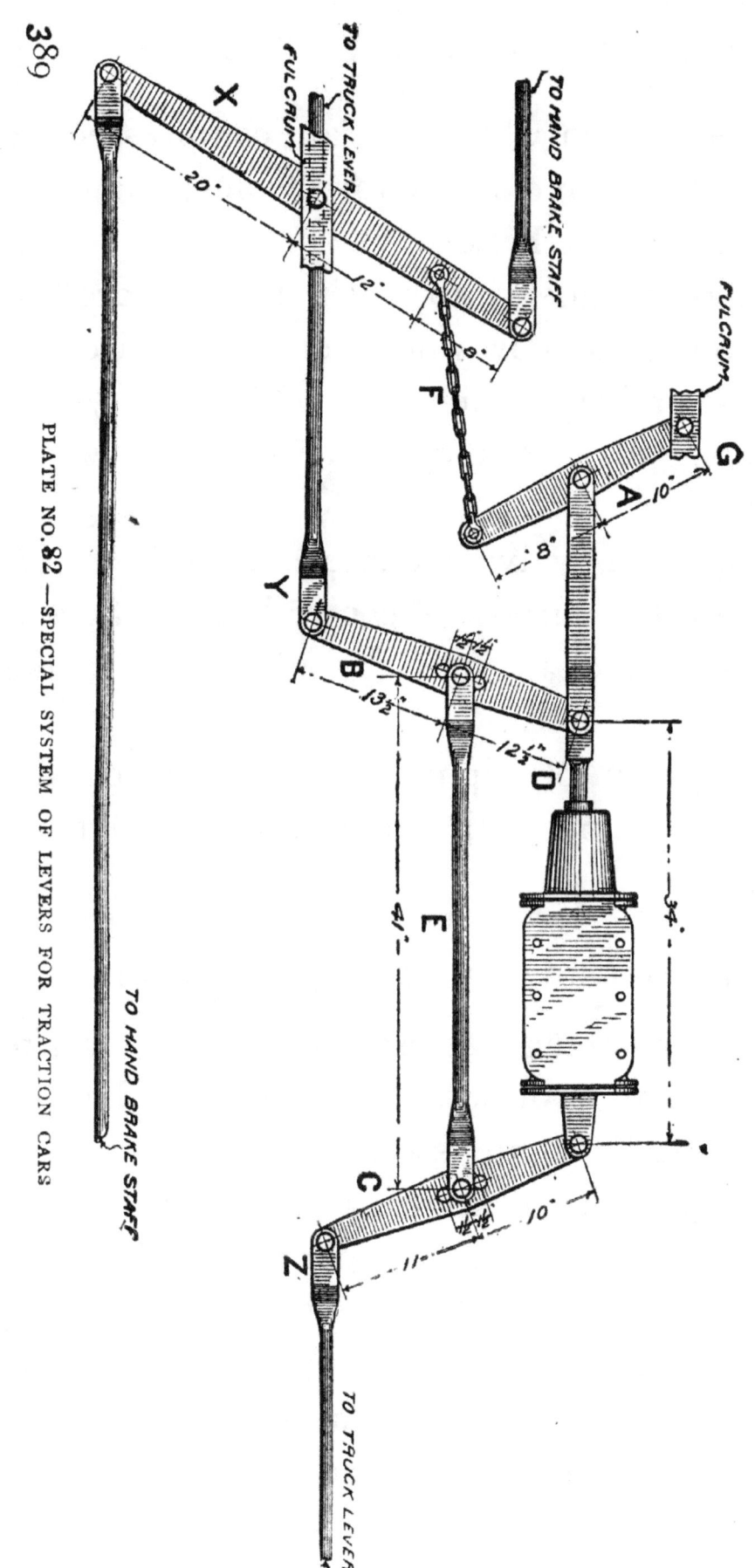

PLATE NO. 82 — SPECIAL SYSTEM OF LEVERS FOR TRACTION CARS

MODERN AIR-BRAKE PRACTICE

A reservoir, in which the air compressed by the pump is stored ready for the instantaneous application of the brakes.

A brake cylinder, into which the compressed air is allowed to flow whenever it is desired to apply the brakes, connected to which is a system of levers, rods and brake shoes, as shown on Plate 82.

An operating valve mounted at each end of the car for the purpose of controlling the flow of the compressed air into or out of the brake cylinder, as desired.

The system of piping with various cut-out cocks, etc., connects the above-mentioned parts together.

THE GEARED AXLE-DRIVEN COMPRESSOR

As the first thing necessary to a power brake is the generating of power, we will begin by considering the air compressor.

Plate 83 is what is known as a "ghost" cut, the white lines representing the parts through which you are looking, in order to get an internal view of the compressor. The compressor is double acting, and has horizontal axis of cylinder at right angles to the car axle; in other words, the car axle operates through the bearing, which, in

ITS USE AND ABUSE

cut 40, is shown to be empty. There are two discharge valves on the compressor and they are located on top of the cylinder, one at either end, and the discharge port is located midway between the discharge valves, as shown in Plate 83. Now, if you will look lower down, you will notice that by the side of the cylinder there are two other valves, one of which is shown in "ghost" outline; these two valves are known as suction valves, or, as we say in the automatic equipment, receiving valves. Leading from

PLATE NO. 83—AXLE-DRIVEN AIR COMPRESSOR

each of these valves are cylindrical chambers connected to each other by a passage that arches over the crank-shaft end-bearing, and between the two receiving valves is the receiving port, through which the atmospheric air is taken into the compressor.

These four valves, together with their renewable seats, are interchangeable, and have no springs to wear out, or gum up.

The cylindrical chambers beneath the receiving valves are connected by means of a fitting, either end of which is piped to the automatic regulator, or governor.

The compressor piston is made of a single casting in the form of two disks connected at the top and bottom. Each disk is provided with a spring packing ring, and carries on its inner side a rectangular surface parallel to the ends of the piston, thus combining a piston and slotted cross-head in one piece. The axis of the crank shaft, which runs parallel with the axle of the car, intersects the axis of the cylinder at its middle point; thus the crank shaft passes between the disks of the piston and imparts to them a reciprocating motion by means of the crank-brass slide between the parallel faces on the interior of the piston disks.

As the compressor operates very rapidly at times, it is necessary that all of these parts run in a *bath of oil*. On the other side of the cylin-

ITS USE AND ABUSE

der there is a flange by which it is bolted to the oil-tight housing that encloses the gear on the crank shaft, as well as the driving gear secured to the axle of the car. This housing is provided with bearings on the axle, which serve to keep the two gears meshed together; one end of the compressor is supported by these bearings and the other end is supported by brackets mounted upon it and the truck frame, respectively, with a rubber cushion between them to deaden the vibration. This type of compressor is especially adapted for mounting on the same axle with the car motor, as the axle gear and its bearings take up only a small space on the axle, and the balance of the housing and the pump cylinder occupy the space back of the motor. There are no stuffing boxes whatever in the entire equipment, and all parts that need lubrication are provided with oil wells and grease pockets. The manufacturers of the geared axle-driven compressor make seven different styles, in order to meet the varying conditions of the different kinds of service. But in all cases the diameter of the axle and pump gear is so proportioned that the piston speed never exceeds the safety limit. All of the compressors are made with a capacity double that required for a motor car with one trailer, running under the most severe conditions of service for which it is designed, so that a large proportion of the time the compressor is running with the pump automatically cut out of operation.

The automatic regulator or governor consists of a chamber which is always in direct communication with the reservoir, one wall of which is formed either by a diaphragm or piston, on one side of which is the reservoir pressure and on the other atmospheric pressure and a graduated spring. As the pressure in the reservor increases, the piston or diaphragm moves outwardly, which causes a D slide valve in the regulator chamber to move outwardly and uncover a port in the slide-valve seat, which admits compressed air to closed chambers, and when the pressure is changed the slide valve moves inwardly and connects this port to one leading to the atmosphere. The closed chambers just referred to are located in the body of the pump cylinder, directly beneath the receiving valves, and are connected by hose to the regulator. Each of the two chambers is provided with an air-tight piston, so that when the reservoir pressure reaches the desired amount the compressed air is admitted beneath them, thereby causing the

ITS USE AND ABUSE

receiving valves to be lifted, so that they are cut out of operation and the pump is thereby thrown out of action. When the reservoir pressure has fallen below the standard at which the governor is set, the D slide valve is forced to its inward position by the spring, thereby allowing the air to escape from trip-piston chambers, and as the pistons are forced down by springs provided for the purpose, the receiving valves seat themselves, and the pump is again in operation until it is cut out again by the regulator. This style of regulator is undoubtedly the best that can be devised, for the reason that regulators which allow the pressure to be discharged into the atmosphere when the pump is cut out usually leak very badly, and have the further disadvantage of keeping the pump valves in service all the time.

THE OPERATING VALVE

For the purpose of controlling the flow of the air either into or out of the brake cylinder, there are ordinarily two operating valves on each car, one at each end, and they are made in two forms. Form OVT, shown in Plate 84, has the valve proper placed upon the platform with the operating head directly above it at the level of the motorman's hand. On the top of this head is a double gauge, the red hand of which shows the reservoir pressure and the black hand the pressure in the brake cylinder. Owing to this convenient location the motorman cannot fail to know at all times the pressure in his reservoir and just how much brake power he is using. The gauge is protected by heavy plate glass and is in practically no danger of being broken. In the head, directly below the gauge, is a revolvable casting provided with a horizontal, cylindrical socket and a latch, so that when the handle is inserted in the socket it lifts the latch and the handle may be rotated, but when the handle is withdrawn the casting is automatically locked in its place. The shell of the head is made so that the handle can be inserted or removed only when it is in one particular position, usually that of lap position, when all ports in the valve are closed. This prevents any one tampering with the valve at the rear end of the car. The revolvable casting in the head is

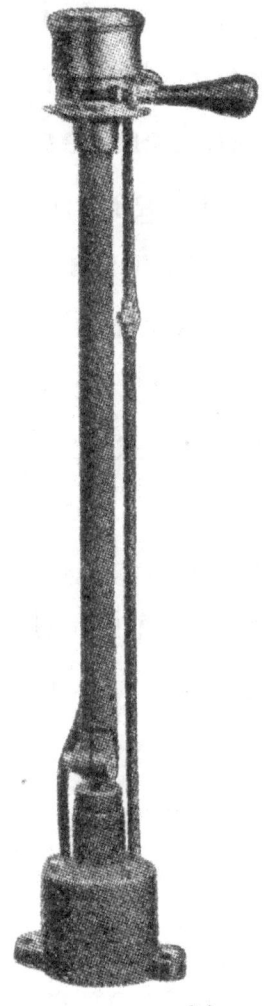

PLATE NO. 84—
OPERATING VALVE
O V T

ITS USE AND ABUSE

connected to the stem of the valve proper by means of a vertical shaft enclosed in a pipe shield and provided with a flexible coupling. This stem is provided with a pinion, which engages with a

PLATE NO. 85—OPERATING VALVE O V G

rack mounted on the slide valve, so that when the handle is moved the valve slides from side to side between its guides.

When it is inconvenient to place the valve on

the floor, type OVG operating valve, as shown in Plate 85, is used; with this valve the head is mounted upon and forms a part of the valve casing, but in all other respects it is identical with valve OVT.

Just below the slot in which the handle of these valves move is a little shelf, on which is clearly marked the different positions to which the brake-valve handle may be moved. There are four positions of the operating valve handle, viz., release, lap, service, and emergency. Release position is marked "off," meaning that the brakes are off or released. Lap position means that the brake cylinder is cut off from communication with both the atmosphere and the reservoir. Service position means that the valve is in position so that a small port is open for the purpose of allowing the reservoir pressure to be gradually admitted to the brake cylinder for making a service stop. Emergency position is used when the full brake power is required and when the handle is in this position the largest port in the brake valve is open, permitting the brakes to be applied instantly.

A very little practice will enable a motorman to handle his brake in such a manner as to avoid the shocking of passengers and the sliding of car

ITS USE AND ABUSE

wheels. Emergency position should never be used except in case of supposed or actual danger. But whenever it is required to use the emergency the handle of the operating valve should be immediately thrown to that position and held there until the car stops or the danger is past.

Plate 86, illustrates the arrangement of the straight air-brake equipment on a motor car with trailer car attached. As each part is plainly marked on the illustration, no further explanation is necessary to describe the plate.

It will be noticed that in this illustration there are two reservoirs; one is known as the preliminary reservoir and the other as the main reservoir. This system of having two reservoirs applies only where the geared axle-driven compressor is used, as when the motor-driven compressor is used only one reservoir is required.

With the equipment illustrated on Plate 86 the passage of the compressed air is as follows: from axle-driven compressor through the check valve to the preliminary reservoir, from the preliminary reservoir through trainpipe to the operating valves on either end of the car. And when the handle of the brake valve is moved to service or emergency position the air passes through the brake valve into the brake-cylinder pipe and from thence into the brake cylinder. When the air leaves the preliminary reservoir and while it is filling the trainpipe it is also passing to the automatic regulator or governor.

You will notice a duplex check valve on the end of the main reservoir. This is placed there for the purpose of allowing thirty-five pounds of air to be accumulated quickly in the preliminary reservoir in order to make a stop within a very short distance after the car first starts out. When the pressure rises to thirty-five pounds the duplex check valve is opened and the main reservoir is then filled along with the rest of the equipment up to the standard pressure at which the automatic regulator is set, which is 45 pounds.

In releasing the brake, the handle of the operating valve is thrown to the position marked "off," which allows the air in the brake cylinder to flow back through the brake valve and out to the atmosphere through the muffler beneath the car platform.

Again referring to Plate 82, I wish to call your attention to the small equalizing lever, marked GA. This is a patented modification of the well-known steam railroad leverage, which is especially adapted to the requirements of traction service. By this system of leverage greater

ITS USE AND ABUSE

brake power is obtained by the hand brake without the use of an excessively long cross lever. In the ordinary construction the point of attachment of the chain to the cross lever is located in line with the axis of the brake cylinder, and the pull is therefore directly at the end of the lever B, and to obtain proper hand-brake power on heavy cars the lever marked X, in Plate 82, would have to be longer than the width of the car. As this would be impracticable, instead of having a long lever X in order to get the proper hand-brake power, the lever A is introduced, thereby overcoming the trouble completely, besides giving a much better clearance between the pull rods and the wheels.

QUESTIONS AND ANSWERS TO SECTION 8

COVERING THE OPERATING AND MAINTENANCE OF THE STRAIGHT AIR-BRAKE EQUIPMENT ON ELECTRIC TRACTION CARS

1. What is the principal difference between the automatic air brake and the straight air brake?

Ans.—The automatic air brake requires the action of a triple valve in order to charge, set and release brakes, whereas with the straight air brake the operating of the brake valve regulates the flow of the air into and out of the brake cylinder.

2. As compressed air is the power by which the brakes are applied, what is it that compresses the air?

Ans.—A geared axle-driven air compressor.

3. Can you explain the operation of the axle-driven compressor?

Ans.—There is a gear attached to the axle of the car, which is meshed into a gearing to which is connected a crank shaft, which passes through the middle of the pump cylinder, in which are

ITS USE AND ABUSE

two pistons which are connected to the crank shaft in such a manner that as the shaft revolves it gives to the pistons a reciprocating movement, and as there are two receiving valves which permit the atmospheric pressure to enter the pump cylinder the motion of the pistons compresses the atmospheric air and forces it through two discharge valves into suitable reservoirs, in which the air is stored ready for use in applying the brakes.

4. What controls the action of the air compressor?

Ans.—An automatic regulator, or governor.

5. At what pressure should the regulator cut out the pump?

Ans.—At forty-five pounds.

6. How does the regulator control the action of the pump?

Ans.—When forty-five pounds has been accumulated in the reservoir the piston in the regulator is forced outwardly by the reservoir presure. As there is a D slide valve attached to this piston, it is moved so that the port in the valve seat is uncovered, which admits compressed air to the under side of the trip pistons below the receiving valves, causing them to unseat the receiving valves so that the pump cannot compress any more air. When the reservoir pressure falls slightly below forty-five pounds the graduating spring in the regulator forces the piston and slide valve inward, so that the air contained in the chamber below the trip pistons can escape to the atmosphere, and the pressure having thus left the under side of the trip pistons the spring on the opposite side forces the trip pistons down and allows the receiving valves to again seat themselves and thereby put the pump again into action.

7. At what pressure is the duplex check valve set between the preliminary and main reservoirs?

Ans.—At thirty-five pounds.

8. Why is it set at thirty-five pounds?

Ans.—In order that sufficient brake power may be accumulated in as short a run as only one hundred yards, and also to enable sufficient brake pressure to be maintained on the inter-urban cars when running at slow speed through cities.

9. What amount of pressure should there be in the **brake** cylinder in making a service stop?

Ans.—In making a service stop the brake cylinder should maintain from twenty-five to thirty

ITS USE AND ABUSE

pounds, as indicated by the black hand of the gauge.

10. What pressure should there be in the brake cylinder in making an emergency action?

Ans.—Forty pounds, which is also indicated by the black hand of the gauge.

11. In what position should the handle of the brake valve be carried in running along?

Ans.—If the brake valve is tight the handle should be carried on lap position, but if the valve leaks slightly the handle should be carried in release or at the position marked "off."

12. If the valve leaks slightly and the handle was carried on lap, what effect would it have?

Ans.—It would cause the brakes to gradually creep on.

13. What attention should be paid to the lubricant in the housing of the axle-driven compressor?

Ans.—The oil should never be allowed to get below the pump shaft.

14. How often should the oil be replenished?

Ans.—This depends upon the service that the car is in and the condition of the bearings on the axle, so that it is not possible to say just how often it will be necessary to replenish the oil. But when a car is first put into service, the cover should be removed from the gear housing at least once a week and enough grease be added to bring the level well above the pump shaft. By noting the amount found remaining in the housing each time, it can be readily seen if it needs grease oftener or if it will run for a longer period without replenishing.

15. What kind of a lubricant should be used?

Ans.—A grease about the consistency of vaseline. A very heavy West Virginia crude oil is the best for the cylinder, and it should be kept at the level of the crank shaft. It should be poured in through the opening on the top of the cylinder, or extension of the housing. It is very important that the lubrication be carefully looked after.

16. What other points should be looked after in maintaining the straight air-brake equipment?

Ans.—Other than attending to the proper supplying of lubricant there is little to do besides keeping the brake shoe-slack taken up and seeing that no nuts have become loosened; this latter inspection should be made at least once a day, if possible, and need take but a minute, as all nuts and bolts that can loosen are on the outside.

ITS USE AND ABUSE

17. How often should the compressor be taken off the axle and cleaned and examined thoroughly?

Ans.—This should be done at regular intervals of three months, if the car is in hard service. The bearings on the axle should then be replaced and the old ones re-babbitted for the next one. As this is practically the only place where oil can escape from the compressor it is necessary to keep these bearings close to the axle.

18. If the pressure cannot be raised in the reservoir, what should you do?

Ans.—Disconnect the discharge hose union, and while the car is running hold the hand over the opening, and if for each revolution of the axle there are two equally sharp spurts of air, the pump is all right, but should you not feel these sharp spurts of air, the discharge and receiving valves should be examined, as they may be stuck. A large leak is somewhat difficult to locate, as with the axle compressor the car must be in motion to do any pumping. For this reason roads having a large number of air-brake equipments should have a stationary compressor, either belt or motor driven, which with two reservoirs make a very convenient testing outfit. If the discharge and receiving valves are found to be all right, the lack of pressure may be caused by the air escaping through the operating valve, as dirt may have gotten between the valves on its seat. If this is found to be all right, the pipes should be examined to see if they have cracked anywhere, or if a fitting has broken.

19. If the compressor fails to pump, what should be done?

Ans.—Remove the fitting under the suction valves and see if the little trip pistons are free, if the suction or regulating pipes were not properly cleaned, dirt may cause one of the pistons to stick and hold the suction valve open. It is also possible to feel from below whether the valves are seating properly.

20. If one suction valve sticks and the other one does not, what is the effect on the pump?

Ans.—The pump will attain maximum pressure, but it will take twice as long to do it.

21. If one discharge valve sticks open, what effect will it have?

Ans.—The pump will only raise the pressure to about twenty pounds.

22. As the pump valves are all interchangeable, what precaution should be taken after cleaning them?

ITS USE AND ABUSE

Ans.—You must be sure to put them back in their old seats. Otherwise they are liable to leak, as no ground valves are interchangeable without re-grinding on the new seats.

23. Should the pump fail to cut out at the point at which the regulator is set, what should you do?

Ans.—Take down the trip fitting and see that the trip pistons are free; instances have occurred of a long trip-piston packing leather being caught between the trip fitting and cylinder body when bolting the fitting on.

24. When the compressor valves are all in good order and the operating valve is tight, what might cause the compressor to pump slowly?

Ans.—A kink in the suction hose by which it is doubled over on itself will cause the compressor to pump slowly, owing to the diminished supply passage.

25. In removing the cover of the housing to oil the compressor, what should you be particular to notice?

Ans.—That nothing is allowed to drop into the housing; the lodging of a stray bolt or nut between the gears will destroy the whole machine.

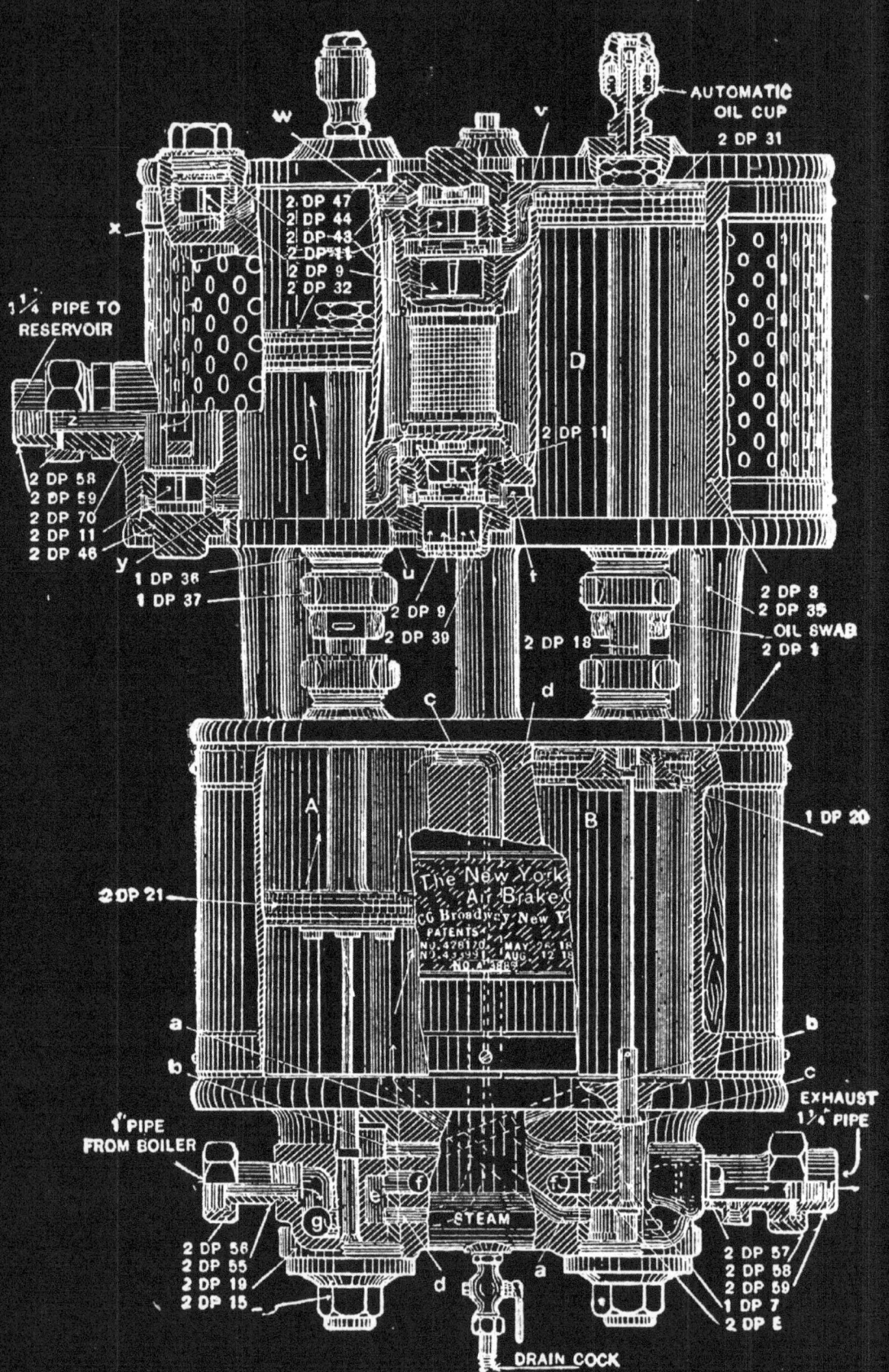

ET LOCOMOTIVE BRAKE EQUIPMENT.

The new locomotive equipment illustrated and described in this article is designated by the symbol ET. It differs materially from the present combined automatic and straight air brake in that it consists of considerably less apparatus. In operation it possesses all the advantages of the latter type of brake equipment and several other important ones which are necessary in modern locomotive brake service to produce satisfactory results.

The design of the principal valves comprising the ET equipment is such that it may be applied to any locomotive regardless of the service in which it is employed without change or modification in any of its parts; and the locomotive so equipped may be used in any kind of service, such as high speed passenger, double-pressure control, all ordinary passenger and freight, and in all kinds of switching service, without change or special adjustment of the brake apparatus. All principal valves are so designed that they may be removed for repairs and replacement without disturbing the pipe joints.

In operation its important advantages are: The locomotive brakes may be controlled with or independently of the train brakes and this without regard to the position of the locomotive in the train, whether coupled to another, as in double heading, or used as a helper and assigned to any position in the train.

MODERN AIR BRAKE PRACTICE

They may be applied with any desired **pressure** between the minimum and the maximum attainable, and this pressure will be automatically maintained in the locomotive brake cylinders regardless of leakage and variation in piston travel, undesirable though these defects are, until released by the brake valve.

They can be perfectly graduated on or off either **in** the automatic or in the independent application; hence, in all kinds of service the train may be handled without shock or danger of parting, and in passenger service especially smooth, accurate stops can be made with **greater ease than** was heretofore possible.

MANIPULATION.

The instructions for manipulating the ET equipment are practically the same as those given for the combined automatic and straight air brake; therefore, no radical departure from present methods of brake manipulation is required to get the desired results.

The **necessary** instructions are briefly as follows:

When not in use, carry the handles of both brake valves in running position.

To apply the locomotive and train brakes, move the handle of the automatic brake valve to the service position, making the required brake-pipe reduction, then back to lap position, which is the one for holding brakes applied.

To release the train brakes, move the handle to the release position and hold it there until all train brakes are **released**; then, move it to holding position, graduat-

ET BRAKE EQUIPMENT

ing off the locomotive brakes by short, successive movements between running and holding positions, aiming to have the locomotive brakes entirely released as the train stops.

To apply the brakes in an emergency, move the handle of the automatic brake valve quickly to emergency position and leave it there until the train stops or the danger is passed.

To make a smooth and accurate two-application passenger stop, make the first application sufficiently heavy to bring the speed of the train down to about 15 miles per hour at a convenient distance from the stopping point, then release train brakes by moving the handle to release position, then the locomotive brakes by moving it to running position for two or three seconds before re-applying. A little experience with the ET equipment will enable the engineer to make smooth and accurate stops with much greater ease than was heretofore possible.

When using the independent brake only, the handle of the automatic brake valve should be carried in running position. The independent application may be released by moving the independent-brake-valve handle to running position. Release position is for use when the automatic brake valve handle is not in running position.

While handling long trains of cars, in road or switching service, the independent brake should be operated with care and judgment, to prevent damage to the cars and lading, caused by running the slack in or out too hard. In cases of emergency arising while the independent brake is applied, apply the automatic brake instantly. The safety valve will restrict the brake cylinder pressure to the proper maximum. The brakes on the locomotive and on the train should be alternated in heavy grade service to prevent overheating of driving-wheel tires and to assist the pressure retaining valves in holding the train while the auxiliary reservoirs are being recharged.

After all brakes are applied automatically, to graduate off or entirely release the locomotive brakes only, use release position of the independent brake valve.

The cylinder gauge will show at all times the pressure in the locomotive brake cylinders, and this gauge should be observed in all brake manipulation.

Release Position of the *Independent Brake Valve* will release the locomotive brakes under any and all conditions.

The train brakes should invariably be released before detaching the locomotive, holding with hand brakes where necessary. This is especially important on a grade as there is otherwise no assurance that the car, cars, or train so detached will not start when the air brakes leak off, as they may in a short time where there is considerable leakage.

The automatic brakes should never be used to hold a standing locomotive or a train even where the locomotive is not detached, for longer than ten minutes, and not for such time if the grade is very steep or the condition of the brakes is not good. The safest method is to hold with hand brakes only and keep the auxiliary reservoirs fully charged so as to guard against a start from brakes leaking off, and to be ready to obtain any part of full braking power immediately on starting.

The independent brake is a very important safety feature in this connection, as it will hold a locomotive with a leaky throttle or quite a heavy train on a fairly

ET BRAKE EQUIPMENT

steep grade if, as the automatic brakes are released, the slack is prevented from running in or out, depending on the tendency of the grade, and giving the locomotive a start. Illustrating the best method by a descending train, apply the independent brake heavily as the stop is being completed, thus bunching the train solidly; then, when stopped, place and *leave* the handle of the independent brake valve in application position, release the automatic brakes and keep them charged. Should the train start through inability of the independent brakes to hold it, the automatic brakes will have been sufficiently recharged to make an immediate stop, and in which case enough hand brakes should be applied to render the necessary aid to the independent brakes.

Many runaways and some serious wrecks have resulted through failure to comply with the foregoing instructions.

When leaving the engine while doing work about it, or when it is standing at a coal chute or water plug, always leave the independent brake valve handle in application position.

In case the automatic brakes are applied by a bursted hose, a break-in-two or the use of a conductor's valve, place the handle of automatic brake valve in lap position.

Where there are two or more locomotives in a train, the double cut-out cock in the brake pipe under the automatic brake valve should be turned to close the brake pipe, and the automatic-brake-valve handle should be placed on lap on each except the one from which the brakes are being handled.

Before leaving the round house, the engineer should try the brakes with both brake valves, and see that no serious leaks exist. The pipes between the distributing valve and the brake valves should be absolutely tight.

PARTS OF THE EQUIPMENT.

1. THE AIR PUMP to compress the air.
2. THE MAIN RESERVOIRS, in which to store the air and collect water and dirt.
3. A DUPLEX PUMP GOVERNOR to control the pump when the pressures are attained for which it is regulated.
4. A DISTRIBUTING VALVE, and small double-chamber reservoir to which it is attached, placed on the locomotive to perform the functions of triple valves, auxiliary reservoirs, double check valve, high-speed reducing valves, etc.
5. TWO BRAKE VALVES, the AUTOMATIC to operate locomotive and train brakes, and the INDEPENDENT to operate locomotive brakes only.
6. A FEED VALVE to regulate the brake-pipe pressure.
7. A REDUCING VALVE to reduce the pressure for the independent brake valve, and for the air signal system when used.
8. TWO AIR GAUGES; one, a DUPLEX to indicate brake-pipe and main-reservoir pressures; the other, a SINGLE POINTER to indicate locomotive brake-cylinder pressure.
9. DRIVER, TENDER, and TRUCK-BRAKE CYLINDERS, CUT-OUT COCKS, AIR STRAINERS, HOSE COUPLINGS, FITTINGS, etc., incidental to the piping, for purposes readily understood.

ET BRAKE EQUIPMENT

The piping hereafter referred to is named as follows:

RESERVOIR PIPE: Connects the main reservoir to the Automatic Brake Valve, Distributing Valve, Feed Valve, and Reducing Valve.

FEED-VALVE PIPE: Connects the Feed Valve to the Automatic Brake Valve.

REDUCING-VALVE PIPE: Connects the Reducing Valve to the Independent Brake Valve, and to the Signal System, when used.

BRAKE PIPE: Connects the Automatic Brake Valve with the Distributing Valve and all Triple Valves on the cars in the train.

BRAKE-CYLINDER PIPE: Connects the Distributing Valve with the Driver, Tender and Truck-Brake Cylinders.

APPLICATION-CHAMBER PIPE: Connects the Application Chamber of the Distributing Valve to the Automatic Brake Valve through the Independent Brake Valve.

DOUBLE-HEADING PIPE: Connects the Application Chamber exhaust port of the Distributing Valve to the Automatic Brake Valve through the Double Cut-Out Cock.

ARRANGEMENT OF APPARATUS.

A piping diagram of the ET equipment is shown in Fig. 266

Air compressed by the pump passes as usual to the main reservoirs and the reservoir pipe. The main-reservoir cut-out cock is to cut off the supply of air when removing any of the apparatus except the governor. The end toward the main reservoir is tapped for a connection

MODERN AIR BRAKE PRACTICE

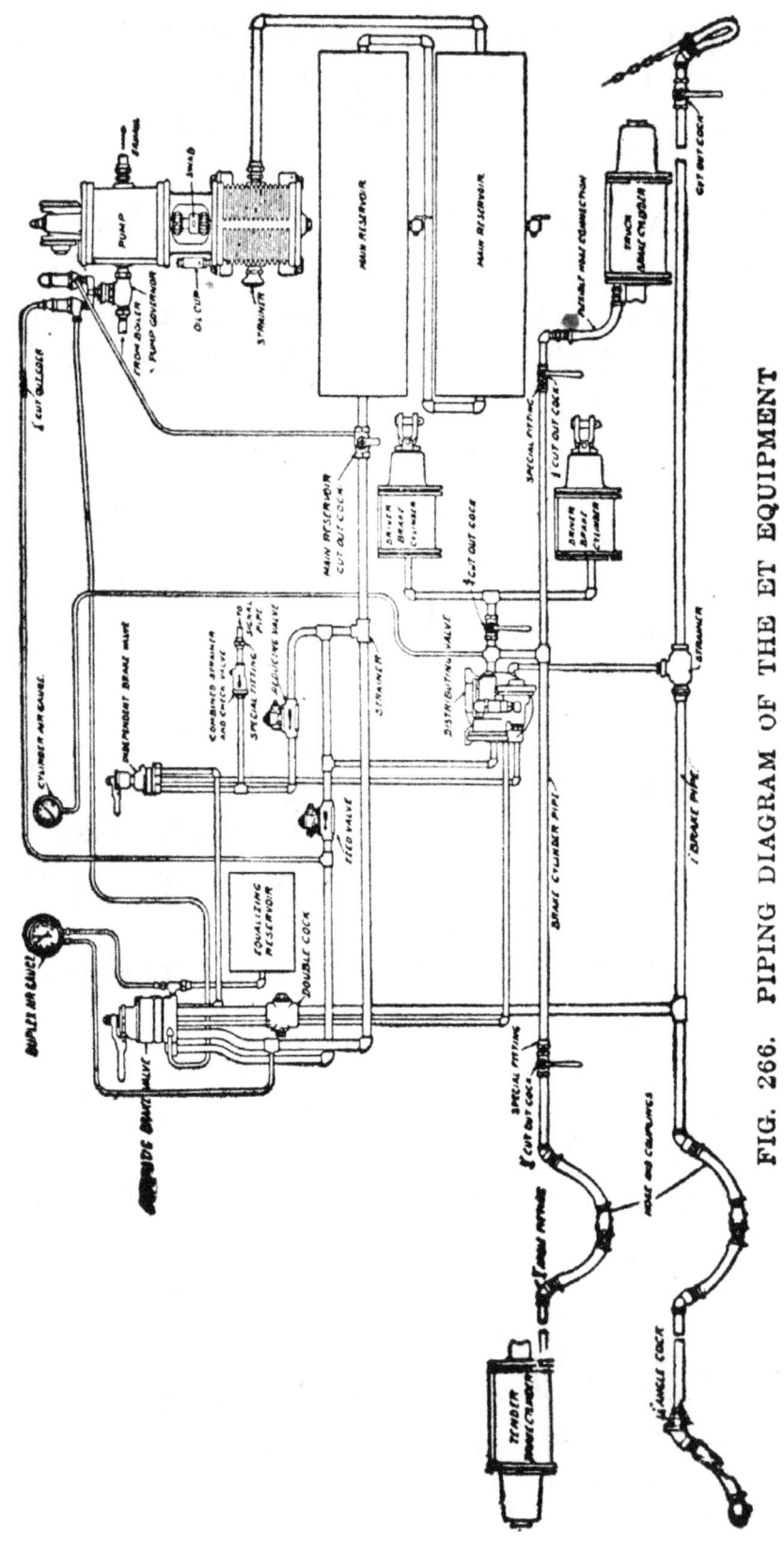

FIG. 266. PIPING DIAGRAM OF THE ET EQUIPMENT

ET BRAKE EQUIPMENT

to the maximum pressure head of the Pump Governor. When closed it discharges the air from the pipe between it and the automatic brake valve.

Beyond the main-reservoir cut-out cock, the reservoir pipe has four branches, one of which runs to the automatic brake valve, one to the feed valve, one to the reducing valve, and one to the distributing valve. As a result, the automatic brake valve receives air from the main reservoir in two ways, one direct and the other through the Feed Valve.

The Feed-Valve Pipe from the feed valve to the automatic brake valve has a branch to the top of the excess-pressure head of the duplex pump governor.

The third branch of the reservoir pipe connects with the reducing valve. Air at the pressure for which this valve is set (45 pounds) is thus supplied to the independent brake valve through the reducing-valve pipe. When the signal system is installed, it is connected to the reducing valve pipe, in which case the reducing valve takes the place of the signal reducing valve usually employed to supply the train air-signal system. In the branch pipe supplying the signal are placed a combined strainer and check-valve, and a special choke fitting. The former prevents any dirt from reaching the check valve and choke plug. The check valve prevents air from flowing back from the signal pipe when the independent brake is applied. The choke plug prevents the reducing valve from raising the signal-pipe pressure so quickly as to destroy the operation of the signal.

The distributing valve has five pipe connections, made through the double-chamber reservoir, three on the left and two on the right. Of the three on the left, the upper is the supply from the main reservoir; the intermediate is the double-heading pipe, leading through the double cut-out cock, when turned to cut out the brake valve from the brake pipe, to the automatic brake valve; and the lower is the application-chamber pipe, leading through the independent-brake valve, when the handle is in running position, to the automatic brake valve. Of the two on the right, the lower is the brake-pipe-branch connection, and the upper is the brake-cylinder pipe branching to all brake cylinders on the engine and tender. In this pipe are placed cocks for cutting out the brake cylinders when necessary, and in the engine truck and tender brake cylinder cut-out cocks are placed special choke fittings to prevent serious loss of main-reservoir air and the release of the other locomotive brakes during a stop, in case of burst brake cylinder hose connection. The cylinder gauge is connected with the brake cylinder pipe.

The automatic-brake-valve pipe connections, other than already mentioned, are the brake-pipe branch through the double cut-out cock, the main-reservoir, the equalizing reservoir, the duplex gauge and the lower connection to the excess-pressure head of the pump governor.

THE DISTRIBUTING VALVE.

This valve is the important feature of the ET equipment. Fig. 267 is a photographic view of the left side of the valve and its double-chamber reservoir. The three pipe connections, as previously referred to, are plainly shown. Fig. 268 is a similar view of the right side, showing the pipe connections there and the two chambers of the reservoir; also the safety valve, 34, which is an essential part of the distributing valve. To simplify the tracing of the ports and connections, the various positions of this valve are illustrated in ten diagrammatic drawings; that is, the valve is distorted to show the parts differently than actually constructed with the object of explaining the operation clearly instead of showing exactly how they

FIG. 267. DISTRIBUTING VALVE AND DOUBLE-CHAMBER RESERVOIR

CONNECTIONS:
SUP—Main-Reservoir Pipe; ABV—Double-Heading Pipe; SBV—Application-Chamber Pipe

are designed. The chambers of the reservoir are for convenience indicated at the bottom as a portion of the valve itself. In Fig. 279, equalizing piston 26, graduating valve 28, and equalizing slide valve 31, are shown as actually constructed. But as there are ports in the valves which cannot thus be clearly indicated, the diagrammatic illustrations show each slide valve in two parts, one below and the other above the piston stem, with similar division of parts in the bush.

FIG. 268. DISTRIBUTING VALVE AND DOUBLE-CHAMBER RESERVOIR
PIPE CONNECTIONS:
Upper—Brake-Cylinder Pipe; Lower—Brake Pipe

ET BRAKE EQUIPMENT

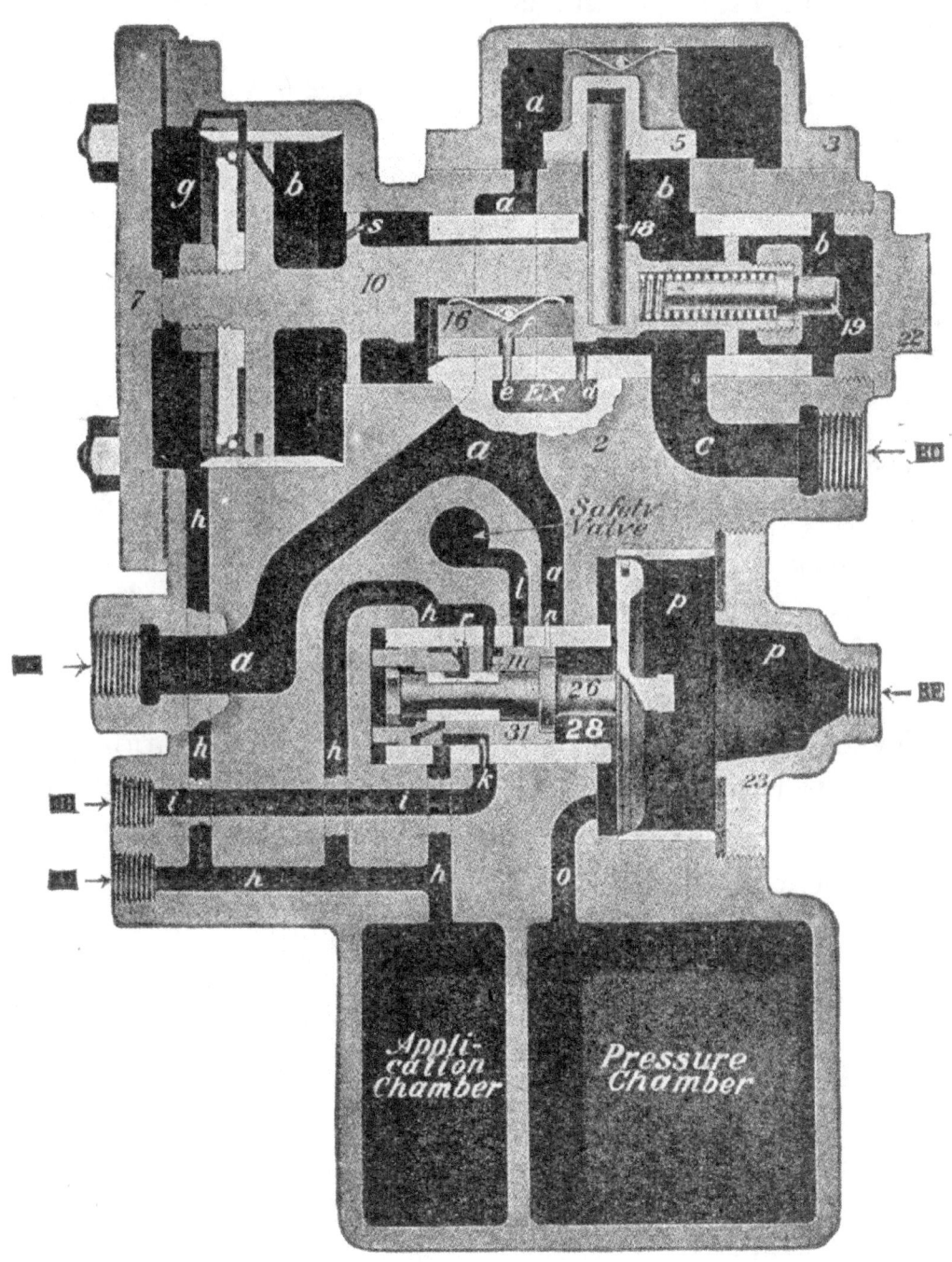

FIG. 269. THE DISTRIBUTING VALVE, DIAGRAMMATIC
CONNECTIONS:
MR—Main-Reservoir Pipe; DH—Double-Heading Pipe; AC—Application-Chamber Pipe; BC—Brake Cylinder Pipe; BP—Brake Pipe

Fig. 269 shows the operative parts in the same position as in Fig. 270 and is used merely for the sake of greater clearness. Referring to these figures it will be seen that main-reservoir pressure is always present in the chamber surrounding application valve 5 by its connection through passage $a, a,$ to the main-reservoir pipe. Chambers b to the right of application piston 10 are always in free communication with the brake cylinder through passage c and brake-cylinder pipe. Chamber g at the left of application piston 10 is a portion of the application chamber, being always connected with it by passage h, and is also connected to the brake valves through the application-chamber pipe.

INDEPENDENT APPLICATION. When the handle of the Independent Brake Valve is moved to the application position, air from the main reservoir, limited by the reducing valve to a maximum of 45 pounds, is allowed to flow to the application chamber, forcing application piston 10 to the right as shown in Fig. 271. We will assume that 45 pounds is so admitted and maintained. This movement of application piston 10 causes exhaust valve 16 to close exhaust ports e and d, and the graduating stem 19 to compress its spring; also open application valve 5 by its connection with the piston stem by pin 18. Main reservoir air then flows through port b and passage c to the brake cylinders until their pressure and that in chamber b equals the application-chamber pressure, in this case 45 pounds. The graduating spring then forces the application piston 10 to the left until application valve 5 closes port b, but without moving exhaust valve 16. This position shown in Fig. 272, is known as INDEPENDENT LAP.

ET BRAKE EQUIPMENT

From the above description it will be seen that application piston 10 has application chamber pressure on

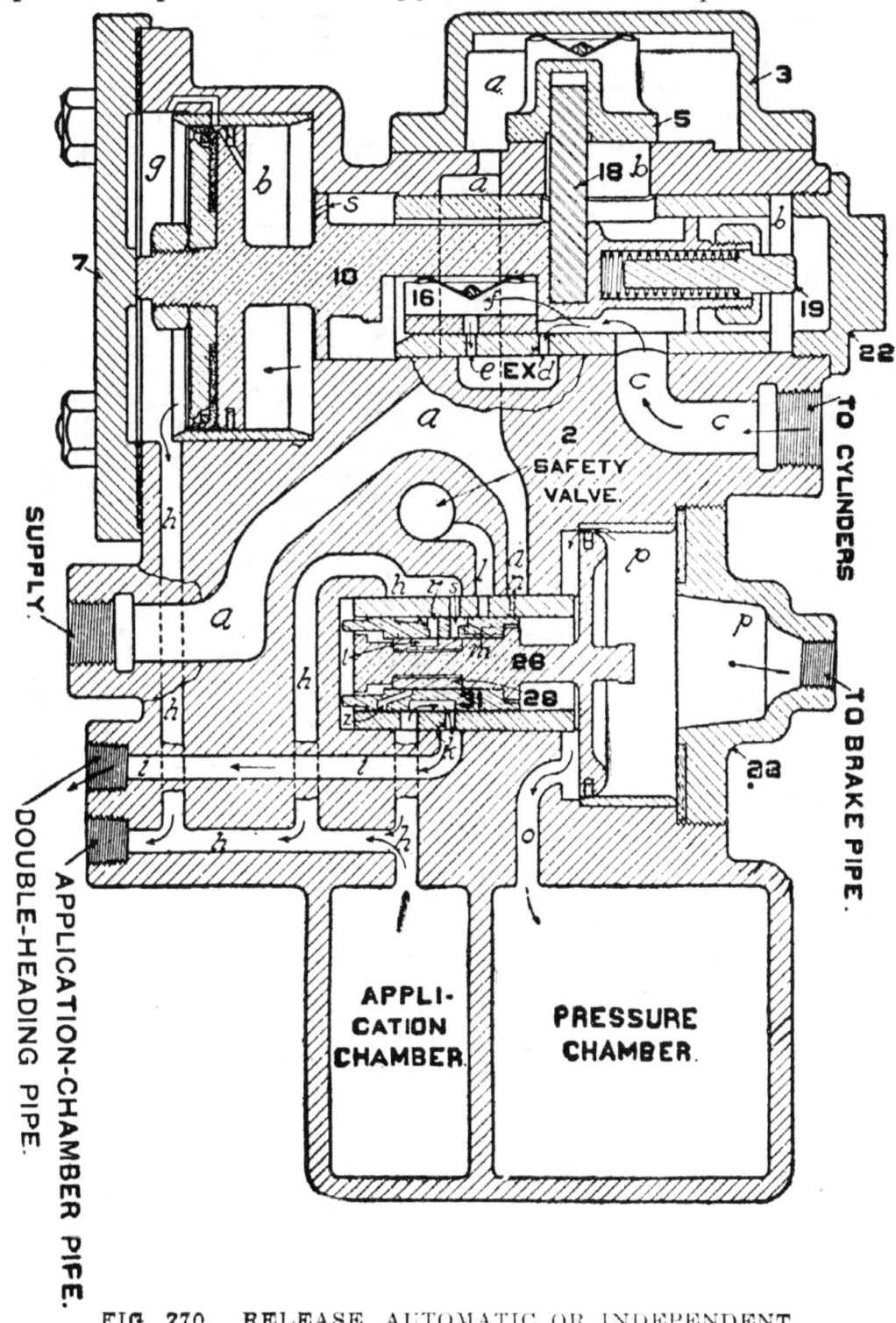

FIG. 270. RELEASE, AUTOMATIC OR INDEPENDENT

one side and brake-cylinder pressure on the other. When either pressure varies, the piston will move toward the

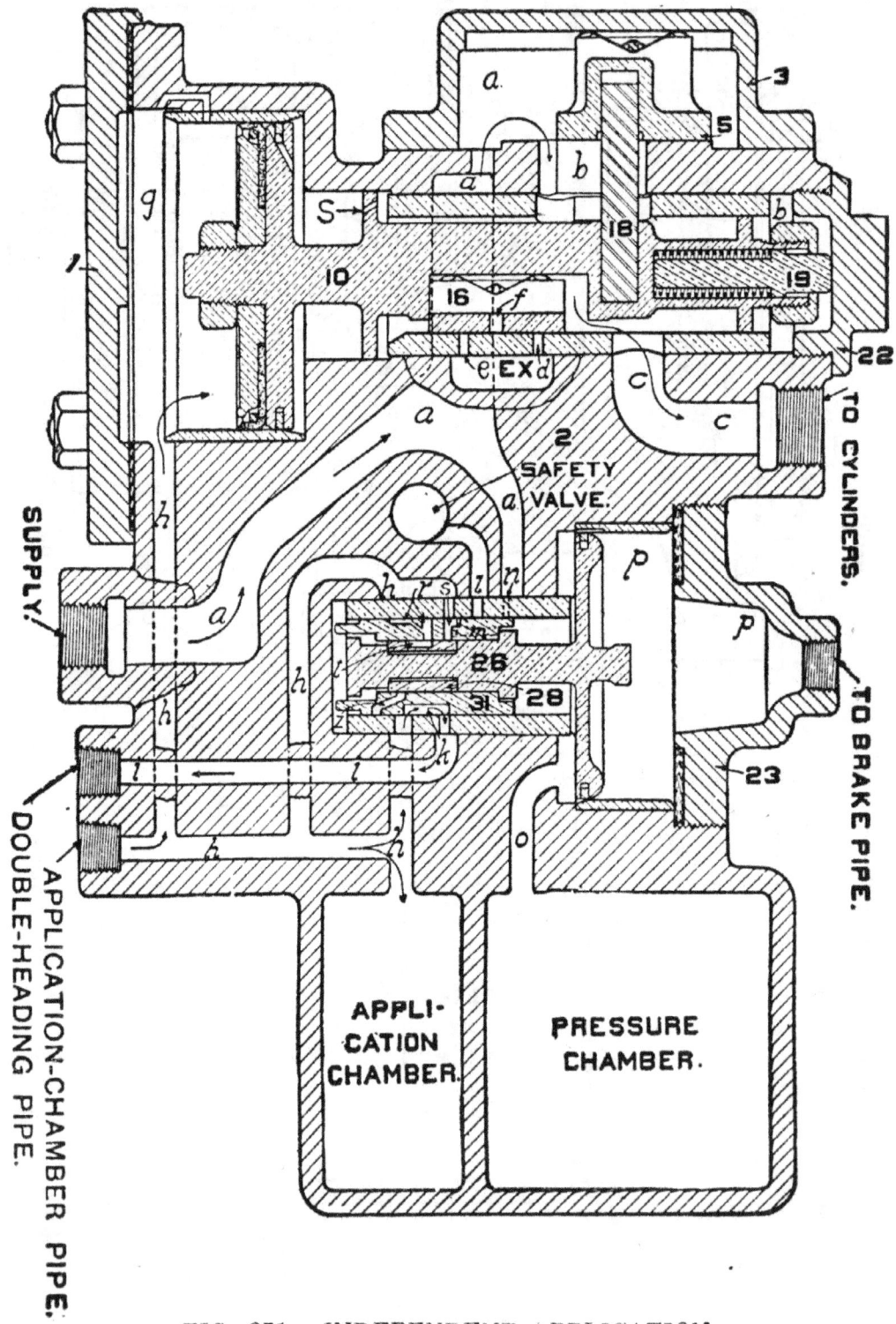

FIG. 271. INDEPENDENT APPLICATION

ET BRAKE EQUIPMENT

lower. Consequently if that in chamber b is reduced, by brake-cylinder leakage, the pressure maintained in the ap-

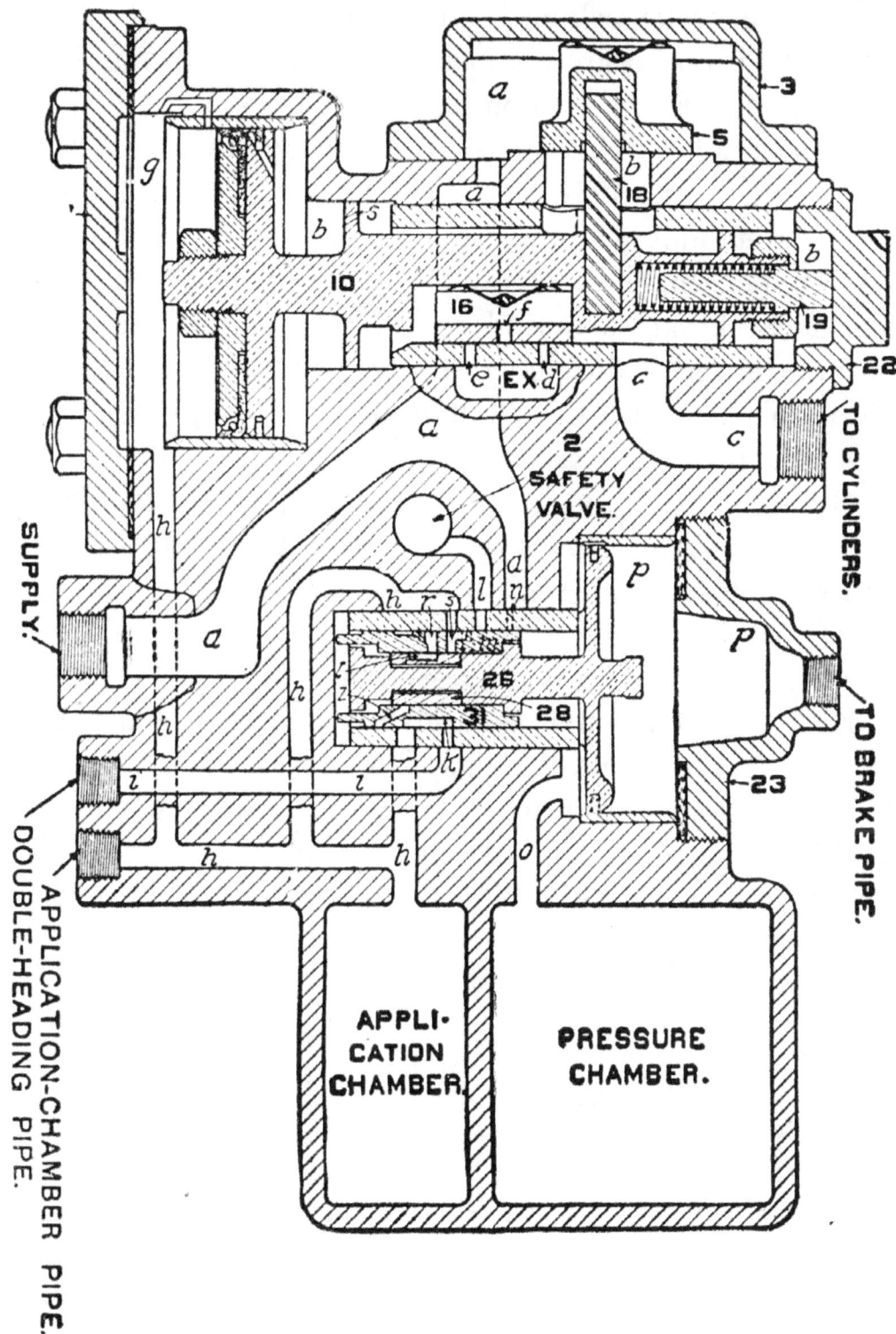

FIG. 272. INDEPENDENT LAP

plication chamber will force piston 10 to the right, opening application valve 5 and again admitting main reservoir air to the brake cylinders until the pressures on both sides of piston 10 are again equal, when the graduating spring will force the piston back to lap position. In this way the brake-cylinder pressure is always maintained to that in the application chamber. This is called the pressure maintaining feature.

INDEPENDENT RELEASE. When the handle of the independent brake is moved to release position, a direct opening is made through the rotary valve from the application chamber to the atmosphere. This permits the pressure in the application chamber to escape; therefore, as this pressure is being exhausted, brake-cylinder pressure in chamber *b* moves application piston 10 to the left, causing exhaust valve 16 to open exhaust ports *e* and *d* as shown in Fig. 270, thereby allowing brake-cylinder pressure to escape to the atmosphere.

If the independent brake valve is returned to lap, before all of the application-chamber pressure has escaped, the application piston 10 will return to independent lap position as soon as the brake-cylinder pressure is reduced a little below that remaining in the application chamber.

AUTOMATIC OPERATION.

During automatic operation of the brakes, the lower movable parts, known as the equalizing parts, are brought into action.

AUTOMATIC RELEASE. Referring to Fig. 270, which shows the movable parts of the valve in the release position, it will be seen that as chamber *p* is connected to

ET BRAKE EQUIPMENT

the brake pipe, brake-pipe air flows through the feed groove around the top of piston 26 into the chamber

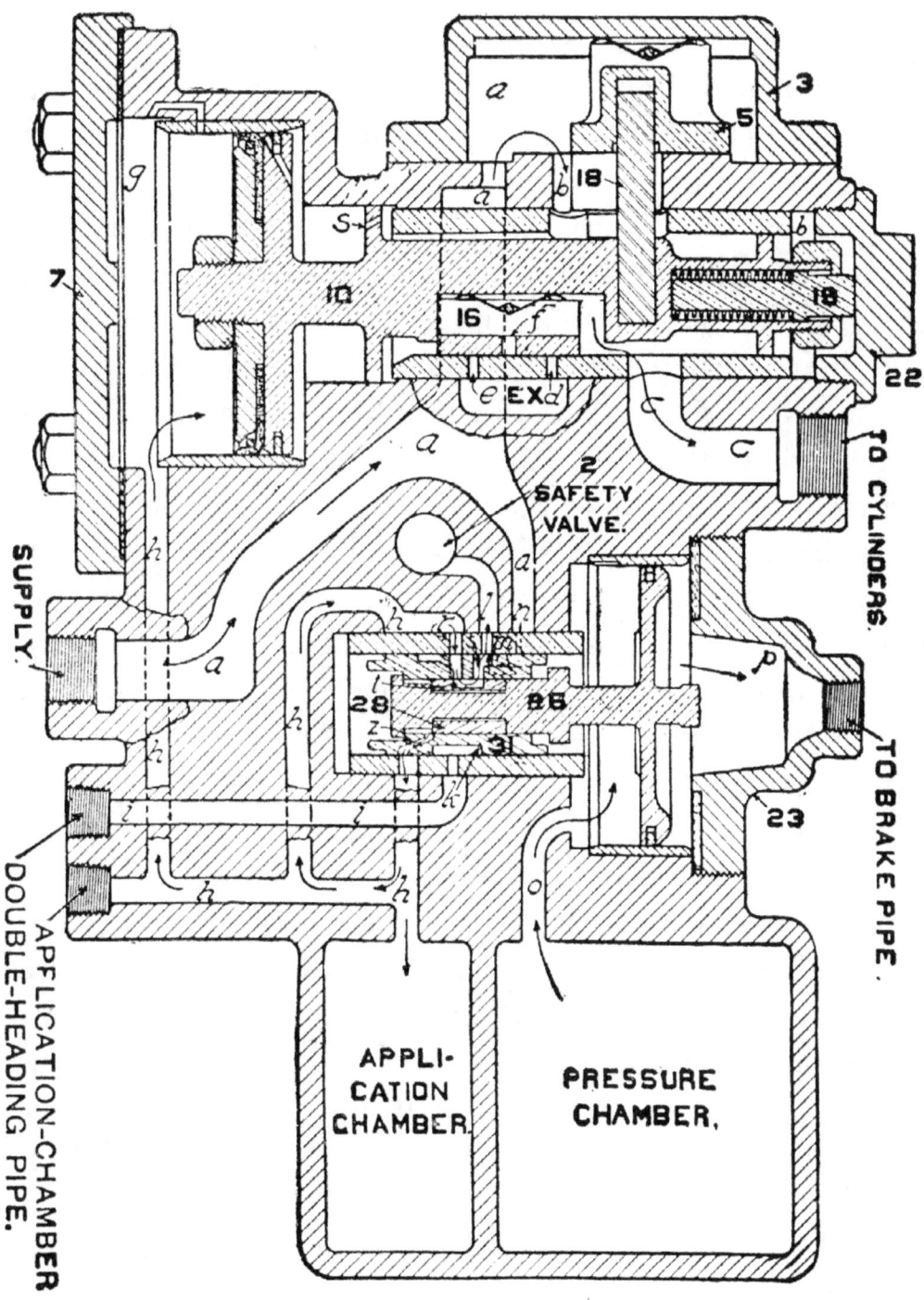

FIG. 273. AUTOMATIC SERVICE

above the slide valve 31, and through port *o* to the *pressure chamber,* until the pressures on both sides of the piston are equal.

SERVICE. When a service application is made with the automatic brake valve, the brake-pipe pressure in chamber *p* is reduced, causing a difference in pressure on the two sides of this piston, which results in the piston moving toward the right. The first movement of the piston closes the feed groove, and at the same time moves the graduating valve until it uncovers the upper end of port *z* in the equalizing slide valve 31. As the piston continues its movement, the shoulder on the end of its stem engages the slide valve, which is then also moved to the right until port *z* in the slide valve registers with port *h* in the seat. As the slide valve chamber is always in communication with the pressure chamber, air can now flow from it to the application chamber. This pressure forces application piston 10 to the right, as shown in Fig. 273, causing application valve 5 to uncover port *b* and allow main reservoir air to flow to the brake cylinders through port *c,* as in an independent application.

During the movement just described, cavity *t* in the graduating valve connects ports *r* and *s* in the equalizing slide valve, and by the same movement ports *r* and *s* are brought into register with ports *h* and *l* in the seat, thus establishing a communication from the application chamber to the safety-valve, which being set at 53 pounds, limits the brake-cylinder pressure to this amount during a full service application.

The amount of pressure resulting in the application chamber for a certain brake-pipe reduction, depends on the comparative volumes of the application and pressure

ET BRAKE EQUIPMENT

chambers. These volumes are such that with 70 pounds in the pressure chamber and nothing in the application

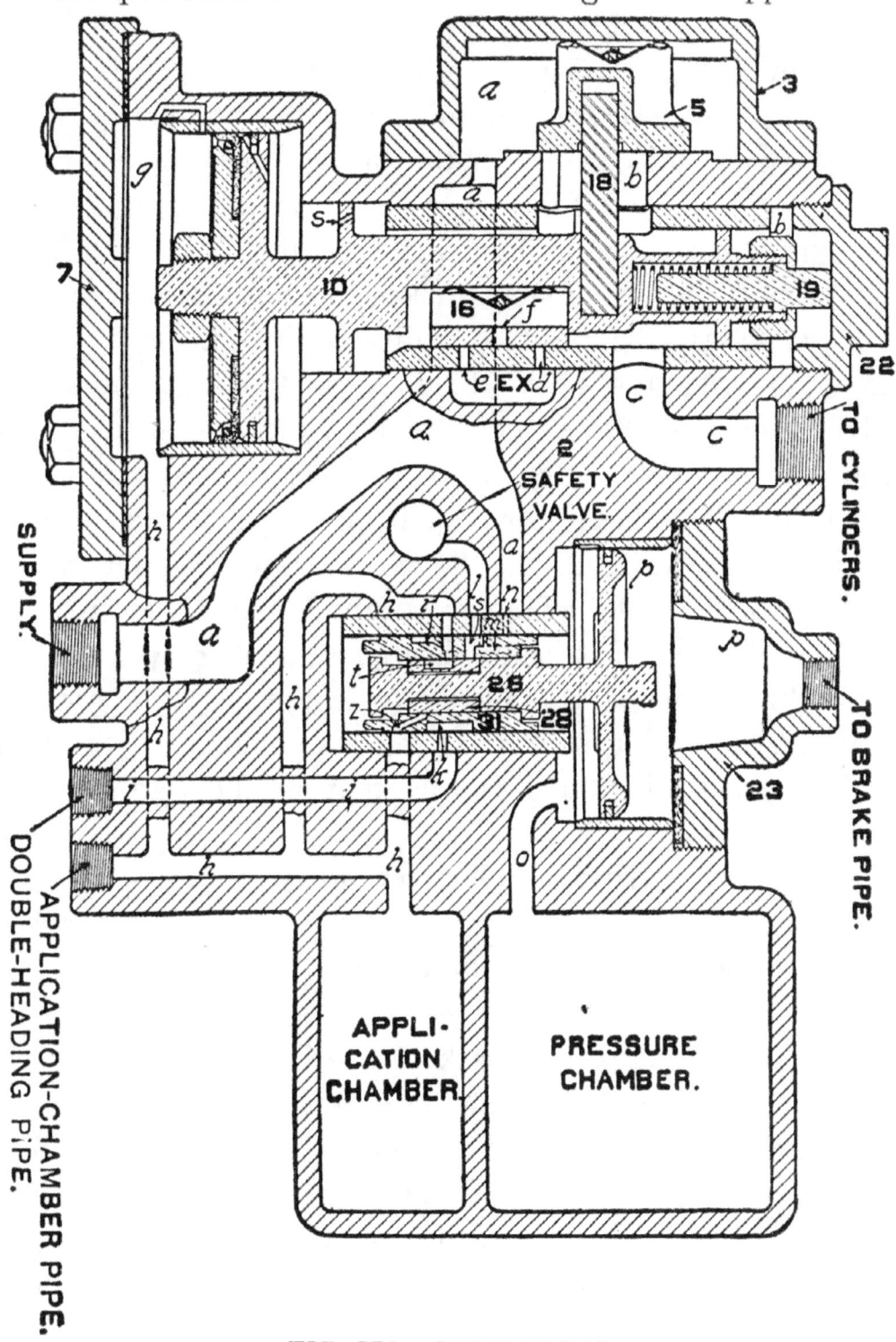

FIG. 274. SERVICE LAP

chamber, if they are allowed to remain connected by the ports in the slide valve, they will equalize at about 50 pounds.

SERVICE LAP. The conditions just described continue until the pressure in the *pressure chamber* is reduced enough below that in the brake pipe to cause the difference in pressure on the two sides of piston 26 to force it and graduating valve 28 to the left until the shoulder on the piston stem strikes the right-hand end of slide valve 31, the position indicated in Fig. 274, and known as SERVICE LAP. In this position, graduating valve 28 has closed port z so that no more air can flow from the *pressure chamber* to the application chamber; and it also has closed port s, cutting off communication to the safety valve. The flow of air past application valve 5 to the brake cylinders continues until their pressure equals that in the application chamber when the graduating spring forces piston 10 to the position shown in Fig. 274, closing port b. The brake-cylinder pressure is then practically the same as that in the *application chamber*.

It will be seen that whatever pressure exists in the *application chamber* will be maintained in the brake cylinder by the "pressure maintaining" feature already described.

When the automatic brake valve is placed in release position, and the brake-pipe pressure in chamber p is increased above that in the pressure chamber, equalizing piston 26 moves to the left, carrying with it equalizing slide valve 31 and graduating valve 28 to the release position as shown in Fig. 270. The feed groove now being open permits the pressure in the pressure chamber

ET BRAKE EQUIPMENT

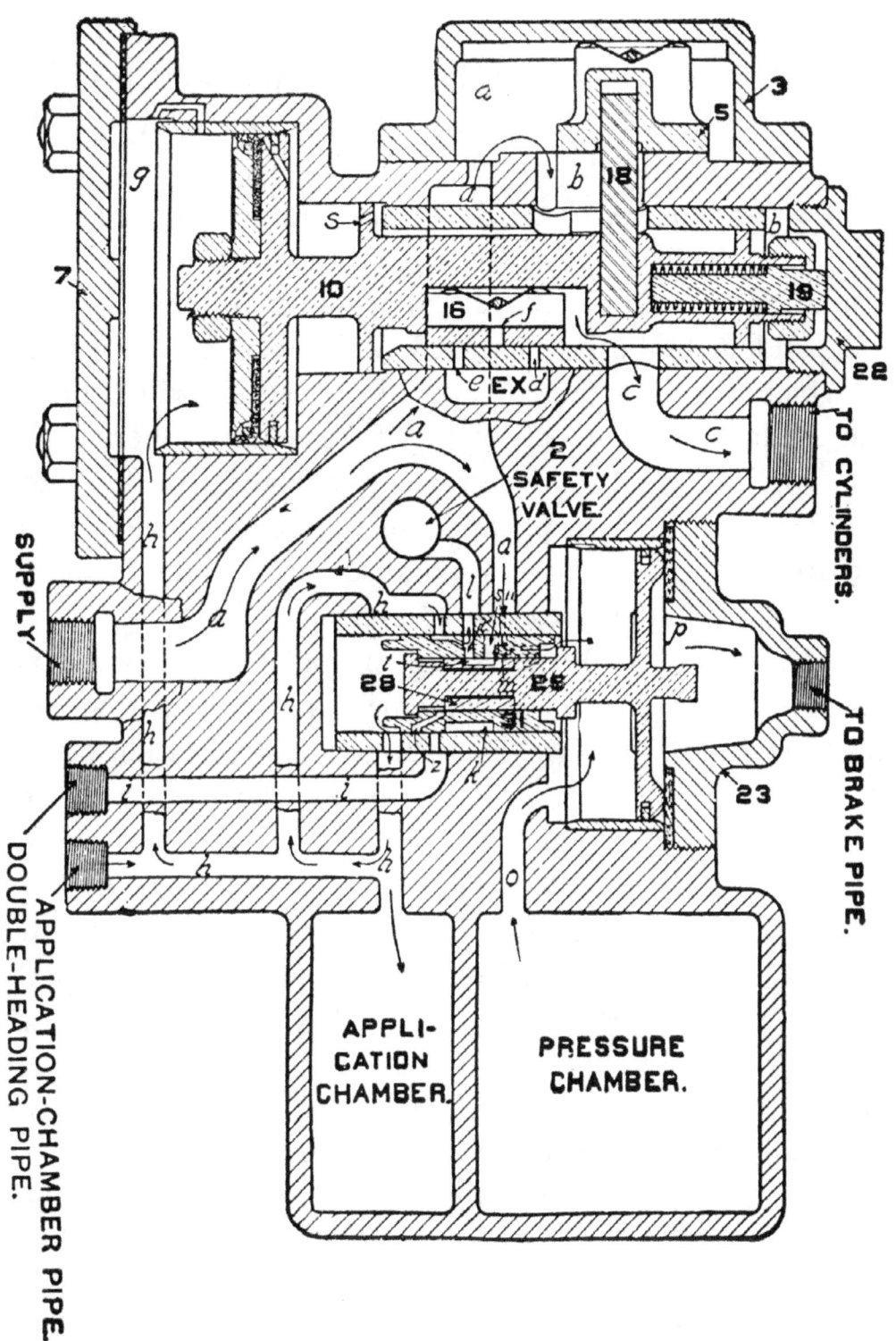

FIG. 275. EMERGENCY

to equalize with that in the brake pipe as before described. This action does not release the locomotive brakes because it does not discharge application chamber pressure. The double-heading pipe is closed at the double cut-out cock, and the application chamber pipe is closed by the rotary valve of the automatic brake valve. Therefore, to release the locomotive brakes, the automatic brake valve must be moved to running position, or the independent brake valve must be held in release position, in which positions the rotary valve of either will connect the application chamber pipe with the atmosphere. As the application chamber pressure escapes, the cylinder pressure will force application piston 10 to the left until exhaust valve 16 uncovers exhaust ports d and e, allowing brake-cylinder pressure to escape.

EMERGENCY. When a sudden and heavy brake-pipe reduction is made, as in an emergency application, the pressure in the pressure chamber forces application piston 26 to the right until it strikes against the leather gasket beneath head 23 as shown in Fig. 275. This movement causes slide valve 31 to uncover port h in the bush, making a large opening from the *pressure chamber* to the *application chamber,* so that they quickly become equalized. In the emergency position of the automatic brake valve, the volume of the equalizing reservoir is connected to that of the application chamber. This reservoir volume together with that of the pressure chamber at 70 pounds pressure equalizes into the application chamber at about 60 pounds. The dotted port m in the slide valve registers with port n in the seat connecting with supply passage a, allowing air from the main reservoir to enter the slide valve and application chambers. A cavity in

ET BRAKE EQUIPMENT

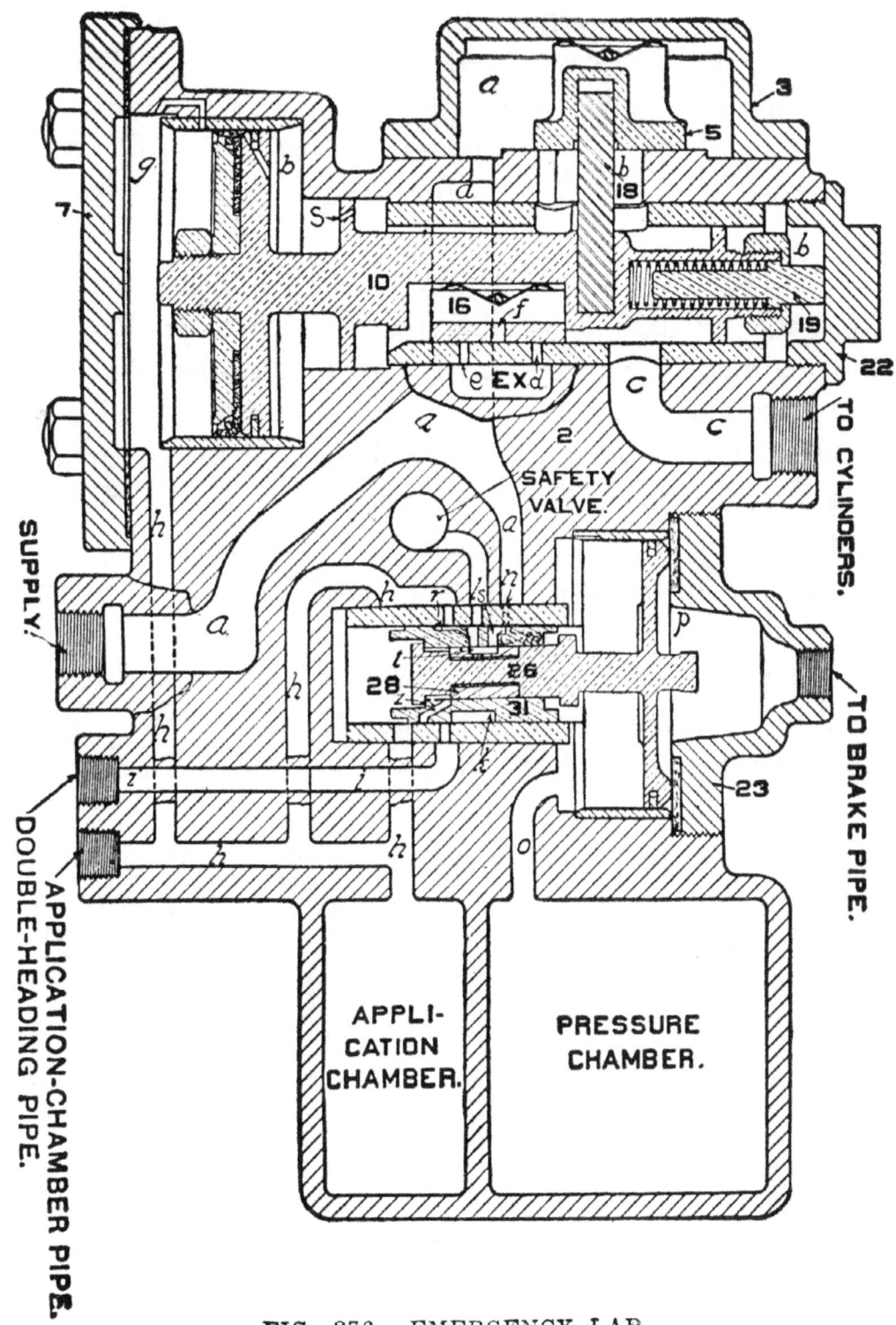

FIG. 276. EMERGENCY LAP

the slide valve registers with port h in the seat. Port r in the slide valve registers with port l leading to the safety valve. The cavity and port r in the slide valve are connected by a small port, the size of which permits the air in the application chambers to escape a little faster than ports m and n can supply it, preventing the pressure from raising above the amount desired.

In High-Speed Brake Service, the feed valve is regulated for 110 pounds brake-pipe pressure instead of 70, and main-reservoir pressure is 130 or 140 pounds. Under these conditions an emergency application raises the *application chamber* pressure to about 85 pounds, but the area of the small passage to port r is so proportioned that the flow of *application-chamber* pressure to the safety valve is just enough greater than the supply through m, to decrease that pressure in practically the same time and manner as is done by the high-speed reducing valve, until it is approximately 60 pounds. The application portion operates similarly to, but more quickly than, in the service application.

EMERGENCY LAP. The above conditions continue until the brake cylinder pressure equals the *application-chamber* pressure, when parts of the valve assume the position known as the *Emergency Lap* and shown in Fig. 276.

The release after an emergency is the same as that following service applications.

Fig. 277 shows the position the distributing valve parts will assume, if the application-chamber pressure is discharged by the independent brake valve during an automatic application. This results in the upper movable portion going to the release position and relieving brake-

ET BRAKE EQUIPMENT

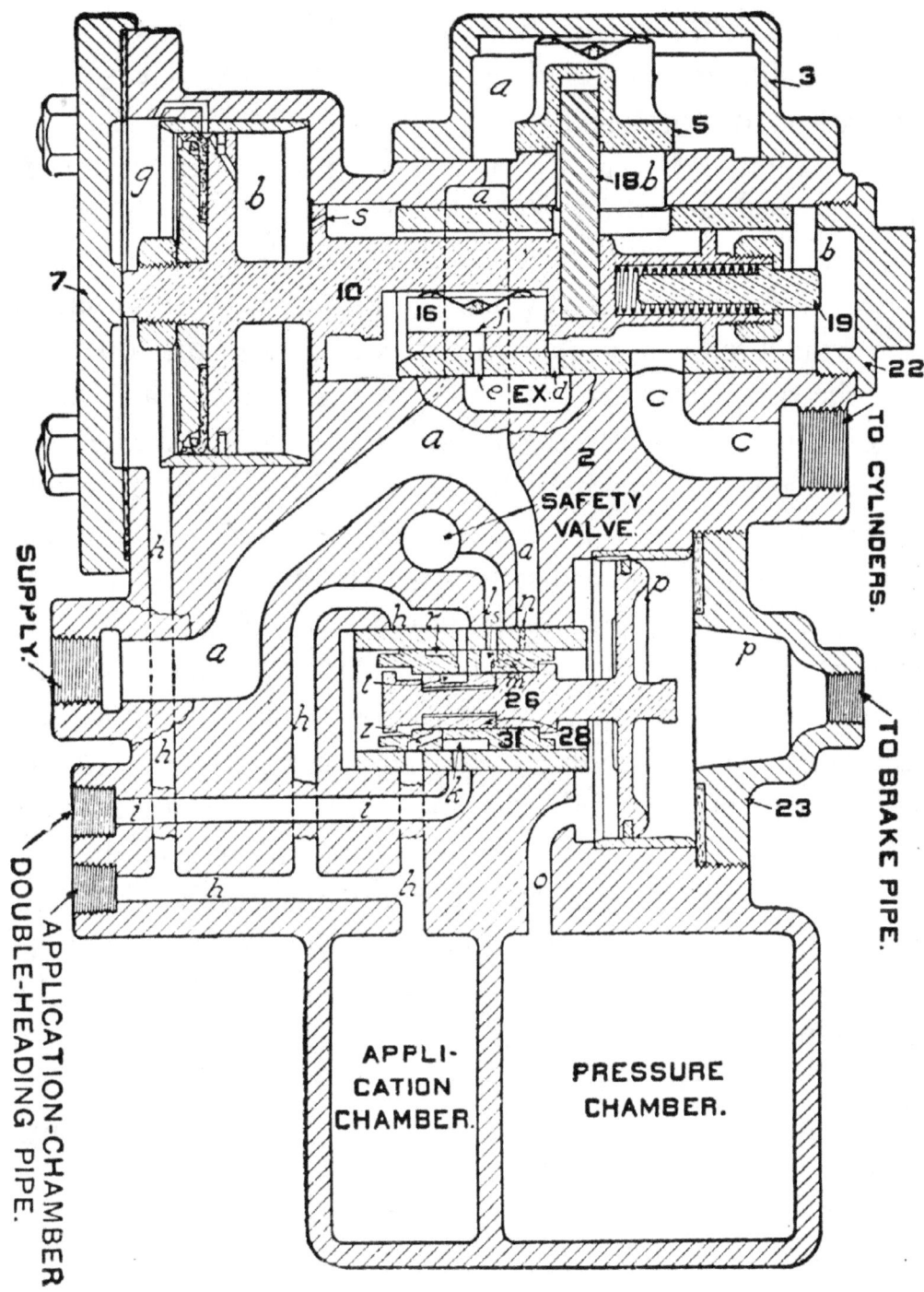

FIG. 277. RELEASE POSITION

When Locomotive Brake is released by Independent Brake Valve **after** an application by Brake Pipe Reduction

cylinder pressure, without changing the conditions in either the *pressure-chamber* or chamber *p*; consequently, the equalizing portion does not move, until released by the automatic brake valve.

DOUBLE HEADING. It will be noted that in all of the above descriptions of the distributing valve, no reference has been made to the double-heading pipe connection. This is only used when the engine does not control the train brakes, and it then becomes an exhaust opening for the distributing valve when the automatic brake valve is on lap and cut off from the brake pipe by the double cut-out cock. This will be better understood from the description of the pipe connections as already explained. The operation of the distributing valve is similar to that described during automatic brake applications with the exception of the release, which is brought about by the equalizing piston 26 moving to the release position and causing exhaust cavity in the equalizing slide valve 31 to connect ports *i* and *h* in the slide valve seat, thereby permitting the pressure in the application chamber to escape to the atmosphere through the double heading pipe, the double cut-out cock and the automatic brake valve. In double heading, therefore, the release of the distributing valve is similar to that of a triple valve.

To remove piston 10 and slide valve 16, it is absolutely necessary to *first* remove cover 3, slide valve 5 and valve pin 18.

Referring to Figs. 268 and 279, the proper names of parts of this apparatus are as follows: 2, Body; 3, Application-Valve Cover; 4, Cover Screw; 5, Application Valve; 6, Application-Valve Spring; 7, Application-Cylinder Cover; 8, Cylinder-Cover Bolt and Nut; 9,

ET BRAKE EQUIPMENT

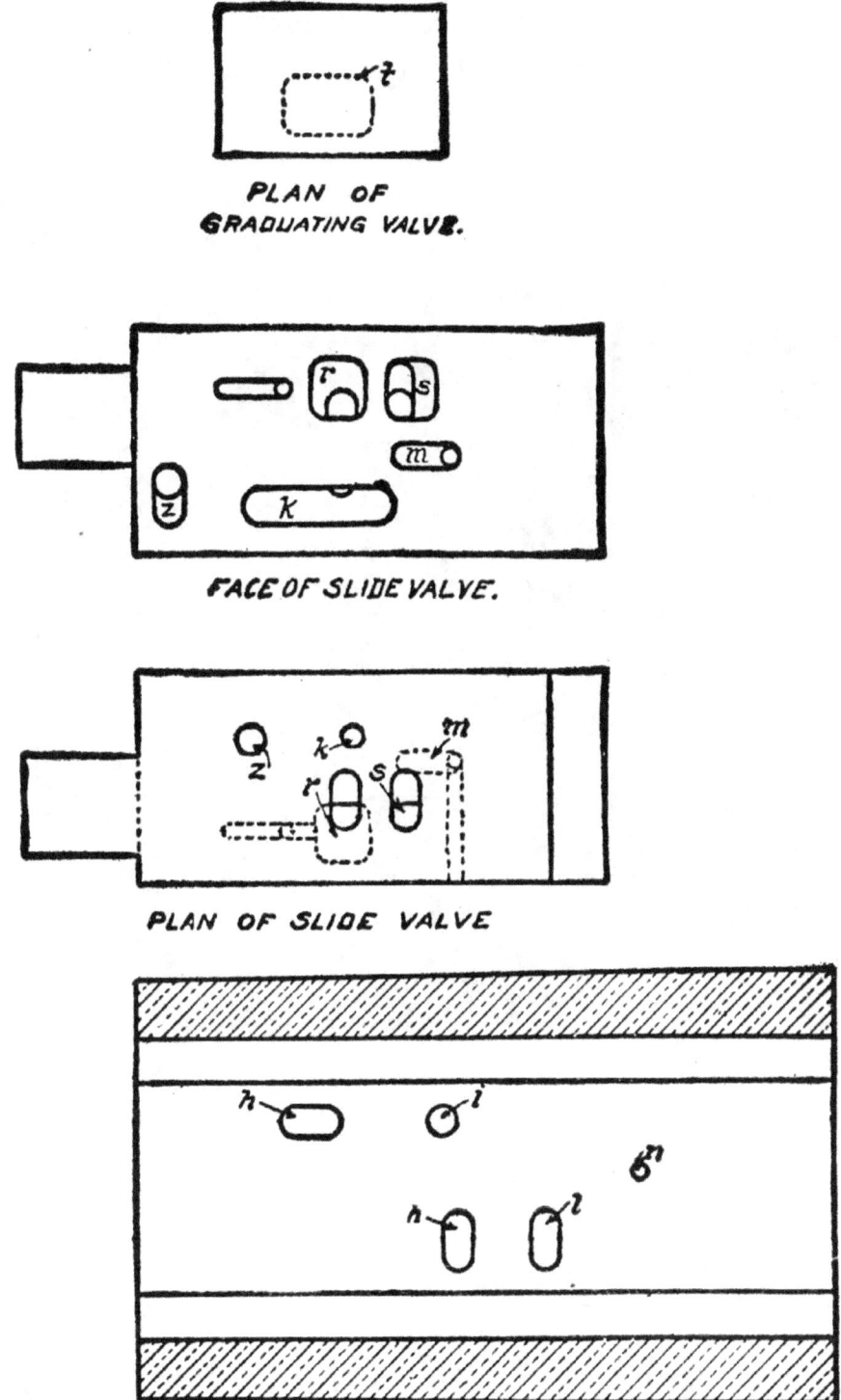

FIG. 278. GRADUATING VALVE, EQUALIZING SLIDE VALVE, AND SLIDE VALVE SEAT

Cylinder-Cover Gasket; 10, Application Piston; 11 Piston Follower; 12, Packing-Leather Expander; 13 Packing Leather; 14, Application-Piston Nut; 15, Applica-

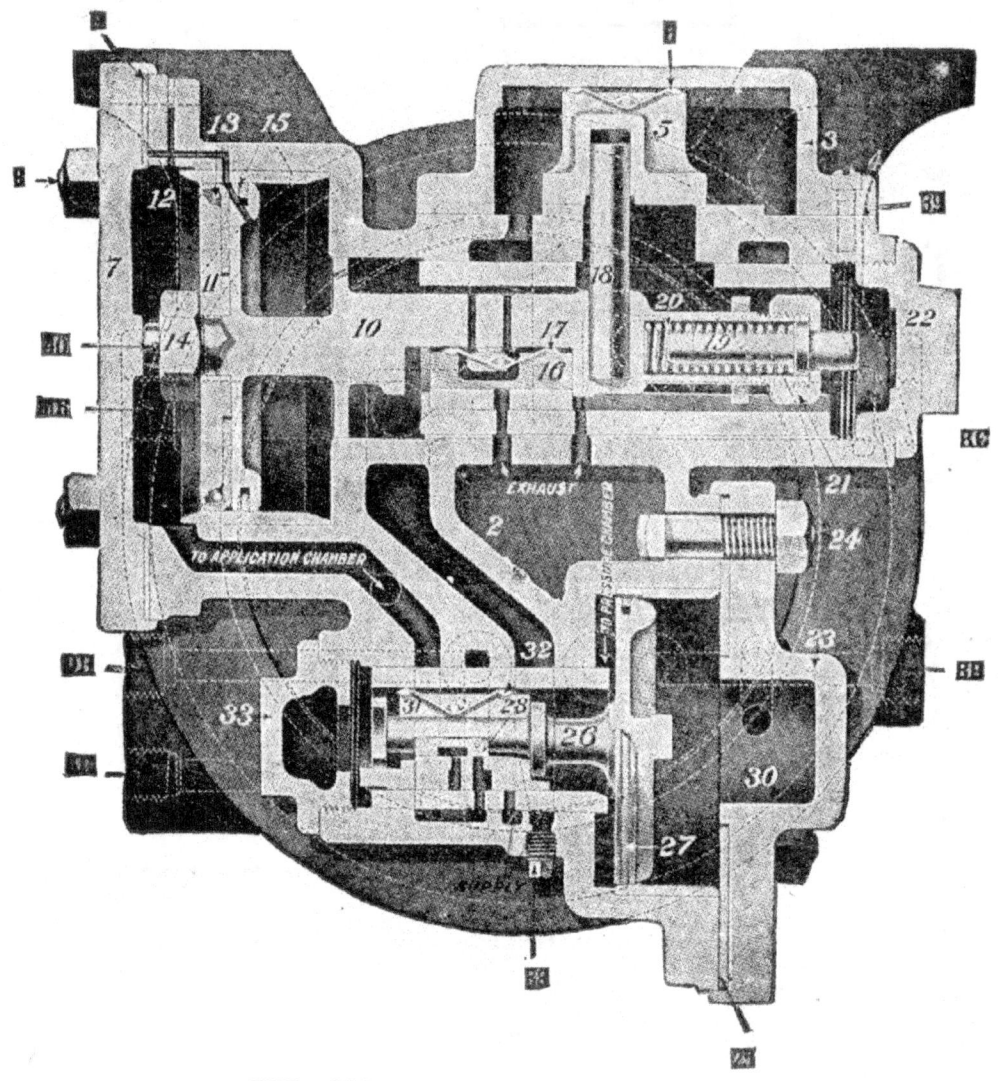

FIG. 279. DISTRIBUTING VALVE
CONNECTIONS:
MR—Main Reservoir Pipe; DH—Double-Heading Pipe; AC—Application-Chamber Pipe; BC—Brake-Cylinder Pipe; BP—Brake Pipe

tion-Piston Packing Ring; 16, Exhaust Valve; 17, Exhaust-Valve Spring; 18, Application-Valve Pin; 19, Graduating Stem; 20, Graduating Spring; 21, Graduat-

ET BRAKE EQUIPMENT

ing-Stem Nut; 22, Upper Cap Nut; 23, Equalizing Cylinder Cap; 24, Cylinder Cap Bolt and Nut; 25, Cylinder-Cap Gasket; 26, Equalizing Piston; 27, Equalizing-Piston Packing Ring; 28, Graduating Valve; 29, Graduating-Valve Spring; 31, Equalizing Slide Valve; 32, Equal-

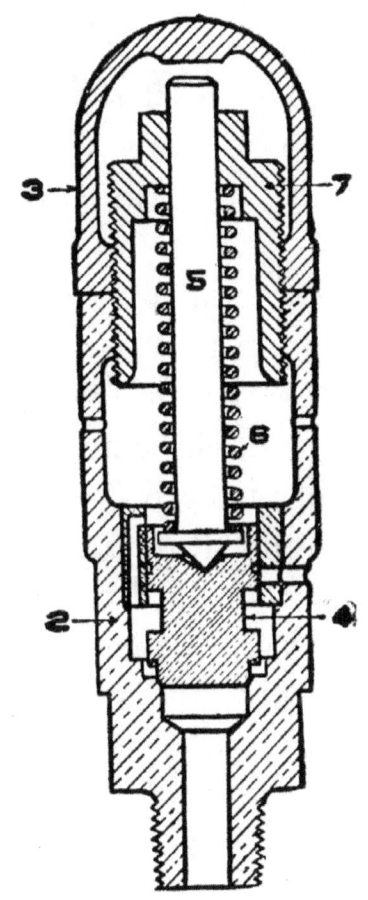

FIG. 280. SAFETY VALVE

izing-Slide-Valve Spring; 33, Lower Cap Nut; 34, Safety Valve; 35, Double-Chamber Reservoir; 36, Reservoir Stud and Nut; 37, Reservoir Drain Plug; 38, Distributing-Valve Drain Plug; 39, Application-Valve-Cover

Gasket; 40, Application Piston Cotter; 41, Distributing-Valve Gasket.

Fig. 280 is a sectional view of the safety valve which is a necessary part of the distributing valve. It is of an improved type, which insures reliability of operation. It is unlike the ordinary safety valve, as its construction is such as to cause it to close quickly with a "pop" action, insuring its seating firmly. It is very sensitive in operation and responds to very slight differences of pressure.

The names of the parts are: 2, Body; 3, Cap Nut; 4, Valve; 5, Stem Valve; 6, Adjusting Spring; 7, Adjusting Nut.

Valve 4 is held to its seat by the compression of the spring 6 between the stem and adjusting nut 7. When the pressure below valve 4 is in excess of the force exerted by the spring, it raises, being guided in its movement by the brass bush in the body 2. Ports are drilled in this bush; one outward through the body to the atmosphere, and the other upward to the spring chamber. Although only one of each of these is shown in the cut, there are eight of the first and two of the second. As the valve moves upward, its lift is determined by the stem 5 striking the cup nut 3. It closes the vertical ports connecting the valve and spring chambers and opens the lower ports to the atmosphere. As the air pressure below valve 4 decreases, and the tension of the spring forces the stem and valve downward, the valve gradually closes the lower ports to the atmosphere and opens those between the valve and spring chambers. The discharge air pressure then has access to the spring chamber. This chamber is always connected to the atmosphere by two small holes through the body, 2; the air from the valve

ET BRAKE EQUIPMENT

chamber enters more rapidly than it can escape through these holes, causing pressure to accumulate above the valve and close it with the "pop" action before mentioned.

The adjustment of this safety valve is accomplished

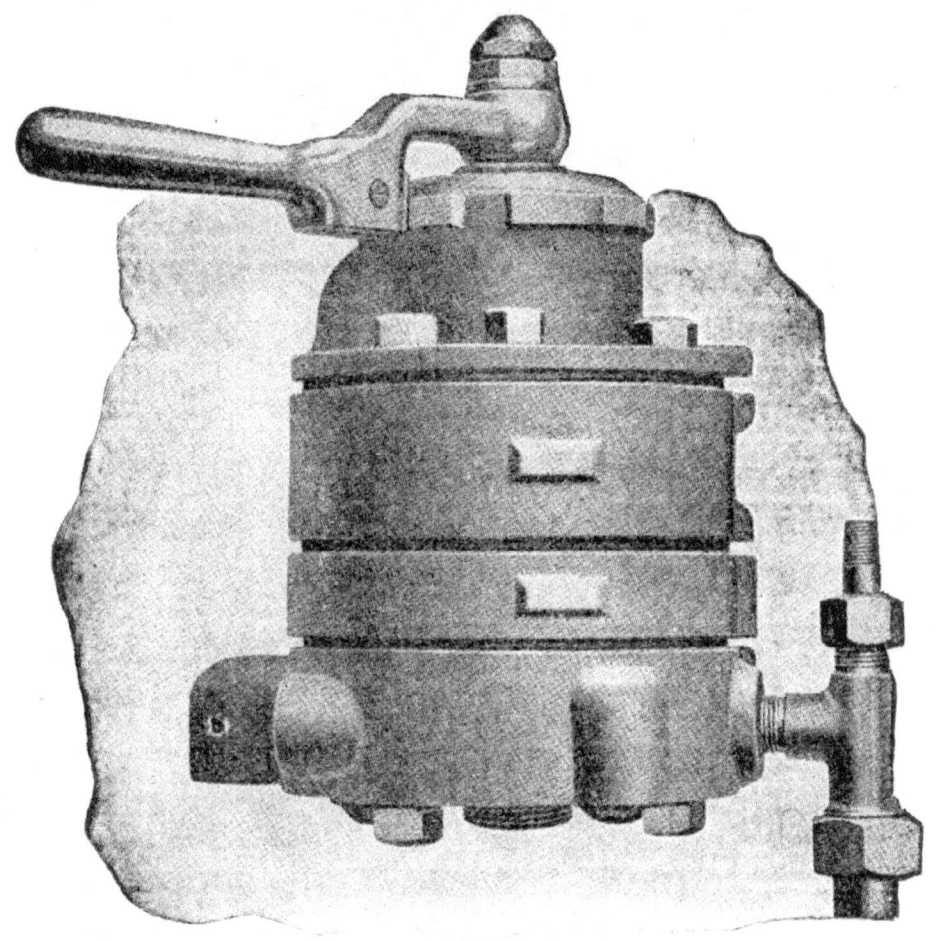

FIG. 281. TYPE H BRAKE VALVE

by removing cap nut 3, and screwing up or down on adjusting nut 7. After the proper adjustment is made, cap nut 3 must be replaced and securely tightened, and the valve operated a few times. Particular attention must be given to the holes in the valve body to see that they are

ROTARY-VALVE SEAT.

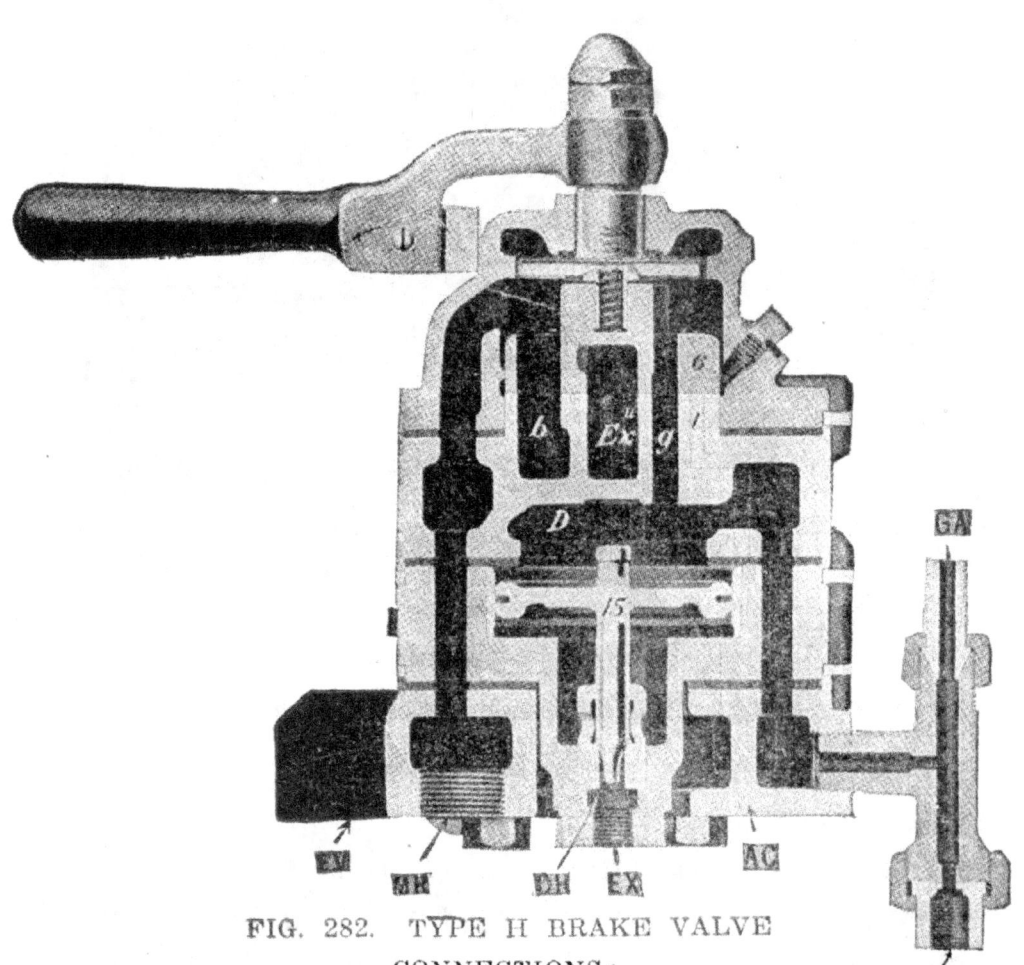

FIG. 282. TYPE H BRAKE VALVE

CONNECTIONS:
FV—Feed-Valve Pipe; **MR**—Main-Reservoir Pipe; **GO**—To Governor; **DH**—Double-Heading Pipe; **EX**—Exhaust; **AC**—Application-Chamber Pipe; **BP**—Brake Pipe; **GA**—Duplex Air Gauge; **ER**—Equalizing Reservoir.

open, and that they are of the proper size, especially the two upper holes.

This safety valve should be adjusted for 53 pounds.

THE TYPE H AUTOMATIC BRAKE VALVE.

This Brake Valve, although modelled to a considerable extent upon the principles of previous valves, is necessarily different in detail, since it not only performs all the functions of such types but also those absolutely necessary to obtain all the desirable operating features of the Distributing Valve.

Fig. 281 is taken from a photograph of this brake valve, while Fig. 282 shows two views, the upper one being a plan view with section through the rotary-valve chamber, the rotary valve being removed; the lower one a vertical section. In these views the pipe connections are indicated.

Fig. 283 is a top view, showing the six positions of the brake-valve handle, which are, beginning at the extreme left, Release, Running, Holding, Lap, Service and Emergency.

Fig. 284 shows two views of this valve similar to those of Fig. 282, with the addition of a plan or top view of the rotary valve. Referring to the latter, a, j and s are ports extending directly through it, the latter connecting with a groove in the face; f and k are cavities in the valve face; o is the exhaust cavity; x is a port in the face of the valve connecting with o; h is a port in the face which passes over cavity k and connects with exhaust cavity o; n is a groove in the face. Referring to the ports in the rotary-valve seat, d leads to the feed-valve pipe; b and c

lead to the brake pipe; *g* leads to chamber D; ex is the exhaust opening; *e* is the preliminary exhaust port leading to chamber D; *r* is the warning port leading to the exhaust; *p* is the port leading to the pump governor; *l*

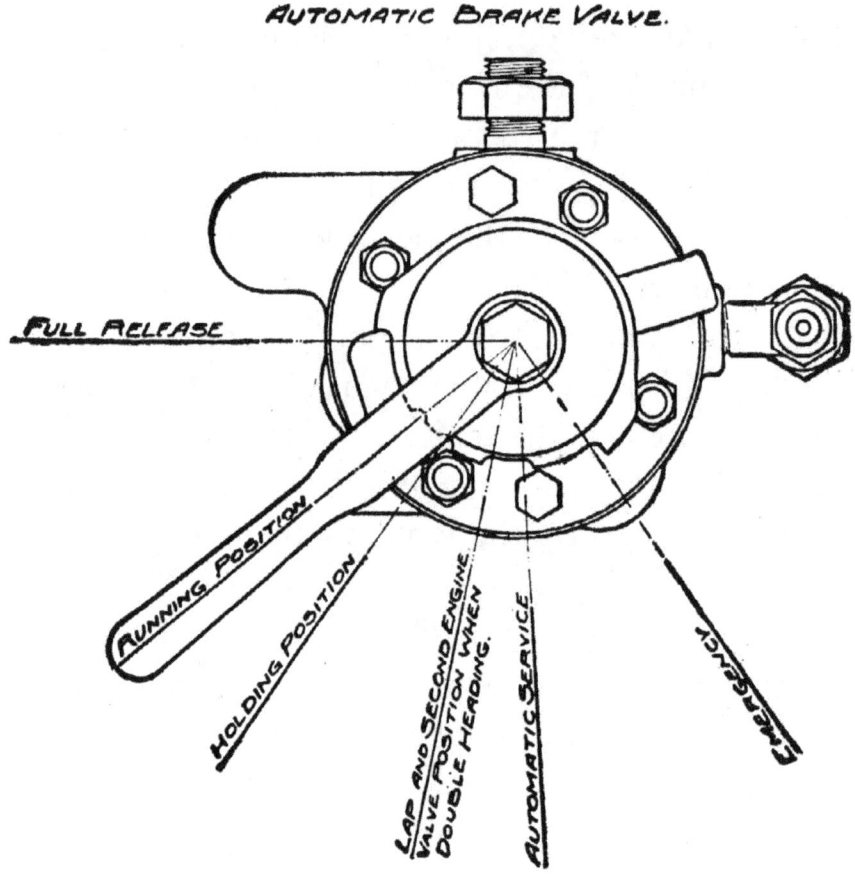

FIG. 283

leads to the application-chamber pipe; *n* leads to the double-heading pipe.

In describing the operation of the brake valve, it will be more readily understood if the positions are taken up in the order in which they are most generally used, rather than their regular order as mentioned above.

ET BRAKE EQUIPMENT

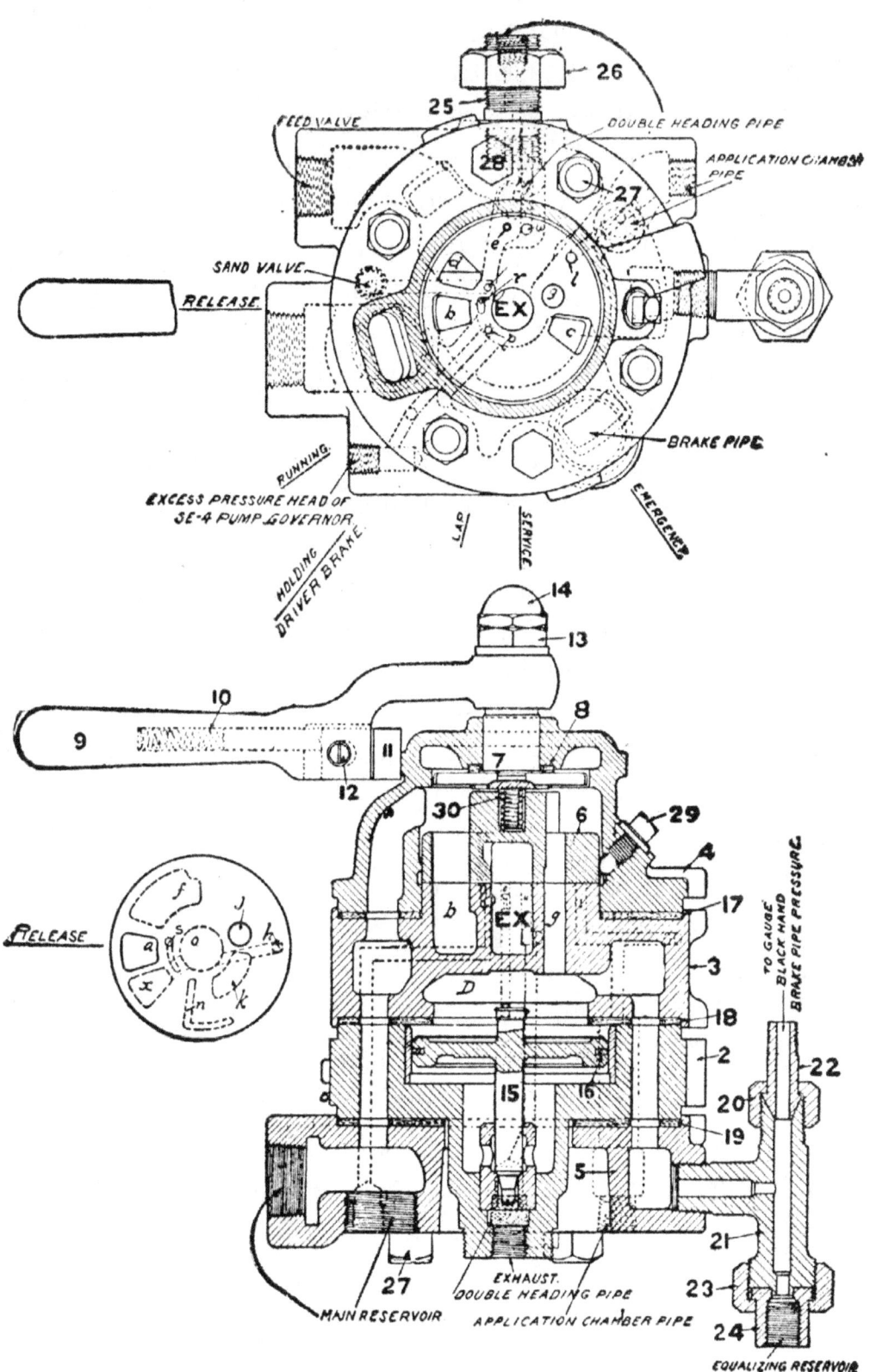

FIG. 284. AUTOMATIC BRAKE VALVE

MODERN AIR BRAKE PRACTICE

RUNNING POSITION. This is the proper position of the handle to release the engine and tender brakes; also when the brakes are not being used and the system is charged and ready for an application. In this position, cavity f in the rotary valve connects ports b and d in the valve seat, affording a large direct passage from the feed valve to the brake pipe, so that the latter will charge up as rapidly as the feed valve can supply the air, but cannot attain a pressure above that for which the feed valve is adjusted. Cavity k in the rotary valve connects ports c and g in the valve seat, so that chamber D and the equalizing reservoir charge uniformly with the brake pipe, keeping the pressures on the two sides of the equalizing piston equal. Port s in the rotary valve registers with port p in the valve seat, permitting main-reservoir pressure, which is present at all times above the rotary valve, to pass to the excess-pressure head of the pump governor. Port h in the rotary valve registers with port l in the seat, connecting the application-chamber pipe to the exhaust cavity EX.

SERVICE POSITION. This position gives a gradual reduction of brake-pipe pressure to cause a service application. Port h in the rotary valve registers with port s in the valve seat, allowing air from chamber D and the equalizing reservoir to escape to the atmosphere through cavities o in the rotary valve and EX in the valve seat. Port e is restricted so as to make the pressure in the equalizing reservoir and chamber D fall gradually. As all other ports are closed, the fall of pressure in chamber D allows the brake-pipe pressure under the equalizing piston to raise it, and unseat the discharge valve, allowing brake-pipe air to flow to the atmosphere. When the pressure in chamber D is reduced the desired amount, the handle is moved to the *lap position,* thus stopping any further reduction in that chamber. Air will continue to discharge from the brake-pipe until its pressure has fallen to an amount a trifle less than that retained in chamber D, permitting the pressure in this chamber to force the piston downward and stop the discharge of brake-pipe air. It will be seen, therefore, that the amount of reduction in the equalizing reservoir determines that in the brake pipe, regardless of the length of the train.

LAP POSITION. This position is used while holding the brakes applied after a service application until it is desired either to make a further brake-pipe reduction, or to release them; also to prevent loss of main-reservoir pressure of the release of the brake in the event of a burst hose, a break-in-two, or the opening of the conductor's valve. Lap position is also used on all engines in a train that are not controlling the train brakes, as, with the handle in this position, port h in the rotary valve connects with port u in the seat. Therefore, when the double cut-out cock is turned to the position which cuts out the brake pipe, it makes a direct opening from port i in the distributing valve through the double-heading pipe to the atmosphere, and is the passage through which the air escapes from the application chamber when the automatic brakes are being released.

RELEASE POSITION. The purpose of this position is to provide a large and direct passage from the main reservoir to the brake pipe, to permit a rapid flow of air into the latter to insure a quick release and recharging of the train brakes, but without releasing the engine and tender brakes.

MODERN AIR BRAKE PRACTICE

Air at main-reservoir pressure flows through port *a* in the rotary valve to port *b* in the valve seat and to the brake pipe. At the same time, port *j* in the rotary valve registers with the equalizing port *g* in the valve seat, permitting main-reservoir pressure to enter chamber D above the equalizing piston.

In this position, port *s* in the rotary valve registers with warning port *r* in the seat and allows a small quantity of air to escape into the exhaust cavity EX, which makes sufficient noise to attract the engineer's attention to the position in which the valve handle is standing. If the handle is allowed to remain in this position, the brake system would be charged to main-reservoir pressure. To avoid this, the handle must be moved to Running or Holding Positions. The small groove in the face of the rotary valve which connects with port *s*, extends to port *p* in the valve seat, allowing main-reservoir pressure to flow to the excess-pressure head of the pump governor.

HOLDING POSITION. This position is so named because the locomotive brakes are held applied, as they are in release position, while the train brakes feed up to the feed-valve pressure. All ports register as in running position, except port *l*, which is closed.

Therefore, the only difference between Running and Holding Positions is that in the former the application chamber is open to the atmosphere, while in the latter it is not.

EMERGENCY POSITION. This position is used when the most prompt and heavy application of the brakes is desired. Port *x* in the rotary valve registers with port *c* in the valve seat, making a large and direct communication between the brake pipe and atmosphere through

ET BRAKE EQUIPMENT

cavity o in the rotary valve and EX in the valve seat. This direct passage causes a sudden and heavy discharge of brake-pipe pressure, causing the triple valves and distributing valve to go to the emergency position and apply the brake in the shortest possible time.

In this position the groove n in the rotary valve connects ports g and l in the valve seat, thereby allowing equalizing reservoir air to flow into the application chamber.

The oil plug 29 is placed in the top case 4, at a point to fix the level of the oil surrounding the rotary valve. Leather washer 8 prevents air in the rotary valve chamber from leaking past the rotary valve key to the atmosphere. Spring 30 keeps the rotary valve key firmly pressed against washer 8 when no main-reservoir pressure is present. The handle 9 contains a latch 11, which fits into notches in the top case, so located as to indicate the different positions of the brake valve handle. The spring 10 back of the latch forces the latter against the body with sufficient pressure to distinctly indicate when the handle arrives at each position.

To remove the brake valve take off nuts 27, thus allowing it to come away without disturbing the pipe bracket or breaking any pipe joints. To take the valve proper apart, remove cap screws 28.

The brake valve should be located so that the engineer can operate it from his usual position, while looking forward or back out of the side-cab window, and in such a manner that the handle will not meet with any obstruction throughout its entire movement.

The oil around the rotary valve furnishes thorough lubrication. Valve oil should be used for this purpose.

MODERN AIR BRAKE PRACTICE

Fig. 284 shows all the principal parts, the proper names of each being as follows: 2, Bottom Case; 3, Rotary-Valve Seat; 4, Top Case; 5, Pipe Bracket; 6, Rotary Valve; 7, Rotary-Valve Key; 8, Key Washer; 9, Handle; 10, Handle-Latch Spring; 11, Handle Latch; 12,

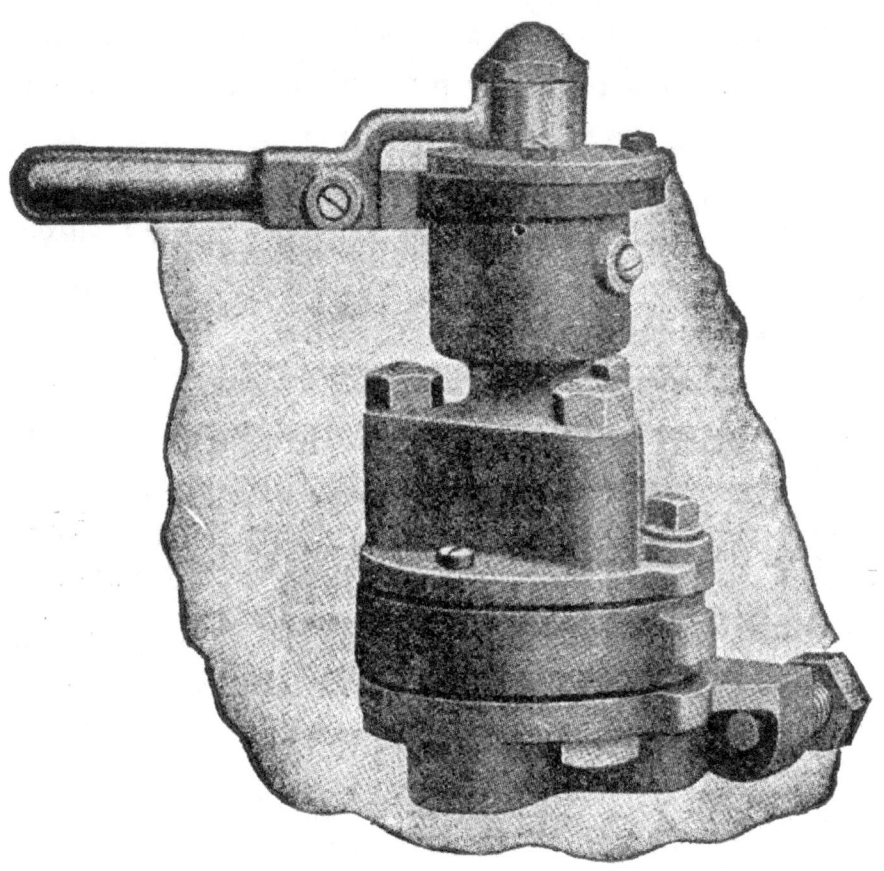

FIG. 285. THE INDEPENDENT BRAKE VALVE

Handle-Latch Screw; 13, Handle Nut; 14, Handle Lock Nut; 15, Equalizing Piston; 16, Equalizing-Piston Packing Ring; 17, Valve-Seat Upper Gasket; 18, Valve-Seat Lower Gasket; 19, Pipe-Bracket Gasket; 20, Small Union Nut; 21, Brake-Valve Tee; 22, Small Union Swivel; 23, **Large** Union Nut; 24, Large Union Swivel; 25, Bracket

ET BRAKE EQUIPMENT

Stud; 26, Bracket-Stud Nut; 27, Bolt and Nut; 28, Cap Screw; 29, Oil Plug; 30, Rotary-Valve Spring.

THE INDEPENDENT BRAKE VALVE.

Fig. 285 illustrates this valve, which is of the rotary type. Fig. 286 shows a vertical section through the center of the valve, and a horizontal section through the valve body, with the rotary valve removed, showing the rotary valve seat. Fig. 287 shows this valve similarly to Fig. 286, with the addition of a top view of the rotary valve. In these views, the pipe connections and positions of the handle are indicated. Port b in the seat leads to the supply connection from the main reservoir through the Reducing Valve. Port c leads to that portion of the application-chamber pipe which connects to the automatic-brake valve. Port d leads to that portion of the application-chamber pipe which connects the distributing valve. Port h, in the center, is the exhaust port leading directly to the atmosphere. Exhaust cavity g in the rotary valve is always in communication with exhaust port h. Groove e in the face of the valve communicates at one end with a port through the valve. This groove is always in communication with supply port b, and through the opening just mentioned air is admitted to the chamber above the rotary valve, thus keeping it to its seat. Port f in the rotary valve consists of two circular openings in the face joined by a cylindrical passage over the top of cavity g.

RUNNING POSITION. This is the position that the independent brake valve should be carried in at all times when the independent brake is not in use. Port f in the

MODERN AIR BRAKE PRACTICE

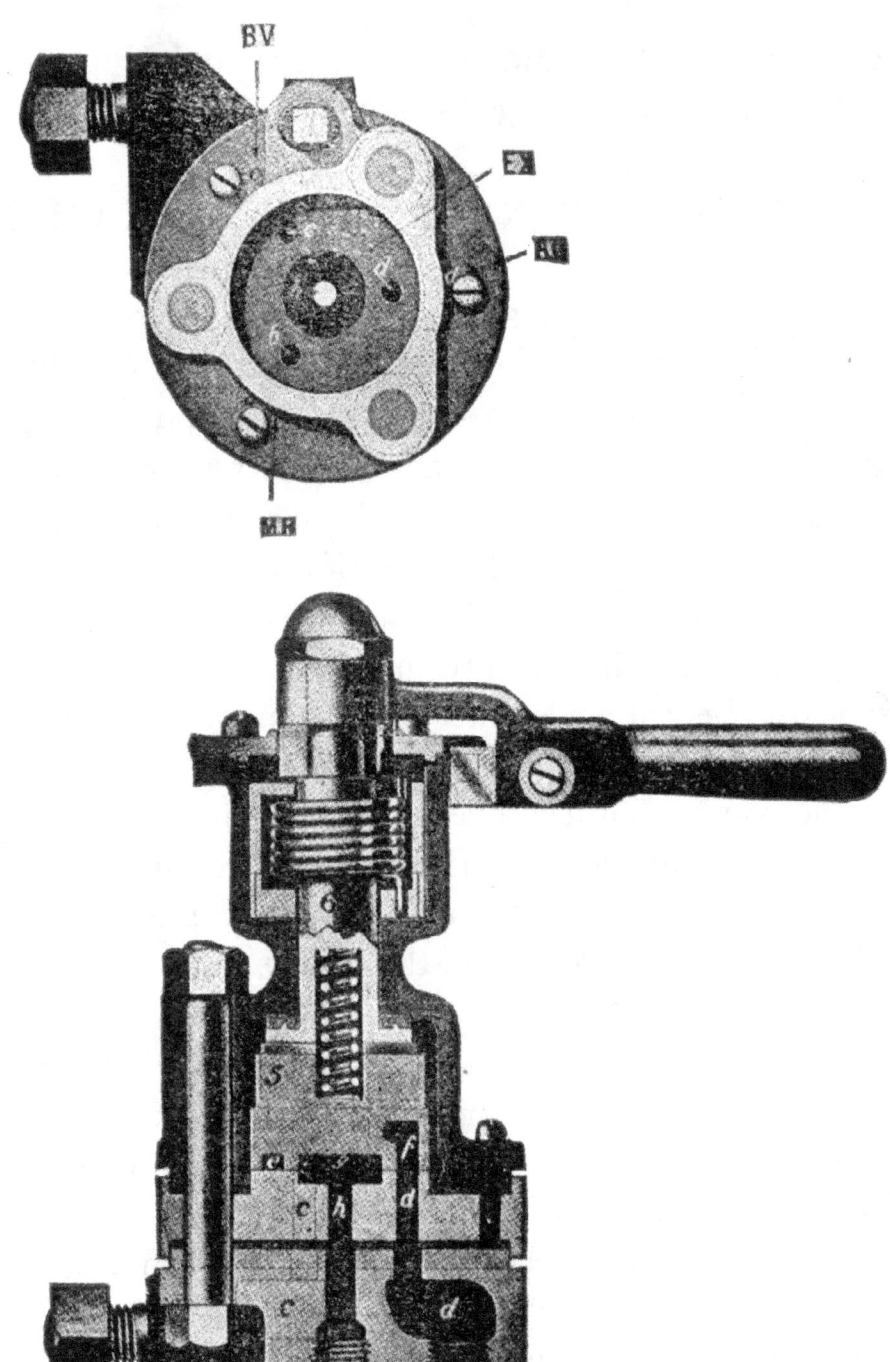

FIG. 286. INTERIOR VIEW OF THE INDEPENDENT
BRAKE VALVE

CONNECTIONS.

BV—Application-Chamber Pipe to Automatic Brake Valve;
EX—Exhaust; AC—Application-Chamber Pipe to Distributing Valve; MR—Reducing-Valve Pipe

ET BRAKE EQUIPMENT

rotary valve connects ports c and d in the valve seat, thus establishing communication between the application chamber of the distributing valve and port l of the automatic brake valve. Therefore, it will be seen that if the automatic brake valve is in running position, and the independent brakes applied, they can be released by returning the independent valve to running position.

SERVICE POSITION. To apply the independent brakes, move the brake valve to the application position; groove e connects ports b and d, allowing air to flow to the application chamber of the distributing valve. Since the supply pressure to this valve is fixed by the regulation of the reducing valve to 45 pounds, this is the maximum cylinder pressure that can be obtained.

LAP POSITION. This position is used to hold the independent brakes applied after the desired cylinder pressure is obtained, at which time all communication between operating ports is closed.

RELEASE POSITION. This position is used to release the pressure from the application chamber when the automatic brake valve is not in running position. In this position, the offset in cavity g registers with port d, allowing pressure in the application chamber to flow through ports d, g and h to the atmosphere.

In order to prevent leaving the handle in the release position and thereby make it impossible to operate the locomotive brakes by the automatic brake valve, spring 9 automatically returns handle 15 from the release to the running position.

The purpose of the oil plug 20 is the same as that described in the automatic brake valve.

The location of this valve should be governed by the

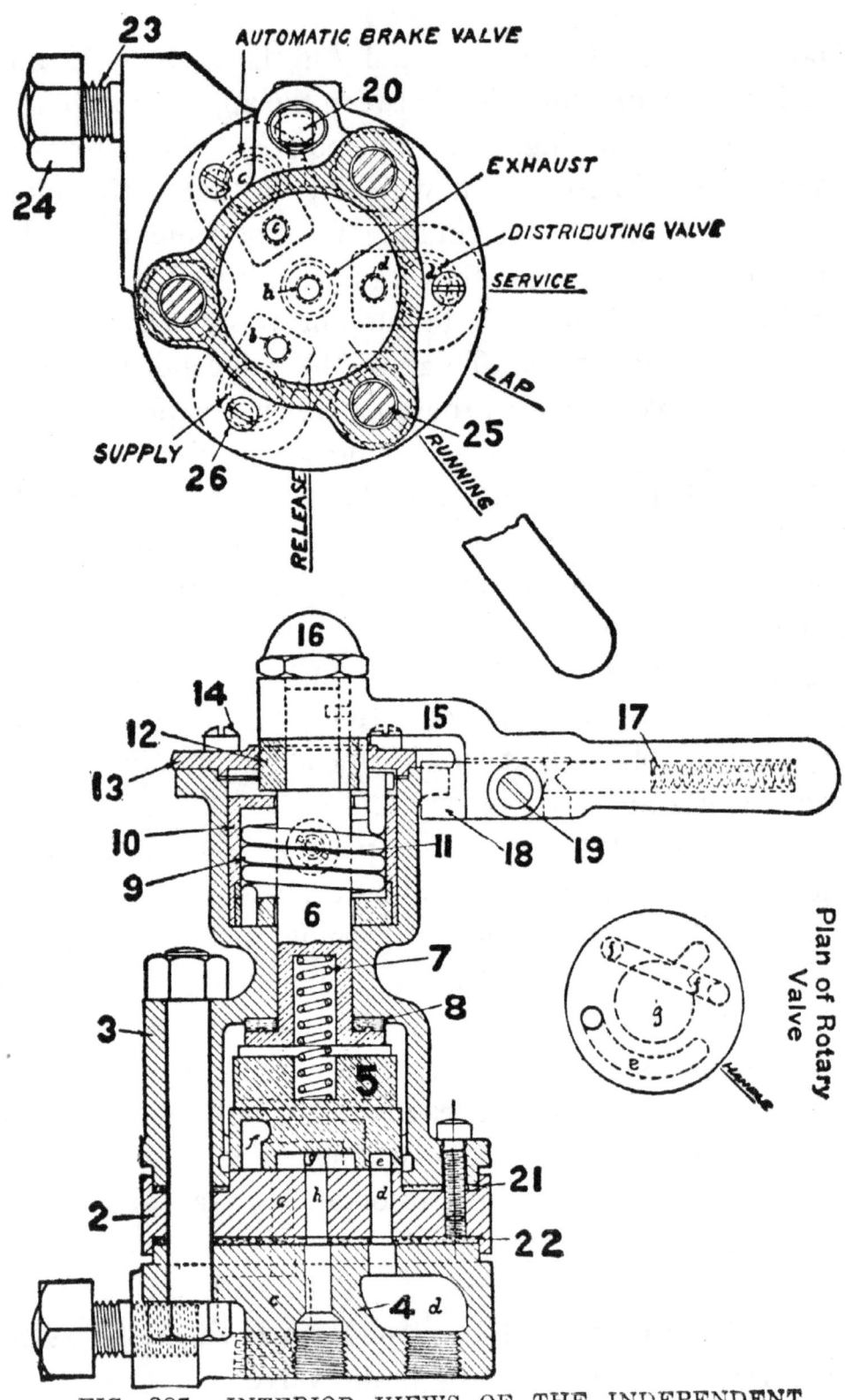

FIG. 287. INTERIOR VIEWS OF THE INDEPENDENT BRAKE VALVE

ET BRAKE EQUIPMENT

same considerations as those mentioned concerning the automatic brake valve. The names of its parts are as follows, referring to Fig. 287.

2, Rotary-Valve Seat; 3, Valve Body; 4, Pipe Bracket; 5, Rotary Valve; 6, Rotary-Valve Key; 7, Rotary-Valve Spring; 8, Key Washer; 9, Return-Spring; 10, Return-Spring Casing; 11, Casing Screw; 12, Return-Spring Clutch; 13, Cover; 14, Cover Screw; 15, Handle; 16, Handle Nut; 17, Latch Spring; 18, Latch; 19, Latch Screw; 20, Oil Plug; 21, Upper Gasket; 22, Lower Gasket; 23, Bracket Stud; 24, Bracket-Stud Nut; 25, Bolt and Nut; 26, Cap and Screw.

THE FEED VALVE.

This valve, Fig. 288, is a slide-valve feed valve of an improved type, and with this equipment is connected to a pipe bracket located in the piping between the main reservoir and the automatic brake valve, receiving its supply of air from the main-reservoir pipe, and delivering it into the feed-valve pipe. It is for the purpose of controlling brake-pipe pressure when the automatic brake valve handle is in running or holding positions.

Figs. 289 and 290 are diagrammatic views of the valve and pipe bracket having the ports and operating parts in one plane to facilitate description. It consists of two sets of parts, the supply and regulating. The supply parts, which control the flow of air through the valve, consist of the supply valve 10, and its spring 11; the supply-valve piston 8 and its spring 6. The regulating parts consist of the regulating valve 13, regulating-valve spring 14, diaphragm 15, diaphragm spindle 17, regulating handle 23. Main-reservoir air enters through port *a, a* to the supply-valve chamber B, forcing supply-valve piston 8 to the left, compressing piston spring 6 and causing supply valve 10 to open port *c*, permitting the air to

FIG. 288. FEED VALVE

pass through ports *c* and *d* to the feed-valve pipe at delivery, and through port *e* to diaphragm chamber L. At the same time air flows through port *f* in supply-valve piston 8 to chamber G, and through port *h* to regulating-valve chamber H. As regulating-valve 13 is raised from its seat, it will flow through port

ET BRAKE EQUIPMENT

k to chamber L. When the feed-valve-pipe pressure, which is always present in chamber L against the diaphragm, exceeds the pressure of regulating spring 18,

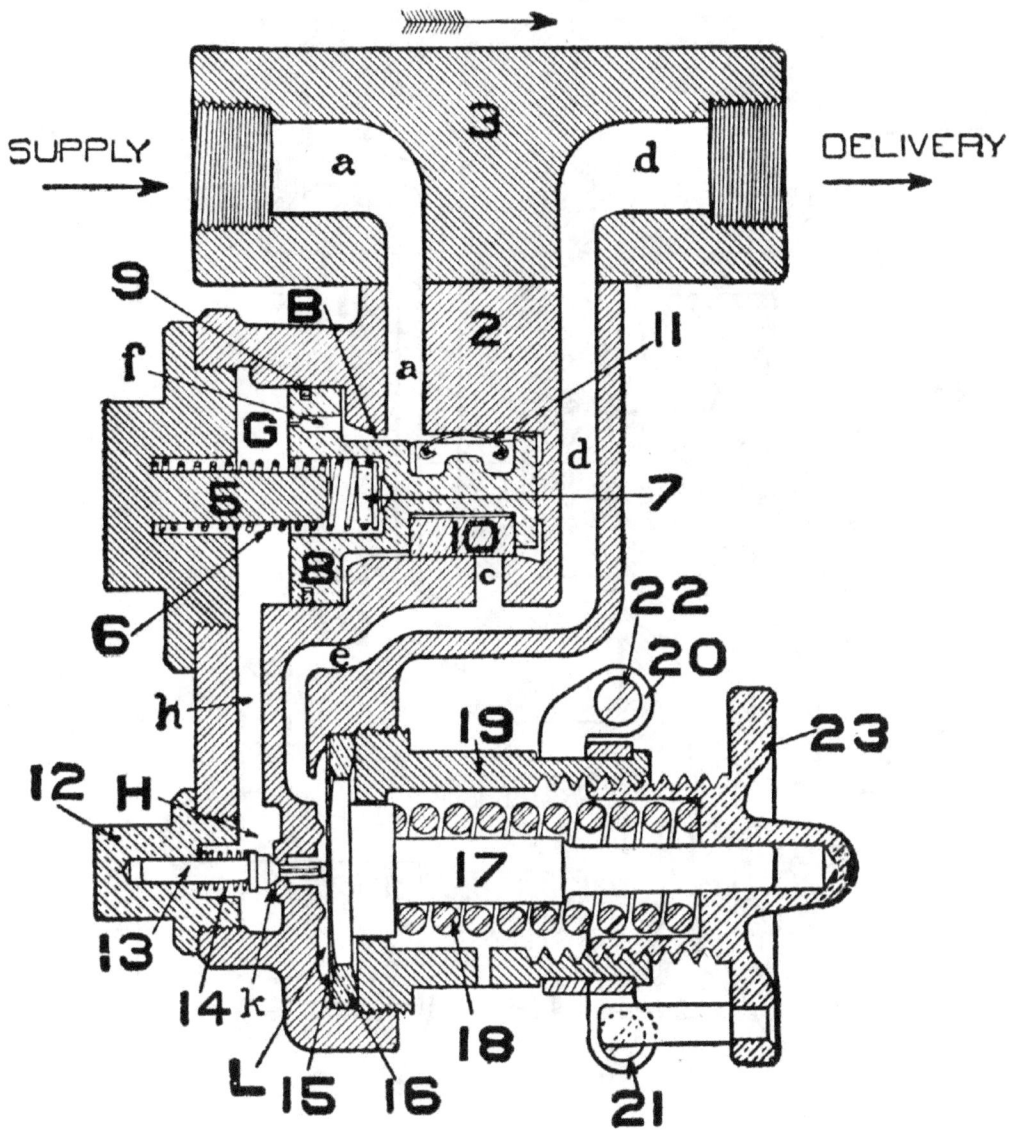

FIG. 289. DIAGRAM OF FEED VALVE, CLOSED

the diaphragm will yield and permit the regulating valve 13 to be forced to its seat, closing port k and cutting off any further flow of air from chamber G. As the air which continues to flow through port f will quickly equal-

MODERN AIR BRAKE PRACTICE

ize the pressure on both sides of piston 8, spring 6 will force the piston to the right, moving supply valve 10 and

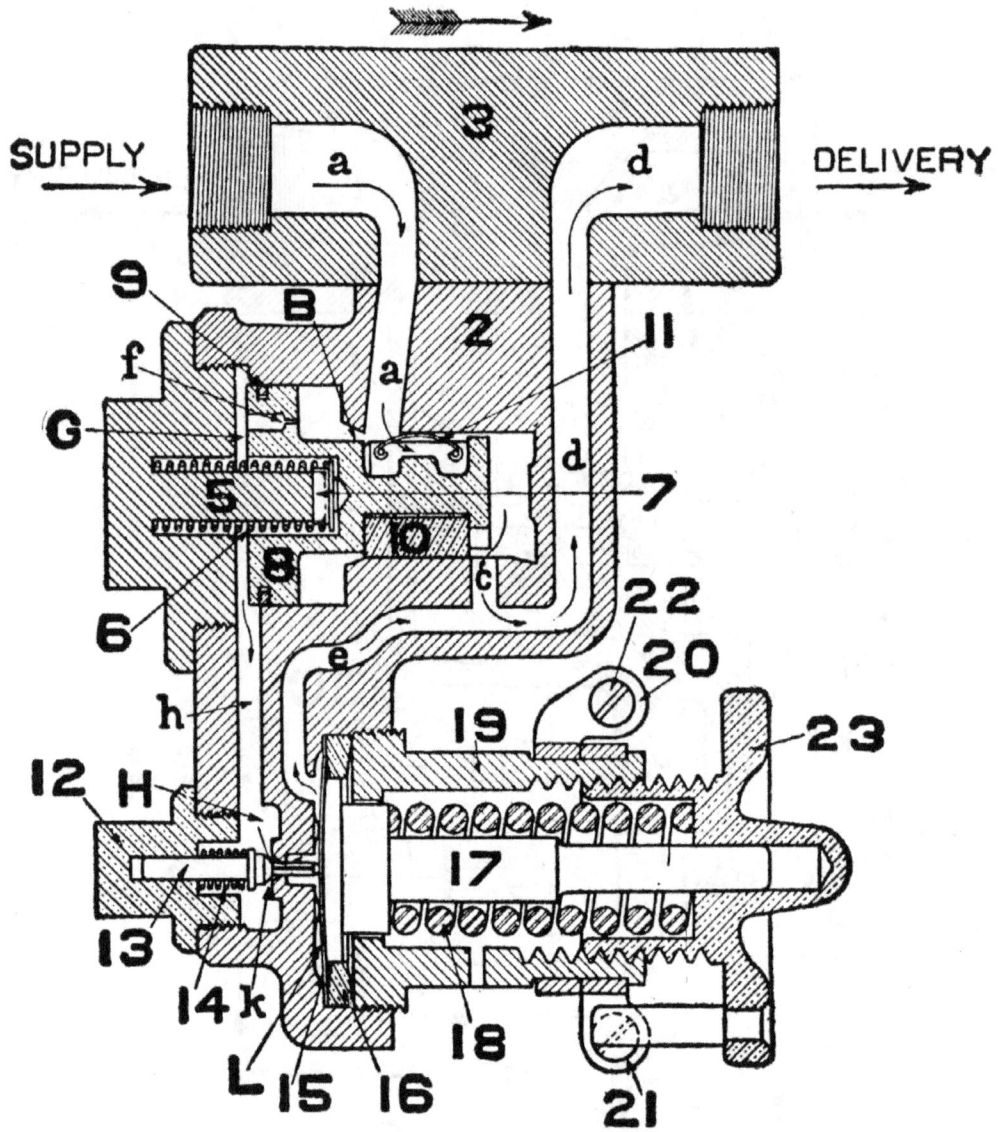

FIG. 290. DIAGRAM OF FEED VALVE, OPEN

closing port *c*, thereby cutting off communication between the supply and the feed-valve pipe.

When the pressure in the feed-valve pipe falls below that for which the valve is adjusted, regulating spring 18

ET BRAKE EQUIPMENT

will force the diaphragm forward, unseat regulating valve 13 and permit the accumulated air pressure in chamber G to escape through port h, chamber H and port k to chamber L. This allows main-reservoir pressure in chamber B to force the supply valve piston 8 to left, and open port c, which again permits air to pass to the feed-valve pipe until its pressure has been restored to the proper amount. Since this feed valve has a duplex adjusting arrangement, it eliminates the necessity of the two feed valves in high and low pressure service, as the turning of handle 23 until its pin strikes stops 20 or 21 changes the regulation from one predetermined brake-pipe pressure to another.

To adjust this valve, slacken screw 22, which allows stops 20 and 21 to turn around spring box 19. Adjusting handle 23 should be turned until the valve closes at the lower brake pipe pressure desired, when stop 21 should be brought in contact with the handle pin, at which point it should be securely fastened by tightening screw 22. Adjusting handle 23 should then be turned until the higher adjustment is obtained, when stop 20 is brought in contact with the handle pin and securely fastened.

The names of the parts shown in the diagram, Figs. 289 and 290, are as follows: 2, Valve Body; 3, Pipe Bracket; 5, Cap Nut; 6, Piston Spring; 7, Piston-Spring Tip; 8, Supply-Valve Piston; 9, Piston Packing Ring; 10, Supply Valve; 11, Supply-Valve Spring; 12, Regulating-Valve Cap; 13, Regulating Valve; 14, Regulating-Valve Spring; 15, Diaphragm; 16, Diaphragm Ring; 17, Diaphragm Spindle; 18, Regulating Spring; 19, Spring Box; 20, Upper Stop; 21, Lower Stop; 22, Stop Screw; 23, Adjusting Handle.

MODERN AIR BRAKE PRACTICE

REDUCING VALVE.

Fig. 291 is a photograph of the exterior of this valve connected to its pipe bracket, the construction and operation of which is the same as the feed valve just described,

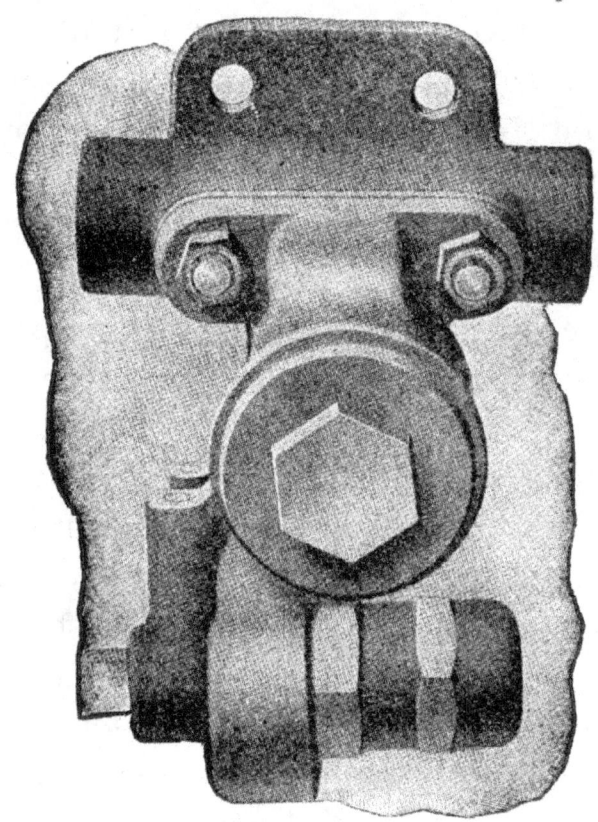

FIG. 291. REDUCING VALVE

with the exception of the adjusting feature, this valve being designed for single adjustment only.

THE PUMP GOVERNOR.

Fig. 292 shows a sectional view of this governor in its normal position. By reference to the piping diagram in Fig. 66 it will be noted that the connection B leads to the boiler; P to the air pump; MR to the main reservoir; ABV to the automatic brake valve; FVP to the feed-valve pipe; W is the waste-pipe connection. Steam enters at B and passes by steam valve 26 to the connection

ET BRAKE EQUIPMENT

P and to the pump. Air from the main reservoir flows through the automatic brake valve to the connection marked ABV into chamber d below diaphragm 52. Air from the feed-valve pipe enters at the connection FVP

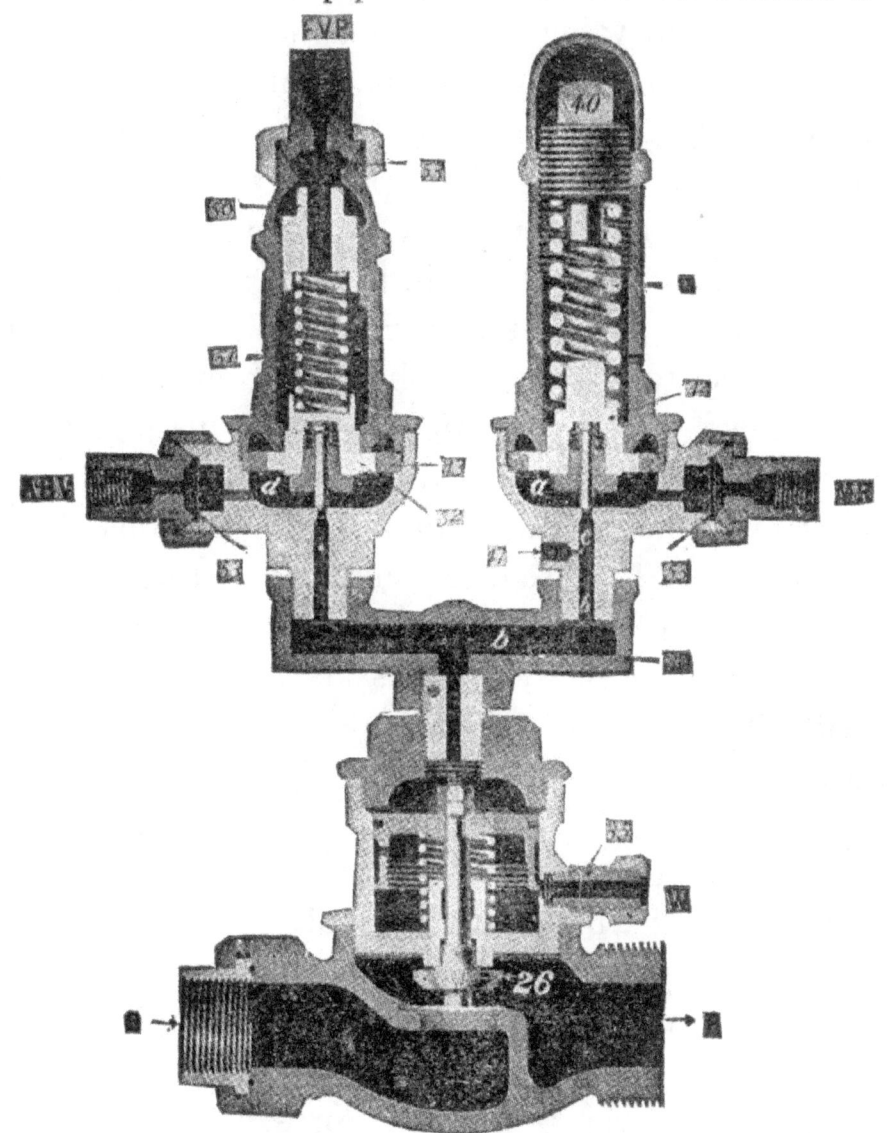

FIG. 292. PUMP GOVERNOR

and passes to the chamber above the diaphragm, adding to the pressure of regulating spring 51 in holding it down. As this spring is adjusted to a compression of about 20 pounds, the diaphragm will be held down until the main reservoir-pressure in chamber d exceeds the feed-valve pipe pressure by this amount. At such time, diaphragm 52 will raise, unseating its pin valve, and allow air to flow through port *b* to the chamber above the governor piston, forcing it downward, compressing its spring and seating steam valve 26. When main-reservoir pressure in chamber *d* is reduced, the combined spring and air pressures above the diaphragm force it down, seating its pin valve. The pressure in port *b* and the chamber above the governor piston, which is always able to escape a little from the vent port *c,* will then escape to the atmosphere and allow the piston spring, and steam pressure below valve 26, to raise it and the governor piston to the position shown. The connection from the main reservoir to chamber *d* is open only when the automatic brake valve handle is in release, running or holding positions; in the other positions it is closed, at which times this governor head is cut out of action. The connection marked MR in the maximum pressure head is always in communication with the main reservoir, so that when the excess pressure head is cut out by the brake valve, this head controls the pump. When main-reservoir pressure in chamber *a* exceeds the compression of adjusting spring 41, diaphragm 42 will rise its pin valve and allow air to flow through port *b* to the chamber above the governor piston, controlling the pump as above described.

As each governor head has a vent port *c,* from which air escapes whenever pressure is present in port *b,* to avoid an unnecessary waste of air, one of these should be plugged.

To adjust this governor, remove the cap nut and turn adjusting nut 50 until the compression of spring 51 is equal to the excess of pressure desired.

QUESTIONS.

904. In what respect does the new ET. Locomotive brake equipment differ materially from the standard automatic and straight air brake?

905. Can the ET. equipment be applied to any locomotive; no matter what kind of service?

906. Mention one of the principal advantages in connection with the design of the valves.

907. What are three important advantages connected with its operation?

908. What can be said regarding the manipulation of the ET. brake, and the automatic and straight air brake?

909. In what position should the handle be, when not in use?

910. When it is desired to apply the locomotive and train brakes, how must the automatic brake valve handle be moved?

911. Describe the method of releasing the train brakes.

912. Describe the method of making an emergency application.

913. In order to make a smooth and accurate, two application passenger stop, what should be done?

914. Describe the proper method of making, and, releasing an independent application.

915. How should the independent brake be operated, when handling long trains on the road, or in switching service?

916. How may the engine brakes only be released, and still leave the train brakes applied?

917. What should be done with the train brakes before detaching the engine?

918. Should the automatic brakes be used to hold a train or engine standing on a grade, for any length of time?

919. What is the safest way to hold a train standing on a grade?

920. Describe the proper method of stopping a descending train, and then holding it with the independent brake.

921. In what position should the independent brake valve handle be left when the engine is standing at a coal chute or water plug?

922. In what position should the handle of the automatic brake valve be placed in case of an accidental application—such as a bursted hose, or break in two of train?

923. In case there are two or more engines in a train, what should be done with the air equipment of each one, except the one from which the brakes are being handled?

924. What should be done with the brake equipment on the engine, before leaving the round house?

925. Name the various parts of the equipment, and the purpose of each.

926. Name the different parts of the piping appertaining to the ET equipment.

927. What are the functions of the main reservoir cut-out cock?

QUESTIONS

928. In what two ways does the automatic brake valve receive air from the main reservoir?

929. How many connections has the distributing valve?

930. What are the functions of the first three of these connections?

931. For what purpose are the other two?

932. What is the important feature of the ET brake equipment?

933. Describe briefly what takes place within the distributing valve during an independent application.

934. What causes independent lap?

935. Between what two pressures does the application piston of the distributing valve vibrate?

936. How is the pressure maintained in the brake cylinders during an application?

937. How is independent release accomplished?

938. How may independent lap be again resumed?

939. What parts are brought into action during automatic operation?

940. Describe in general terms what occurs within the distributing valve during a service application of the automatic brake.

941. Upon what does the amount of pressure in the application chamber of the distributing valve depend?

942. What causes the position known as service lap?

943. Does the pressure maintaining feature also apply to automatic application?

944. Are the engine brakes released when the automatic brake valve is placed in release position?

945. What, then, must be done in order to release the engine brakes?

946. What takes place within the distributing valve during an emergency application?

947. What pressures are carried for high speed brake service?

948. Under these conditions, what pressure is attained in the application chamber?

949. What conditions are necessary to cause the parts of the valve to assume the position known as emergency lap?

950. What is the purpose of the double-heading pipe connection?

951. Describe the action of the safety valve.

952. How is this valve adjusted, and for what pressure?

953. In what respect does type H automatic brake valve differ from previous types?

954. When should the handle be kept in running position?

955. What does service position give?

956. For what purposes is lap position used?

957. What is the purpose of release position?

958. What results follow, when the valve is placed in holding position?

959. Explain the difference between running and holding position.

960. When is emergency position used?

961. Of what type is the independent brake valve?

962. In what position should this valve be carried when not in use?

963. Describe the service position of this valve.

QUESTIONS

964. When is lap position used for the independent brake valve?

965. When is release position used?

966. What type of valve, and for what purpose is the feed valve?

967. Describe the construction and operation of the reducing valve.

968. Describe the construction and operation of the pump governor.

OTHER BOOKS FROM CGR PUBLISHING AT CGRPUBLISHING.COM

1915 San Francisco World's Fair in Color: The Grandeur of the Panama-Pacific...

San Francisco 1915 World's Fair: The Panama-Pacific International Expo.

1904 St. Louis World's Fair: The Louisiana Purchase Exposition in Photographs

Chicago 1933 World's Fair: A Century of Progress in Photographs

19th Century New York: A Dramatic Collection of Images

The American Railway: The Trains, Railroads, and People Who Ran the Rails

The Aeroplane Speaks: Illustrated Historical Guide to Airplanes

The World's Fair of 1893 Ultra Massive Photographic Adventure Vol. 1

The World's Fair of 1893 Ultra Massive Photographic Adventure Vol. 2

The World's Fair of 1893 Ultra Massive Photographic Adventure Vol. 3

Henry Ford: My Life and Work - Enlarged Special Edition

History of the Crusades Volumes 1 & 2: Gustave Doré Restored Editions

Ethel the Cyborg Ninja Book 1

The Complete Ford Model T Guide: Enlarged Illustrated Special Edition

How To Draw Digital by Mark Bussler

Best of Gustave Doré Volume 1: Illustrations from History's Most Versatile...

OTHER BOOKS FROM CGR PUBLISHING AT CGRPUBLISHING.COM

Ultra Massive Video Game Console Guide Volume 1

When Trolleys Ruled the Earth: A Photographic History of Streetcars...

Complete Works of Edgar Allan Poe Volumes 1 - 10

Classic Cars and Automobile Engineering: Volumes 1-5

Antique Cars and Motor Vehicles: Illustrated Guide to Operation...

Chicago's White City Cookbook

The Clock Book: A Detailed Illustrated Collection of Classic Clocks

The Complete Book of Birds: Illustrated Enlarged Special Edition

1901 Buffalo World's Fair: The Pan-American Exposition in Photographs

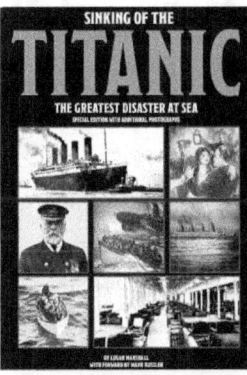
Sinking of the Titanic: The Greatest Disaster at Sea

Gustave Doré's London: A Pilgrimage: Retro Restored Special Edition

Milton's Paradise Lost: Gustave Doré Retro Restored Edition

The Art of World War 1

The Kaiser's Memoirs: Illustrated Enlarged Special Edition

History in the Age of Vikings Series Volumes 1-3

The Complete Butterfly Book: Enlarged Illustrated Special Edition

- MAILING LIST -
JOIN FOR EXCLUSIVE OFFERS

www.CGRpublishing.com/subscribe

www.ingramcontent.com/pod-product-compliance
Lightning Source LLC
Chambersburg PA
CBHW082103230426
43671CB00015B/2594